FROM RESEARCH TO CLINICAL PRACTICE

The Implications of Social and Developmental Research for Psychotherapy

FROM RESEARCH TO CLINICAL PRACTICE

The Implications of Social and Developmental Research for Psychotherapy

Edited by

George Stricker
Adelphi University
Garden City, New York

and

Robert H. Keisner
C. W. Post Center
Long Island University
Greenvale, New York

PLENUM PRESS • NEW YORK AND LONDON

Library of Congress Cataloging in Publication Data

Main entry under title:

From research to clinical practice.

Includes bibliographies and index.
1. Psychotherapy. 2. Social psychology—Research. 3. Developmental psychology—Research. I. Stricker, George. II. Keisner, Robert, H. [DNLM: 1. Psychotherapy. 2. Research. WM 420 F931]
RC480.5.F756 1985 616.89′14 84-24881

ISBN-13: 978-1-4684-4822-1 e-ISBN-13: 978-1-4684-4820-7
DOI: 10.100/978-1-4684-4820-7

Softcover reprint of the hardcover 1st edition 1985
A Division of Plenum Publishing Corporation
233 Spring Street, New York, N.Y. 10013

For Joan and Bonnie

Contributors

JOSEPH ADELSON, PH.D., *Department of Psychology, University of Michigan, Ann Arbor, Michigan*

JONATHAN BLOOM-FESHBACH, PH.D., *Georgetown University Counseling Center and Departments of Psychiatry and Behavioral Sciences and Child Health and Development, George Washington University, Washington, D.C.*

SALLY BLOOM-FESHBACH, PH.D., *American University Center for Psychological and Learning Services and National Research Council Committee on Child Development Research and Public Policy, Washington, D.C.*

BARBARA BENEDICT BUNKER, PH.D., *Department of Psychology, State University of New York at Buffalo, Amherst, New York*

MARTHA DAVIS, PH.D., *Department of Mental Health Sciences, Hahnemann University, Philadelphia, Pennsylvania*

SCOTT DICKMAN, PH.D., *Department of Psychology, Oberlin College, Oberlin, Ohio*

NORBERT FREEDMAN, PH.D., *Research Unit for the Study of Communicative Process, Department of Psychiatry, Downstate Medical Center, State University of New York, Brooklyn, New York*

KATHRYN S. GALVIN, PH.D., *Counseling Center, Texas A & M University, College Station, Texas*

STANLEY GRAND, PH.D., *Research Unit for the Study of Communicative Process, Department of Psychiatry, Downstate Medical Center, State University of New York, Brooklyn, New York*

JOHN H. HARVEY, PH.D., *Department of Psychology, Texas Tech University, Lubbock, Texas*

ELAINE HATFIELD, PH.D., *University of Hawaii at Manoa and King Kalakaua Center for Psychotherapy, Honolulu, Hawaii*

JOHN S. JORDAN, PH.D., *Duke Medical School, Duke University, Durham, North Carolina*

JAMES W. JULIAN, PH.D., *Department of Psychology, State University of New York at Buffalo, Amherst, New York*

ROBERT H. KEISNER, PH.D., *C. W. Post Center, Long Island University, Greenvale, New York*

CHARLES A. KIESLER, PH.D., *College of Humanities and Social Sciences, Carnegie-Mellon University, Pittsburgh, Pennsylvania*

JUDITH F. LASKY, PH.D., *257 Central Park West, New York, New York*

ANNA C. LEE, PH.D., *Rehabilitation Services, Hospital for Special Surgery, New York, New York*

JAMES E. MADDUX, PH.D., *Department of Psychology, Texas Tech University, Lubbock, Texas*

GILEAD NACHMANI, PH.D., *115 East 87th Street, Suite 32F, New York, New York*

KENNETH S. POPE, PH.D., *11747 Sunset Boulevard, Suite 226, Los Angeles, California*

KAISA PUHAKKA, PH.D., *Department of Psychiatry, Medical College of Ohio, Toledo, Ohio*

LEE SECHREST, PH.D., *Department of Psychology, University of Arizona, Tucson, Arizona*

ROBIN B. SHAFRAN, PH.D., *262 Central Park West, Suite 1-B, New York, New York*

JOYCE SLOCHOWER, PH.D., *Department of Psychology, Hunter College, City University of New York, New York, New York*

SUSAN SPRECHER, PH.D., *Department of Sociology, University of Wisconsin, Madison, Wisconsin*

JANICE M. STEIL, PH.D., *Institute of Advanced Psychological Studies, Adelphi University, Garden City, New York*

GEORGE STRICKER, PH.D., *Institute of Advanced Psychological Studies, Adelphi University, Garden City, New York*

RICHARD L. VICKERS, PH.D., *Institute of Advanced Psychological Studies, Adelphi University, Garden City, New York*

GIFFORD WEARY, PH.D., *Department of Psychology, Ohio State University, Columbus, Ohio*

JASON WORCHEL, PH.D., *Virginia Institute of Short-term Dynamic Psychotherapy, Charlottesville, Virginia*

STEPHEN WORCHEL, PH.D., *Department of Psychology, Texas A & M University, College Station, Texas*

Foreword

It is an intriguing feature of human experience that in our present world, amid thousands of indications of the effectiveness of the scientific method, so many of us persist in demonstrably illusory or magical beliefs whether religiously related or simply reflections of long-standing superstitions. At a time when millions can observe on television the first landings of human beings on the moon, when our daily lives in the so-called developed countries are replete with conveniences that reflect scientific advances, we still persist in daily wagers on the state lotteries, in paying astrologers or palmists for their readings, in investing thousands of dollars and hours of our legislators' time in discussing such issues as the value of daily prayer in the elementary schools. The emergence of modern medicine based increasingly on scientific research in chemistry, biology, and physics has considerably reduced people's resort to shamans and witch doctors within the major sectors of our own society, although it has by no means eliminated such practitioners.

The field of psychotherapy has emerged to some degree from the medical practices of the late nineteenth century and its spread in this century is mainly attributable to scientifically trained individuals, most notably Sigmund Freud. The practice of clinical psychology, which has included the development of diagnostic and personality or intellectual assessment instruments and procedures and also direct intervention through psychological means whether in the form of psychotherapy or consultation, grew up from a scientifically based science of psychology that emerged under the aegis of Wundt in Germany, and James and Titchener in the United States. Freud never gave up his own firm belief that psychoanalysis was rooted in basic neurophysiology, and although he might have questioned the usefulness of individual psychological experiments in testing the complexity of information adduced from hundreds of hours of analytic sessions, he never gave up his faith in the continuing link between basic science and psychoanalytic practice.

In the United States a tremendous impetus was given to the spread of psychotherapy practice among clinical psychologists by the client-centered approach developed by Carl Rogers, a university-based psychologist, who also introduced some of the basic methodology for evaluating process and outcome of psychotherapy. A second major influence on the widespread practice of psychotherapy came with the behavior modification techniques introduced by Joseph Wolpe, again on the premise of a link between scientific learning theory and the possibilities of systematic behavior change. Probably 90% of the actual basic research on clinical processes, from evaluation of diagnostic systems through evaluation of the various components of psychotherapeutic process, has been carried out by clinical psychologists linked in some way to universities or to research centers and hospitals and clinics. Given these developments it seems likely that most people who resort to psychotherapy for help with emotional problems rather than consulting a religious figure, a practitioner of voodoo, an astrologer, or some other representative of the historical folk methods of healing must be assuming that psychotherapy as practiced by psychiatrists, psychologists, or social workers is in some way an outgrowth of basic psychological science.

I have personally taken the position that to the extent to which clinical practitioners in psychology have lost track of the scientific roots of their practice and have become adherents of a particular system or school that does not engage in research, they may be betraying the scientific trust of their clients. Although we do not have clear evidence that all clients who resort to psychotherapy with psychologists are expecting them to perform their services directly on the basis of the most recent scientific data, it is certainly true that people who visit physicians would be appalled if they had reason to believe that the physicians were not relating their work to the latest scientific research in chemistry and physiology. As a clinician I am aware that in working directly with an individual client one must have a certain tolerance for ambiguity to function effectively. It is not possible mentally to run through recent articles in the *Journal of Clinical and Consulting Psychology* or the *Journal of Abnormal Psychology* in order to understand what a client is saying or to respond naturally and meaningfully. But I would propose that a clinician who has been keeping abreast, for example, of some of the latest results in research on depression would take a different view of the phenomenon as it is manifested by a client than would someone who approaches the situation primarily in terms of an early version of psychoanalytic theory.

Practitioners are busy people who have made a commitment to direct service, are often personally somewhat isolated from larger groups, and who confront day after day the dilemmas of their clients with a courage and honesty that must be commended. Nevertheless, a natural correlate of this lifestyle (and this is often true of medicine as well) is the tendency to lose touch with the recent scientific developments in one's field. This volume provides a remarkable opportunity for the practicing clinician to read up on developments in a number of crucial areas directly linked to psychological practice. It has been too easy in the past to rationalize away the obligation to read the literature because undoubtedly many specific psychological experiments seem trivial exercises in relation to the complexity confronted by the clinician. Nevertheless, there has been a gradual and impressive buildup of extremely useful knowledge in cognitive psychology, social psychology, developmental psychology, and psychology of personality that has tremendous potential for practical application. Reviews of social psychology research on aggression, attitudes, interpersonal attraction, and so on that are presented in this volume are extremely valuable in helping clinicians overcome some of their own misdirected tendencies, for example, the excessive belief in the inherent value of catharsis as a form in discharging aggression, and also in learning many useful ways of identifying potential areas of legitimate influence on the self-defeating behaviors of clients. There has also been, I believe, an excessive emphasis in clinical practice on a mythology about childhood based to some degree on the theorizing of psychoanalysts who have available only adult memories, which are often proved to be inaccurate or at least subject to gradual accretions of adult perspectives. There has been an extremely valuable growth of direct observational research on children and this literature has impressive possibilities for correcting some of the misinformation clinicians have been employing and also for sharpening their own effective work both with adults and children.

For me this book fulfills a long-standing need for a much closer link between clinical application and basic research in psychology. Drs. Stricker and Keisner are surely to be commended for their efforts in organizing this volume. I hope it will become a part of the basic library of clinical practitioners.

JEROME L. SINGER

Preface

Both of us are university professors with a history of—and commitment to—research activities, and practicing psychodynamic psychotherapists. The role conflict this engenders is frequently troublesome, and we are frequently asked by students on the one hand, and by therapeutic colleagues on the other, how we can reconcile these two functions. This book is an attempt to respond to that question by showing the many ways in which a knowledge of research can be of value to the practicing clinician, and also the ways in which the practice of psychotherapy can be a valuable source of ideas to the research scientist.

In searching for authors to help us with this task, we sought people with a good working knowledge of both areas. In some cases we were able to locate such individuals; in others, a collaboration between two or more authors brought both perspectives to the chapter. Each author was sent the following letter, which shows our intentions in arranging the book:

> This project will focus on the contributions that basic research in social and developmental psychology can make to the practice of psychodynamic psychotherapy. The author(s) must have a knowledge of both a research area and treatment, and be capable of moving back and forth between the two. As this is rarely done, it will require a good deal of creativity to identify connections that are usually overlooked.
>
> Since the primary audience will probably be practicing psychotherapists, we can assume a minimum knowledge of the research area but a ready familiarity with clinical concepts. Therefore each chapter should begin with a clear summary and review of relevant research. This should be followed by an extension of the research concept to a parallel clinical concept (e.g., the implications of experimenter bias for countertransference, or the relation of cognitive dissonance to resistance to change of ego-syntonic behaviors). Case material should be used to illustrate how similar incidents can be interpreted in research or clinical terminology.

> For the chapter to be successful, it should not stop with a simple translation of research language into clinical language. A number of provocative questions may be considered. Does the research literature suggest a different understanding? Does it suggest any modifications in technique? In what ways does it help us to go beyond what we already know? Finally, are there any clinical findings or problems that suggest hypotheses that might be tested in the research arena, and allow us to go beyond the current research position?
>
> If the chapter fulfills the task, it should give new meaning to the term scientist-professional. A feedback loop should be established that shows how research and practice can inform each other, and how each area of inquiry can benefit from the accomplishments in the other.

Some of the authors followed these instructions very closely. Others followed the spirit of the instructions and employed a creative approach to demonstrating a research–practice synergy. Our excitement in developing the project mounted as the chapters came in, providing what we had hoped to find and, in some cases, a good deal more. We are very fortunate in the authors who agreed to write chapters, and grateful to them for their scholarship and creativity.

We also would like to express our gratitude to those who were helpful in assembling the manuscript. The facilities of Adelphi University were available to us, and the support provided by the university was essential to completing the task. Virginia Bruchhauser, in particular, attended to many tiresome but necessary details throughout the project. We are also appreciative of our students who listened and responded to these ideas, and our families who provided active inspiration and willingly endured the long hours while the work was being done.

GEORGE STRICKER
ROBERT H. KEISNER

Contents

I. Basic Issues 1

1. The Relationship between Research and Practice 3
George Stricker and Robert H. Keisner

2. Research on Memory and Clinical Practice 15
Scott Dickman and Lee Sechrest

3. Implications of Nonclinical Research for Mental Health Policy 45
Charles A. Kiesler

II. Social Psychology 63

4. The Leadership Literature and Its Relevance to Psychotherapy 67
Barbara Benedict Bunker and James W. Julian

5. Nonverbal Behavior Research and Psychotherapy 89
Martha Davis

6. Aggression and Psychodynamic Therapy: From Theory to Practice 113
Stephen Worchel and Jason Worchel

7. Attitude Change Theory, Research, and Clinical Practice 139
John H. Harvey, Gifford Weary, James E. Maddux, John S. Jordan, and Kathryn S. Galvin

8. Social Stereotypes 157
Robin B. Shafran

9. Interpersonal Attraction 179
Susan Sprecher and Elaine Hatfield

10. The Experience of Injustice: Social Psychological and Clinical Perspectives 217
Janice M. Steil and Joyce Slochower

III. DEVELOPMENTAL PSYCHOLOGY 243

11. Implications of Infant Development Research for Clinical Practice 245
Richard L. Vickers

12. Psychodynamic Psychotherapy and Research on Early Childhood Development 265
Jonathan Bloom-Feshbach and Sally Bloom-Feshbach

13. Normal and Pathological Gender-Role Development in Children 287
Anna C. Lee

14. Adolescence for Clinicians 313
Joseph Adelson

15. Midlife Change 327
Judith F. Lasky

IV. SPECIAL TOPICS 351

16. Shielding: An Associative Organizer 353
Norbert Freedman and Stanley Grand

17. Fantasy 375
Kenneth S. Pope

18. Psycholinguistics and Psychopathology 401
Gilead Nachmani

19. Religious Experience and Psychoanalytic Psychotherapy 411
Kaisa Puhakka

20. Self-Fulfilling Prophecies in Psychodynamic Practice 427
Robert H. Keisner

Index 447

I

BASIC ISSUES

This book is concerned with the relationship of research to practice, a topic that has been the basis of much theoretical and political debate. Stricker and Keisner address this issue in the first chapter, describing some of the tensions between the two areas of endeavor and exploring some of the possible semantic and psychological barriers to rapprochement. Each of the subsequent chapters in the volume attempts to illustrate the meaning of a body of basic research for the practicing clinician, and by doing so show the inherent complementarity of research and practice.

Failure to recall past events and distortions in memory are key issues in clinical practice. In Chapter 2, Dickman and Sechrest develop hypotheses from an information-processing perspective that focus on nonemotional factors in long-term memory. They demonstrate how patients' recollections and responses, as well as therapists' conclusions and interventions, can be influenced by such cognitive functions as retrieval, cuing, coding, elaboration, and labeling. Clinical examples, including specific alternative approaches, are presented throughout the chapter.

In Chapter 3, Kiesler addresses an issue that is broader than the practice of psychotherapy by the individual clinician. Mental health policy in the United States, which affects the practice of every clinician and the fate of many patients, is an aggregate of laws and practices that are drawn in response to political and social pressures without any clear attempts at articulation or evaluation. These far-reaching policies can be constructed and evaluated more rationally on the basis of empirical evidence, and Kiesler illustrates the ways in which these policies can be better understood if we attend to available information.

1

The Relationship between Research and Practice

GEORGE STRICKER AND ROBERT H. KEISNER

It is clear that research and clinical practice have a synergistic relationship, with each one informing the other. It is clear that effective clinical psychologists will guide their practice on the basis of established research findings. It is clear that contributing research clinicians will be actively involved in practice, deriving hypotheses from clinical experience in order to better their research which, in turn, will influence their practice. And it is clear that each preceding statement is a romantic platitude, more honored in the breach than in performance. More likely, the practicing clinician does not read research journals very often, having discovered, early on, the irrelevance of the findings to practice. More likely, the research clinician has not treated a clinical patient for years, so the lack of apparent relevance of the findings is of little surprise. And most likely, the practitioner will register disdain for the researcher, who will complain about the disregard of the practitioner for the evidence of science.

The tension between research and practice in psychotherapy has a long history. Both activities are difficult and require extensive individual effort, reducing the possibility that any one person can perform both

GEORGE STRICKER • Institute of Advanced Psychological Studies, Adelphi University, Garden City, New York 11530. ROBERT H. KEISNER • C. W. Post Center, Long Island University, Greenvale, New York 11548.

activities competently even if he should wish to. This being the case, synergy is an idealistic goal that is rarely achieved. Yet polarization, the characteristic outcome, seems unnecessary and destructive.

A spectrum of views about the importance of research to practice has been presented and remains unresolved. Barlow (1981), introducing an important series of invited articles on this subject, simply states, "At present, clinical research has little or no influence on clinical practice" (p. 147). He goes on to cite Matarazzo, who has been able to be active in both research and practice yet acknowledges,

> Even after 15 years, few of my research findings affect my practice. Psychological science per se doesn't guide me one bit. I still read avidly, but this is of little direct practical help. My clinical experience is the only thing that has helped me in my practice to date. (p. 148)

In a similar vein, Raush (1974) asserts that "so far as one can see, again with the possible exception of behavior modification approaches, research has not influenced practice" (p. 678). On the other hand, Singer (1980), with equal vehemence, states:

> The practice of psychotherapy . . . is an outgrowth of basic knowledge in psychology, behavioral and social sciences more broadly and, to a somewhat lesser degree, the biological sciences. The ethical practice of psychotherapy must reflect the current status of knowledge in those fields. The practitioner who has not examined recent developments in the research literature or who has not kept abreast of evaluation studies of various forms of treatment may well be violating a central ethic of the profession. (p. 372)

Strupp (1981), in contrast to Barlow and Raush, has considered this issue and concludes,

> Considering its short history, I believe research in psychotherapy has made considerable contributions to clinical practice, mainly in the sense of sharpening clinical observations and forcing clinicians to think more critically, to take empirical data more seriously, to avoid excursions into esoteric theory, and to pursue realistic therapeutic goals. (p. 217)

This impressive litany of behaviors into which the clinician has been forced is all too reminiscent of Meehl's (1954) earlier argument for the unavoidability of statistics in clinical practice. He convincingly maintained that the honest clinician would need validation of his hypotheses and would not be content with the imprecision of clinical experience; hence the need for validation through statistical demonstration. The argument is compelling, but practitioners have spent the subsequent generation demonstrating that statistics can be avoided after all, and will probably spend the next generation eluding the force to which Strupp refers.

The debate about the desirability and feasibility of research utilization is further complicated by the failure to define either *research* or *utilization*. This makes it possible for the argument to proceed with each side referring to a different phenomenon although using the same term, so that agreement is precluded. When *research* is used pejoratively by clinicians, they often think of a narrow, static, methodology-bound, laboratory-based effort that has little potential for generalizability. Researchers, on the other hand, see clinicians as seeking a loose, impressionistic, vague set of speculations that cannot contain any internal validity, making the question of external validity moot.

The difficulty in defining research may be illuminated, as is true of so many issues, by returning to the works of Sigmund Freud. Although he is characterized as the arch-clinician, most well known for his intricate theoretical system built on evidence gleaned from a small group of patients, he always considered himself to be a scientist. Freud was well aware of the distinction between psychoanalysis as a method of treatment and psychoanalysis as a theory of personality and, if he had to choose, always opted for the latter. Luborsky and Spence (1971) cite a remark he made in 1917: "Even if psychoanalysis should show itself as unsuccessful with other forms of nervous and mental diseases, it would still remain justified as an instrument of scientific research" (p. 409). This view is reiterated in a consideration of lay analysis (Freud, 1927/1950) when he states: "I only want to feel assured that the therapy will not destroy the science" (p. 106). Yet Luborsky and Spence also report Freud's classic response to a request to comment on a research study by Rosenzweig:

> I have examined your experimental studies for the verification of the psychoanalytic assertions with interest. I cannot put much value on these confirmations because the wealth of reliable observations on which these assertions rest make them independent of experimental verification. Still, it can do no harm. (p. 408)

This last quotation, often cited as evidence of Freud's disdain for research, does not represent a break from the earlier position or a denigration of research. Rather, it is a statement that the validity of the theory rests on evidence of a different, and to his mind superior, sort ("wealth of reliable observations"). Science can embrace a number of methodologies, and it might be of some value to both clinicians and researchers to recognize that psychological research consists of systematic empirical inquiry. It can range from a tightly controlled laboratory analog to a well-done phenomenological study, as long as the careful explication of the system permits replicability.

Similarly, whereas there are many statements about the failure of clinicians to utilize research, there is little consideration given to the meaning of utilization. Cohen (1981), who has given this matter both speculative and research attention, distinguishes among research awareness, research consideration, and research implementation. Awareness requires some degree of familiarity with research; consideration requires the active contemplation of the implications of that research; and implementation is the actual integration of research and practice. Cohen also notes the failure to consider the time dimension for utilization. The resolution of the apparent contrast between Barlow's pessimism and Strupp's optimism about the impact of research rests on the recognition that Barlow is looking for time-limited research implementation, whereas Strupp is thinking of a time-extended research consideration.

Although Cohen resolves the opposition of Barlow and Strupp on the ground that they mean different things by utilization, it is also possible that they mean different things by research. Barlow is thinking of research as a rather stereotyped, group comparison strategy, and it is this that he dismisses as having little influence on clinical practice. He then goes on to espouse an approach that is modeled after Cronbach's (1975) "intensive local observation." This is an idiographic approach in which individual clinicians attend closely to their impact on patients, describing procedures in careful detail while noting the nature, proportion, and parameters of success and failure. The accumulation of these data would be of enormous heuristic value. Barlow notes that this was the research strategy of both Wolpe and Masters and Johnson, but does not add that it was also that of Freud. Interestingly, Barlow's recommendation following his lament about utilization entails precisely the understanding of research that led Strupp to his more optimistic vision. He believes that "most of what the field has learned about psychotherapy . . . has come from astute and creative clinical observations" (p. 216). He goes on to integrate this point with support for the research endeavor by noting that "each therapeutic dyad constitutes an experiment, albeit an experiment with built-in limitations. The therapeutic situation can be although it is not necessarily, a unique laboratory situation" (p. 216). Thus, both Barlow and Strupp agree as to the most constructive approach to research while disagreeing about the impact of utilization: Barlow sees utilization as a strategy for the future, whereas Strupp recognizes the extent to which it has already been implemented. It should also be noted that Raush (1974), too, agrees "that it is not research that is being rejected, but one kind of research" (p. 679).

There is another dimension that must be considered in order to understand the disagreement about the link between research and practice. Aside from differences in methodology, there are also substantial differences in research goals. Clinical research can be directed either at outcome or process of psychotherapy. Outcome research attempts to answer the question, Does therapy work?—a question that most practicing therapists feel has been answered to their satisfaction. Process research deals with the question, How does it work? and this technique-oriented question is of far more interest to the practitioner. In contrast to clinical research is nonclinical, or basic, research, which is directed at the understanding of people rather than the means of intervening with them. For reasons that may relate to inadequate methodology, technique-oriented research has not been very contributory to the practitioner (Barlow, 1981; also Matarazzo) and is not likely to be unless research strategies are altered (Barlow, 1981; Raush, 1974; Strupp, 1981). On the other hand, there is a rich body of basic research that will help us to understand the patient, but all too often it has been ignored by the practitioner (Singer, 1980).

The distinction between technique-oriented and understanding-oriented research leads to a consideration raised by Mitroff and Featheringham (1974). We are all familiar with the dangers of Type 1 and Type 2 errors in experimentation, but Mitroff and Featheringham raise the possibility of an error of the third kind (E3). This occurs when the wrong experiment is performed when the right one should have been, and is usually based on faulty assumptions about the nature of the problem. If we assume that the most useful research for the practitioner is technique oriented, we are likely to commit an E3, carry out flawed studies, and reduce the likelihood of rapprochement between research and practice. A correction of this assumption could reduce the likelihood of an E3, promote more understanding-oriented research, and raise the probability that clinicians will find the work of value.

This goal-oriented demarcation of research may also help to explain some of the apparent changes in attitudes toward research that have been described. Perl and Kahn (1981) surveyed over 2,000 graduate students in academic and professional schools regarding their attitudes toward research. In the academic departments, 22% of the clinical students indicated that their career goals were in research, about one-third had had research articles accepted for publication, and 27% had presented a research paper at a convention. Their attitudes toward their research training were very positive, and 47% felt that they were more interested in research than when they first entered graduate school, in contrast to 17% who felt they were now less interested. Attitudes of graduate students in professional schools were a good deal less positive. They have less interest in research, less of a history of research productivity, and less career interest in research. Interestingly, almost three-quarters of them indicate an intention to spend 10% or less of their time in research. This is interesting in light of another survey, by Garfield and Kurtz (1974), which looked at a large number of attitudes of almost 1,000 practicing clinical psychologists and found that research-related activities consumed slightly less than 10% of the time they devoted to various activities. The authors note that this imbalance between research and service delivery was consistent with the results of a similar survey over a decade earlier (Kelly, 1961) and conclude, rather conservatively, that "research does not appear as enticing or important to clinical psychologists as does clinical service" (p. 9). These findings are consistent with survey data shown by Cohen (1979), who estimated that clinical psychologists read two to four research articles per month. This figure is in agreement with a survey by Garvey and Griffith (1971), who estimated that within a 2-month period from publication, almost half the research articles published are read by at most a tiny fraction of psychologists, and no article is likely to be read by more than 7% of psychologists.

The pattern, then, is for initial enthusiasm for research to be sharply reduced as students become professionals. This change must be early and precipitous, since Garfield and Kurtz (1974) did not see any relationship between attitudes and years of experience, although their method of categorization obscured any changes over the immediate postdoctoral period. In order to understand this phenomenon, we should return to the distinction based on research goals. We can speculate that students in academic programs are exposed to a good deal of basic research, and they seem to be excited by it. We can also assume that their needs when they go on internship or begin practice are for more technique-oriented research, but that they are disappointed when they turn to it. This produces the disenchantment which is seen in later years. Although the beginning professional needs technique-oriented research, the more experienced professional is more likely to value research that is oriented to understanding. Unfortunately, it appears as though the early disillusionment was sufficient to bar them from a return to research, which would prove far more valuable now that their needs have changed from technique to understanding. No extant data will allow a specific test of this hypothesized course of development, but it is consistent with the various survey findings.

The relationship between research and practice is also affected by potential sources of belief. Factors that may influence the beliefs of clinicians include authority, theory, experience, and evidence. Authority (supervision), theory, and experience are the usual sources of the clinician's beliefs; evidence is presented by the researcher. One would expect that strong evidence would lead to the abandonment of inconsistent beliefs, and weak evidence would be disregarded; but exceptions to this rule of epistemological primacy can be found, such as with ESP. Technique-oriented research often can produce no more than weak evidence because of all the inherent problems of generalizability. Basic research is more capable of producing strong evidence, but is less likely to come to the attention of the clinician, again leaving the weight of beliefs with the experience of the clinician. Interestingly, the "intensive local observation" suggested by both Barlow and Strupp would serve to convert experience into evidence, making technique-oriented research more relevant to practice, and perhaps opening the door to the examination of understanding-oriented research.

If practitioners are to know about and utilize basic psychological research in their clinical practice, it is essential that their attitudes toward research change dramatically. Although one can acknowledge that "a systematic evaluation of the effectiveness of traditional research for its impact on clinical thinking and practice would come off worse than the most damning evaluation of psychotherapy" (Raush, 1974), it seems clear that there is a body of basic research currently available and ignored by clinicians that could be used as another tool in their efforts to promote mental health. Along with the barriers established by researchers, the attitudes and perceptions of clinicians are a major source of the scientist–practitioner schism. These attitudes, learned early in training and reinforced in often dogmatic clinical surroundings, represent major obstacles to the clinician's awareness and utilization of basic psychological research.

Before identifying several implicit attitudes, we believe that there is a fundamental human assumption about similarities and differences that has contributed to the establishment and maintenance of divisive attitudes between researchers and clinicians. For any group, the formation of an identity requires the dialectical process of establishing criteria and rules for inclusion and exclusion. Historically, this process can be seen as an essential ingredient in stereotyping and prejudice toward individuals who are not part of the in-group. Psychologists, like all professional and scientific groups, have been moving more and more in the direction of specialization, more and more toward emphasizing ways in which we are different from other professionals and from one another. In spite of all the outcries against the abuse of classifying people, greater and greater emphasis has been placed on differential diagnosis. Integration of orientations, groups, and ways of classifying patients and therapists represent the exception to the rule. Any effort to integrate basic research with professional practice goes against the tide of identity formation among researchers and practitioners who, in addition to creating inclusionary guidelines, have also established unwritten negative identifications that are attributed to members and activities of the other groups. It is our hope that the awareness of this universal group phenomenon will encourage clinicians to take a good look at what they have in common with researchers and hence approach researchers with a greater sense of collaboration rather than confrontation. In order for this to happen, clinicians must perceive what they do as being more similar to what researchers do.

The first of several fundamental, dubious assumptions of the clinician is that research represents an objective approach to understanding human activity whereas the clinical inquiry is based on a subjective model. The scientist versus artist roles are concomitants of this belief that scientists function objectively and artists subjectively or intuitively. It appears that this attitude distorts the activities that researchers and clinicians actually engage in. To the extent that clinicians perceive the "objectivity" of clinical work and the "subjectivity" of research, the two models could be perceived with increased similarity. The first point, clinical objectivity, is less justified and less relevant to the present argument. Nevertheless there are those (e.g., Racker, 1968) in the analytic community who have argued that analysts' awareness of their own countertransference reactions (subjectivity) ought to enable them to experience the analytic matrix with an increased amount of "objectivity." In fact, recent evidence in the interpersonal psychoanalytic literature suggests the usefulness of knowing how we feel in the presence of a patient (Epstein & Feiner, 1979). This may be seen as a tool to increase objectivity insofar as the clinician would have more insight into the patient's reality.

A high degree of subjectivity in the research process is more easily argued. The literature collectively known as the "social psychology of the psychology experiment" has demonstrated the illusory nature of the belief that psychological research is unaffected by the wishes, intentions, motives, beliefs, and unintentional behavior of experimenters. Experimenters are "participant observers" in their own laboratories. First, experimenters and the experimental setting provide subjects with demand characteristics—implicit and explicit—regarding what is expected of them. Those characteristics influence subjects to be cooperative and sometimes provide the experimenter with data to fit the research hypothesis. Experimenters do not have as much control over the human subjects' environment as they would like to believe (Orne, 1969). Rosenthal (1964,

1966) demonstrated that the experimenter's sex, warmth, anxiety, needs, and experience all influence the behavior of research subjects. Even though other researchers have questioned the validity of these findings (e.g., Barber & Silver, 1968), it still seems appropriate to conclude that "expectancy effects" are an inevitable ingredient in some research outcomes. It seems justified to conclude that research and clinical practice overlap when it comes to the objective-subjective continuum.

A second assumption that perpetuates the research-clinical split is that research is conducted with "normal" populations whereas clinicians are engaged in psychotherapy with "abnormal" individuals. For example, Brehm and Smith (1982) question the validity of applying social psychology to clinical practice because normal subjects participate in the vast majority of social psychology research. How, they ask, can we generalize most research to clinical populations? This assumption may not be as accurate as is generally believed. The behavior of numerous participants in a variety of social psychological studies seems to demonstrate how easy it is to elicit immoral, compliant, irrational, apathetic, and harmful responses in "normal" populations. For example, research on obedience (Milgram, 1974) demonstrated that most people will harm others simply because they are instructed to do so. Subjects who perceived themselves as anonymous, identified by the experimenters as "deindividuated," engaged in the performance of antisocial acts (Zimbardo, 1970). "Well-adjusted" college students, assigned by experimenters to participate as guards or prisoners in a mock prison study, quickly became aggressive and even sadistic, or helpless and self-deprecating (Zimbardo, Banks, Haney, & Jaffe, 1973). Almost half of the subjects who witnessed a simulated emergency (in groups) acted apathetically (Darley & Latané, 1968). Finally, individuals in one study devalued those who suffered for them, evidencing rejection rather than compassion for victims (Lerner & Simmons, 1966). Although these observations can be understood as stemming from compelling situational demands, the behavior of these and numerous other "normal" participants in psychological studies is *not* radically different from the "pathological" actions of psychotherapy patients. In fact, apathy, sadism, self-deprecation, antisocial tendencies, helplessness, and devaluation of others are frequently identified as symptoms, presenting problems, or maladaptive responses to interpersonal situations.

From the clinical perspective, there is the issue of relative degrees of psychopathology characteristic of clinical and nonclinical populations. Practitioners have often found their patients less or at least no more disturbed than friends or members of their families. Notwithstanding conscious and unconscious distortions, this conclusion is based on the assumption that patients are basically different from others in that they recognize the need for treatment. Havens (1982), in discussing the issue of "pathological location," notes that

> one no longer assumes that one knows who is sick. The family member who comes for treatment is often the most flexible and insightful one, and if not insightful, depressed and self-deprecating enough to blame himself. (p. 40)

Psychotherapy patients may not be healthier than the general population, but they may not be any more abnormal. The argument that research is not generalizable to practice because of differences between normal research subjects and pathological psychotherapy patients seems entirely unjustified.

Third, there is the assumption that research in psychology is generally nomothetic whereas clinical practice is idiographic in approach. Researchers are concerned with identifying general laws and principles applicable to everyone, whereas clinicians are oriented to what is unique about the individual patient. Therefore the practitioner is somewhat justified in believing that researchers have a different basic goal. Even this issue is not as black-and-white as it first appears. Dependence on differential diagnosis represents a *subnomothetic* approach within a broader idiographic framework. Individual patients are classified and sometimes treated on the basis of the guidelines available for the diagnostic groupings. For instance, a clinician's treatment of a narcissistic patient may be based on experiences with other similarly classified patients. In addition, the clinician has etiological and treatment guidelines available for particular disorders. Hence, the emphasis is placed not only on a person's uniqueness but also on traits held in common with patients considered to be similar. In fact, some practitioners believe that the emphasis in clinical practice is overly nomothetic thus disregarding the uniqueness of each patient. The nomothetic-idiographic distinction seems to be as exaggerated as the two previously discussed assumptions.

There are two noticeable exceptions to the chasm between basic research and clinical practice. The methods of intervention developed by behavior therapists and behavior modifiers are based on the principles of learning discovered in the psychology laboratory. This research–practice link, as Dickman and Sechrest point out (in Chapter 2), is a fragile one that may be threatened by the "cognitive revolution" in behaviorism. In line with the present effort, there have been several recent attempts to integrate basic research in social psychology and clinical practice (Brehm, 1976; Weary & Mirels, 1982). Brehm's work focused on the theoretical and empirical implications of three existing psychological perspectives—dissonance, reactance, and attribution processes—

for psychotherapy and counseling. Unlike the present text, however, she focused on the implications of laboratory research primarily for cognitively oriented psychotherapy. The theoretical focus in this book is psychodynamic. In addition, a serious effort has been made to address the issue of relevance of research to actual case study observations.

Weary and Mirels brought together articles that explore such important issues as the valid application of social psychology to clinical practice, the components of the clinical process, beliefs about the causes of mental disorders, social psychological processes in the development of maladaptive behaviors, and approaches to treatment. Interestingly, Weary and Mirels conclude their social-clinical integration with the recognition that their approach is not likely to be readily accepted. They identify three major obstacles to reducing the separation between social psychology research and clinical practice: (1) the requirement of an extensive knowledge base, (2) the belief that the kinds of behavior and attitudes of the two disciplines differ widely, and (3) the fact that clinical research for social psychologists requires a special effort, ingenuity, and tolerance for delays and interruptions. Although different obstacles to an integrative endeavor have been identified, there is clearly complete agreement with the view that linking research to practice will contribute to the understanding and reduction of dysfunctional patterns of behavior. And, as Singer has stated,

> if psychotherapy is not simply to be a fad of the century or settle into the role of another folk practice like astrology or phrenology, we must accept its scientific foundations as a fundamental value of practice. (1980, p. 381)

REFERENCES

Barber, T. X., & Silver, M. J. Fact, fiction, and the experimenter bias effect. *Psychological Bulletin Monograph Supplement*, 1968, *6* (Part 2), 1–29.

Barlow, D. H. On the relation of clinical research to clinical practice: Current issues, new directions. *Journal of Consulting and Clinical Psychology*, 1981, *49*, 147–155.

Brehm, S. S. *The application of social psychology to clinical practice*. Washington, D.C.: Hemisphere, 1976.

Brehm, S. S., & Smith, T. W. The application of social psychology to clinical practice: A range of possibilities. In G. Weary & H. Mirels (Eds.), *Integrations of clinical and social psychology*. New York: Oxford University Press, 1982.

Cohen, L. H. The research readership and information source reliance of clinical psychologists. *Professional Psychology*, 1979, *10*, 780–785.

Cohen, L. H. *Research utilization by clinical psychologists*. Paper presented at the meeting of the American Psychological Association, Los Angeles, August 1981.

Cronbach, L. J. Beyond the two disciplines of scientific psychology. *American Psychologist*, 1975, *30*, 116–127.

Darley, J., & Latané, B. Bystander intervention in emergencies: Diffusion of responsibility. *Journal of Personality and Social Psychology*, 1968, *8*, 377–383.

Epstein, L., & Feiner, A. *Countertransference*. New York: Jason Aronson, 1979.

Freud, S. *The question of lay analysis* (J. Strachey, Trans.). New York: W. W. Norton, 1950. (Original work published 1927)

Garfield, S. L., & Kurtz, R. A survey of clinical psychologists: Characteristics, activities and orientations. *The Clinical Psychologist*, 1974 *28*(1), 7–10.

Garvey, W., & Griffith, B. Scientific communication: Its role in the conduct of research and creation of knowledge. *American Psychologist*, 1971, *26*, 349–362.

Havens, L. The choice of clinical methods. *Contemporary Psychoanalysis*, 1982, *18*, 16–42.

Kelly, G. L. Clinical psychology—1960: Report of survey findings. *Newsletter. Division of Clinical Psychology of the American Psychological Association*, 1961, *14*(1), 1–11.

Lerner, M. J., & Simmons, C. H. Observer's reaction to the "innocent victim." *Journal of Personality and Social Psychology*, 1966, 4, 203–210.

Luborsky, L., & Spence, D. P. Quantitative research on psychoanalytic therapy. In A. G. Bergin & S. L. Garfield (Eds.), *Handbook of psychotherapy and behavior change* (1st ed.). New York: Wiley, 1971.

Meehl, P. G. *Clinical versus statistical prediction*. Minneapolis: University of Minnesota Press, 1954.

Milgram, S. *Obedience to authority: An experimental view*. New York: Harper & Row, 1974.

Mitroff, I. I., & Featheringham, T. R. On systematic problem solving and the error of the third kind. *Behavioral Science*, 1974, *19*, 383–393.

Orne, M. T. Demand characteristics and the concept of quasi-controls. In R. Rosenthal & R. I. Rosnow (Eds.), *Artifact in behavioral research*. New York: Academic Press, 1969.

Perl, K. G., & Kahn, M. W. *Psychology graduate students' attitudes toward research: A national survey*. Paper presented at the meeting of the American Psychological Association, Los Angeles, August 1981.

Racker, H. *Transference and countertransference*. New York: International Universities Press, 1968.

Raush, H. L. Research, practice, and accountability. *American Psychologist*, 1974, *29*, 678–681.

Rosenthal, R. The effect of the experimenter on the results of psychological research. In B. A. Maher (Ed.), *Progress in experimental personality research* (Vol. 1). New York: Academic Press, 1964.

Rosenthal, R. *Experimenter effects in behavioral research*. New York: Appleton-Century-Crofts, 1966.

Singer, J. L. The scientific basis of psychotherapeutic practice: A question of values and ethics. *Psychotherapy: Theory, Research and Practice*, 1980, *17*, 372–383.

Strupp, H. H. Clinical research, practice, and the crisis of confidence. *Journal of Consulting and Clinical Psychology*, 1981, *49*, 216–219.

Weary, G., & Mirels, H. (Eds.). *Integrations of clinical and social psychology*. New York: Oxford University Press, 1982.

Zimbardo, P. The human choice: Individuation, reason, and order versus deindividuation, impulse, and chaos. In W. J. Arnold & D. Levine (Eds.), *Nebraska Symposium on Motivation*. Lincoln: University of Nebraska Press, 1970.

Zimbardo, P., Banks, W. C., Haney, C., & Jaffe, D. A Pirandellian prison. *New York Times Magazine*, April 8, 1973, pp. 38–60.

2

Research on Memory and Clinical Practice

SCOTT DICKMAN AND LEE SECHREST

There is a vast body of psychological literature on basic processes in human behavior that ought clearly to have relevance to clinical practice. Nearly 20 years ago, Goldstein, Heller, and Sechrest (1966) attempted to demonstrate the relevance of some of that literature for psychotherapy practice. The rationale for that work was implicit in its title, *Psychotherapy and the Psychology of Behavior Change*. Those authors started with the assumption that the aim of psychotherapy is to change behavior, with behavior being broadly defined, and that, therefore, research bearing on behavior change ought to be relevant to the practice of psychotherapy. Even if therapists reply that their aim is to change the ways people think or feel, research bearing on how that may be brought about should be relevant to psychotherapy practice.

Their youth and enthusiasm, coupled with what may in retrospect be seen as considerable naïveté, doubtless led Goldstein *et al.* to expect that their persuasive demonstrations of the usefulness of basic research in psychology would be followed by rapid and widespread increases in interest in the basic science of psychology among clinicians. That increase, needless to say, appears not to have occurred, either rapidly or widely. We believe that the relevance of psychology to psychotherapy is as great

SCOTT DICKMAN • Department of Psychology, Oberlin College, Oberlin, Ohio 44074. LEE SECHREST • Department of Psychology, University of Arizona, Tucson, Arizona 85721.

as ever, equaled only by the lack of any evident concern for it. Even a casual inspection of reference lists in articles written by and for clinicians, even research articles, will reveal a striking paucity of references to "basic" research.

The one seemingly notable exception to this estrangement between laboratory and clinic is the research and therapeutic efforts of those working within the theoretical framework of radical behaviorism. Models of learning developed in the laboratory have been applied to the amelioration of a variety of psychological difficulties (e.g., Hersen, Eisler, Smith, & Agras, 1972; Lovaas & Newsom, 1976; Wolpe, 1972). Recently, however, behaviorism has undergone what has been termed a "cognitive revolution" (Gomes-Schwartz, Hadley, & Strupp, 1978). Behaviorally oriented clinicians have moved away from treatment approaches based solely on altering the frequency of observable behaviors. Increasingly, they have come to emphasize changes in such nonobservable psychological functions as the deployment of attention and the acquisition of cognitive coping strategies (Beck, 1976; Kanfer, 1977; Meichenbaum, 1976).

Although this theoretical shift seems likely to be a fruitful one, there is a danger that the always fragile alliance between the experimentalist and the clinician will fall victim to it. Laboratory work on such phenomena as reinforcement schedules may come to seem less pertinent to clinical practice than it once was. Even behaviorally oriented clinicians may not be sufficiently attentive to the wide range of experimental developments relevant to their work.

We do not have the space here even to sketch out the implications for psychotherapy of the many exciting recent and current developments in experimental, physiological, social, and other fields of psychology. We have chosen, instead, to focus on one fairly narrow area of research that we think highly relevant, in the hope that our illustrations might demonstrate this relevance and that our approach might serve as an example of how scientific developments in basic psychology can be examined and extrapolated to clinical practice. Although we shall focus on implications for psychotherapy, other clinical problems such as diagnosis are also potential functions of interest.

The importance to clinical practice generally, and to psychotherapy specifically, of cognitive functions is clear. Even radical behaviorists must rely in their clinical practices on cognitions of their patients—not to say on their own! The salience of cognitive processes appears to be increasing with the growing trend toward cognitive behavioral approaches. Clinicians, however, have shown little interest in laboratory research on such basic cognitive functions as memory and perception. This sort of research has been seen as largely irrelevant to the problems encountered in clinical practice. There exists, nonetheless, a substantial body of experimental work that ought to be directly relevant to cognitively oriented treatment approaches. This is work, within an information-processing paradigm, on such aspects of cognitive functioning as perception and memory. This body of research also has considerable relevance for more traditional face-to-face verbal treatment approaches, including psychodynamically oriented ones. In particular, psychodynamically oriented clinicians have relied heavily on assumptions about such cognitive functions as memory (e.g., in theories about repression).

In this chapter we suggest a variety of hypotheses, derived from the information-processing literature, that seem directly relevant to the clinical work of cognitively, behaviorally, or psychodynamically oriented clinicians. These are simply hypotheses; they need to be tested in the clinic. We are not proposing any immediate changes in clinical practice. The importance of such hypotheses as we shall propose is that they suggest alternate explanations for phenomena of considerable interest to clinicians—explanations that few clinicians would be likely to consider without a familiarity with the information-processing literature. The present approach is similar to that taken by Goldstein *et al.* (1966) in their earlier review of the relationship between laboratory research and clinical practice.

It is not possible to cover here the entire body of information-processing research. For illustrative purposes, then, we shall focus on research on memory, particularly long-term memory, a body of research that has implications for both cognitive and psychodynamic treatment approaches.

Psychodynamically oriented clinicians generally place much emphasis on the recovery of memories that are thought to have been rendered inaccessible because of emotional factors. These include memories of both the distant past (i.e., of childhood) and the recent past (e.g., of emotionally disturbing situations). It is, unfortunately, rare that such clinicians entertain sophisticated explanations for recall failure that do not assume the importance of emotional factors; under such circumstances, "repression" might well appear to be a more common cause of recall failure than it actually is.

Research on memory should have a good deal of relevance for cognitively oriented treatment approaches, as well as for psychodynamic ones. A critical element in most cognitively oriented treatments is to get the patient to recall the external circumstances and thought processes that are associated for him with emotional distress; a patient might, for instance, be asked to recall the situation in which he feels anxious, and the thoughts about those situations that play a role in producing his anxiety. Thus, for cognitively oriented clinicians as for psychodynamically oriented ones, a knowledge of the factors that enhance and impair recall could result in increased clinical effectiveness.

To illustrate the clinical problem of reliance on memory, let us give a few examples of questions that might be asked by therapists of different persuasions, the answers to which depend in considerable degree on memory functions.

Psychodynamic therapist

What were your feelings when your mother told you that your father was not going to live with you anymore?

What is the earliest thing you can remember in your life?

When did you first become aware that your mother was pregnant? So then what happened?

Cognitive behavioral therapist

Were there ever any times in your life when things seemed better?

When you were at the party yesterday, did you feel when you first walked in that you had to make everyone like you?

Was there anything that happened at work that would justify your feelings of incompetence when you left?

What thoughts went through your mind yesterday just before you became anxious?

What did your wife do the last time you criticized her?

What happened the last time you had this panic reaction?

Is there anything in particular that seems to bring on your feelings of futility?

What was going on in the evening before your last episode of impotence?

We shall begin our discussion of memory with a very brief overview of current theories of memory. Then we shall discuss a variety of nonemotional factors that have been demonstrated to affect the ability to recall past events. We shall also discuss ways in which nonemotional factors can produce distortions in the recall of past events. Finally, we shall suggest specific ways in which this research bears upon phenomena encountered by clinicians.

MEMORY

Although the evidence is far from conclusive, memory seems generally to operate as if it consisted of two distinct systems. One of these systems, short-term memory (STM), contains that information that is being processed at any given moment. The other system, long-term memory (LTM), is the repository of the bulk of information being stored in memory; this information is in an inactive state. Information enters STM from two sources. One source is the external world, via the sensory receptors. The other source is LTM; previously stored information can be reactivated, and hence enter STM.

In STM, input from the external world is analyzed. The meaning and physical characteristics of that input are determined. The information that results from this encoding process is stored in LTM. There is considerable controversy about the way information is stored in LTM. However, it is generally agreed that LTM is organized in terms of the meaning rather than physical attributes of information.

The process of recalling information that is stored in LTM is thought to involve two steps. First, various "locations" in LTM are searched in an attempt to find the desired information. Then an item of information that has been retrieved in this way is evaluated to determine whether it is in fact the information being sought. It has been proposed that LTM can be subdivided into separate *episodic* and *semantic* memories. In this view, semantic memory contains information of a general nature ("Afghan hounds have long hair"), whereas episodic memory contains information about specific events ("I saw an Afghan hound yesterday that had long hair"). There is some controversy over whether it is necessary to posit two distinct memory stores of this sort. It is possible that general knowledge (which is acquired through discrete learning episodes) is stored like episodic information.

There is general agreement that information can be represented in LTM in nonverbal as well as verbal forms. Information can be stored in visual or auditory form, as motor programs, and perhaps in tactile and olfactory forms as well.

This brief summary of current thinking about memory is intended only to provide a general background for the discussion that follows. More extensive reviews of the memory literature can be found in Posner (1973), Crowder (1976), and Baddeley (1976).

FAILURE OF RECALL

A variety of factors have been demonstrated to affect the ability to recall past events. Some of these factors affect the initial encoding of input; others affect the retrieval of stored information. These factors are ones that clinicians rarely consider in trying to explain a patient's inability to recall clinically important information. Yet, in some cases at least, a knowledge of these factors may aid the clinician in devising a strategy to enhance the patient's recall. Thus, if a patient recollects being punished by a first-grade teacher and feeling humiliated but cannot remember the misdeed that led to the punishment, clinicians would be inclined to regard the problem as one of motivated forgetting, repression. Yet there are other possibilities, and they have implications both for interpreting the incident and for either facilitating recall or deciding not to pursue the matter further.

Inadequate Encoding

One factor that might result in an inability to recall certain information is a failure initially to encode that information in an adequate way. For instance, encoding seems to be impaired by the absence of a schema, a general organizing principle that allows for the integration of new information with previously acquired information. A study by Dooling and Lachman (1971) illustrates this phenomenon. They presented their subjects with the following paragraph, then asked them to recall as much of it as they could:

> With hocked gems financing him, our hero bravely defied all scornful laughter that tried to prevent his scheme. "Your eyes deceive," he had said, "an egg not a table correctly typifies this unexplored planet." Now three sturdy sisters sought proof, forging along sometimes through calm vastness, yet more often over turbulent peaks and valleys. Days became weeks as many doubters spread fearful rumors about the edge. At last from nowhere welcome winged creatures appeared signifying momentous success. (p. 217)

Some of the subjects were informed in advance that the passage was about Columbus's discovery of America; other subjects were not given this information. The subjects who were given a framework within which to interpret the passage were able to recall considerably more of it. Presumably the schema represented by the "Columbus discovers America" framework enabled subjects reading the passage to encode it effectively so that they could later retrieve it. For example, "an egg not a table" can be encoded and assimilated to the idea that "the world is round not flat."

A later study (Dooling & Mullett, 1973) demonstrated that when an integrative schema is lacking at the time information is presented, the lack may not be remedied by providing such a schema at the time of recall. Dooling and Mullett presented their subjects with the same Columbus passage that Dooling and Lachman (1971) had used. Some of the subjects were not informed about the nature of the passage before it was presented, but were told that it was about Columbus just before they were asked to recall it. These subjects did not show significantly better recall than subjects who were never told about the nature of the passage.

The findings of these two studies have several possible clinical implications. For instance, when a patient is unable to recall the details of a childhood incident, psychodynamically oriented clinicians often attribute this failure to motivational factors (i.e., repression). An alternative explanation that should at least be considered is that the incident the patient is trying to recall made little sense to him as a child because he lacked a schema into which to integrate it. If a child, for example, does not have a schema for "spoiling another person's work," he may have difficulty in remembering later that it was scribbling with crayon on another child's paper that led to punishment. Under these circumstances, it may well be that no amount of probing of the emotional implications of the incident will improve recall. And the Dooling and Mullett (1973) study suggests that a therapist's attempts to provide the patient with an explanatory framework for interpreting the childhood events will not improve recall. As we shall discuss later, such interventions by a therapist may actually cause the patient's recall to become distorted.

Retrieval Failure: Verbal versus Nonverbal Codes

There is another factor that might produce impaired recall for childhood events. As has been mentioned, information can be stored in nonverbal as well as verbal forms. There is evidence that young children do not rely on verbal storage to the same extent as adults (Conrad, 1972). It is therefore possible that an adult might fail to retrieve a childhood memory because he was searching for a verbally encoded memory whereas the actual memory was encoded in some other form. One of us once had a client who could remember certain early childhood experiences only when primed with what she described as a visual image of a "blocked door."

It is not clear what types of codes young children rely on. If, however, a patient in therapy is unable to report the details of a childhood event, it is probably worthwhile for the therapist to try to direct the patient's memory search toward nonverbal memories by means of nonverbal prompts. For instance, the therapist might ask the patient to try to imagine the physical sensation he would have experienced during an event of the sort he is trying to recollect. Then the patient might be encouraged to try to recall instances in his childhood in which he actually experienced those physical sensations. Malpass and Devine (1981) have devised a procedure they term "guided memory" that improves memory for faces. The procedure requires subjects to relax and think about the whole situation, their thoughts, feelings, and so on. Probably many skillful clinicians use similar techniques, but perhaps others do not. Wagstaff (1982) found that a similar "revivification" procedure for memory enhancement improved memory of unhypnotized subjects.

Retrieval Failure: Different Types of Verbal Codes

There are likely to be circumstances in which a disparity between encoding and retrieval strategies produces recall failure even when the initial encoding and the memory search are both verbal in nature. Stimuli have a variety of semantic attributes, all of which can be used to guide the memory search process. A disparity between the semantic attributes emphasized in initial encoding of stimulus information and those attributes emphasized in memory search can produce recall failure. A study by Light and Carter-Sobell (1970) illustrates this phenomenon.

Light and Carter-Sobell presented their subjects with a list of words. Each word in the list was polysemous; that is, it had multiple meanings (e.g., JAM). An adjective was presented along with each word; its purpose was to direct the subject's attention to one of the meanings of the word (e.g., strawberry-JAM; traffic-JAM). Later, subjects were presented with the words in the list, along with words that had not been in the list. Subjects were tested to see how many of the words in the original list they were able to recognize. Sometimes these words were presented alone, sometimes with the same adjective that accompanied them originally, and sometimes with an adjective that was associated with a different meaning of the word than that associated with the original adjective.

The best recognition performance was obtained when the word was presented along with the adjective that had originally accompanied it. When the word was presented with an adjective that biased its interpretation toward a different meaning from that associated with the original adjective, recognition performance fell below the level obtained when the word was presented alone. And this was not simply because of the effects of pairing the word with a new adjective. In a second experiment, Light and Carter-Sobell administered a recognition test that involved pairing the words with new adjectives that biased their interpretation toward the same meaning as did the original adjectives (e.g., strawberry-JAM; raspberry-JAM). Recognition performance here was still better than when the word was presented with a novel adjective that also biased its interpretation toward a new meaning (e.g., traffic-JAM). It appears that the original encoding of each word was affected by the adjective that accompanied it; when the word was assigned a different meaning (i.e., by a new adjective), subjects had difficulty recognizing it.

Before discussing the clinical implications of these findings, it is worth noting that the initial encoding of the meaning of a word can affect its recognition in subtle ways. Barclay, Bransford, Franks, McCannell, and Nitsch (1974) presented their subjects with words that were incorporated into sentences. The sentences were constructed so as to emphasize one particular semantic attribute of the target word (e.g., "The man lifted the *piano*"; "The man tuned the *piano*"). Subjects were later asked to recall the target words; for each target word, a recall cue was provided. It was found that when the recall cue was associated with the same attribute of the target word as had been emphasized in the original sentence, recall was better than when the recall cue emphasized a different attribute of the target word (e.g., "Something heavy"; "Something with a nice sound").

Clinically, the findings of the Light and Carter-Sobell (1970) and Barclay *et al.* (1974) studies raise the possibility that a patient might be unable to recall an important event in his past because the meaning originally assigned to the event is different from the meaning assigned to the same event as an adult; this new meaning presumably guides the patient's attempts at recall, but in an inappropriate way. A therapist might, for instance, be concerned with determining whether a patient had been depressed at times during his childhood. When the therapist asks the patient to try to recall such periods of depression, the patient might be unable to do so. It is possible, in this instance, that the patient was never in fact depressed during his childhood. A psychodynamically oriented clinician might wonder whether the patient had actually been depressed, but had repressed his memories of those feelings. The research just cited suggests a third possibility, one that a clinician might fail to consider. It might be that the patient did not label his periods of childhood depression as depression; as a child, he might have thought of them as periods of fatigue, or times when he never seemed to be hungry. The patient's failure to recall periods of depression might, therefore, be because of the use of an inappropriate retrieval cue. A clinician who was aware of this possibility would not assume that because the adult patient could recognize that fatigue and loss of appetite were signs of depression, he would necessarily be able to recall such experiences by searching his memory for times when he was depressed. Instead, the clinician might explicitly ask his patient to try to recall times when he was tired, or lacking in appetite, or unable to keep well. Such retrieval cues might be closer to the way depression was encoded in childhood and, hence, more effective.

Context-Dependent Learning

Another factor that can cause a person to fail to recall information actually stored in memory is a disparity between the context in which the information was originally acquired and the context in which recall is attempted. Greenspoon and Raynard (1957) had their subjects learn a list of nonsense syllables in one of two laboratory rooms. The two rooms were set up to differ from each other in many ways; for example, one room was quiet and dimly lit, the other noisy and bright. Greenspoon and Raynard found that subjects who had learned the list of nonsense syllables in one room were better able to recall that list when tested there than when tested in the other room.

Godden and Baddeley (1975) confirmed these findings using a different set of contexts. Godden and Baddeley's subjects were divers who learned a list of words either on shore or 10 feet underwater. Recall for the word list was found to be better when it was obtained in the same environment as the original learning.

Although it may seem bizarre to try to move from noisy, bright rooms or from underwater to the clinical setting, we believe that these findings have implications that should at least be considered and possibly tested in a clinical context. One clinical implication of findings of this sort is that when a patient cannot recall an event or a set of events that occurred in a particular locale, recall might be improved if the patient returned to that locale and tried to recall the events there. Thus, a patient who had undergone a distressing experience in a particular setting, but could not remember exactly what it was that had caused the distress, might profit from returning to that same setting to try to recall what had transpired there. Wagstaff (1982) notes that police investigators sometimes take witnesses back to scenes of crimes in order to improve their memories. He also reports experimental evidence that witnesses will, in fact, remember a face more accurately if they look at "mug shots" while at the scene of the crime.

A phenomenon similar to context-dependent learning is state-dependent learning. In some instances, information that is acquired when the individual is in a certain physical state can be better recalled when the individual is in that same physical state. Overton (1972) reviews a body of research bearing on the effects of a state of alcoholic intoxication on recall.

It may be that a patient's failure to recall certain events that occurred when drunk, physically exhausted, or in some other unusual physical state is a manifestation of a state-dependent learning effect. Obviously, in that case, a therapist would rarely be justified in advising the patient to reenter that same physical state just for the sake of recalling some additional information. If the information is important, a clinician might attempt to recreate the relevant state through hypnosis or, in exceptional cases, actually induce the state again. A therapist might very well, however, try to enhance memory by having a patient try imagining himself in the critical state or urge the patient to try to recover critical memories and jot them down should the state reoccur for any reason. Wagstaff's (1982) report on revivification of experience in nonhypnotized persons may be relevant here also.

Decision Criterion

Some evidence suggests that one aspect of the process of recall consists of determining whether the information that has been retrieved from memory actually represents the information that is being sought after (Anderson & Bower, 1972). It is possible that in psychotherapy a patient might report that certain events were rare in his life (e.g., having angry thoughts) because when he tries to recall such events he adopts an especially high criterion for accepting his recollections of them as accurate. Note that this differs from the concept of repression in that the patient is able to retrieve the desired information from memory; he fails to report that information only because he is uncertain of its accuracy. Under such circumstances, the therapist could facilitate recall by urging the patient to adopt a lower decision criterion (e.g., "Just describe what memories come to mind even if you are unsure about their accuracy"). Such a tactic could, of course, result in the patient adopting a decision criterion that was too low and recalling events that did not actually occur. The therapist would therefore have to be especially careful in assessing the likelihood of the recalled events on the basis of other available information about the patient.

The Negative Cuing Effect

There are circumstances in which recall may be impaired if a person is provided with additional, relevant, information. In particular, when a research subject is asked to try to recall as many members of a particular category as he can, providing him in advance with some of the members of that category can actually impair his ability to recall the other category members. Brown (1968), for instance, asked his subjects to try to recall as many of the states of the Union as they could. He provided some of his subjects with a partial list of the states. Brown found that the subjects who had been provided with these recall cues were poorer in their recall of the remaining states than were subjects who had not been given any recall cues. It appears that providing a subject with some members of a category disrupts usual strategies for searching for members of that category; the manner in which this disruption is caused is as yet undetermined.

It is possible, then, that a therapist who wants a patient to recall a certain category of events (e.g., all of the circumstances in which anxiety is aroused) might be better off not reminding the patient of some members of that category of events (e.g., mentioning to the patient certain circumstances that are already known to cause anxiety). The therapist may actually be reducing the likelihood that the patient will recall new anxiety-provoking situations by cues involving known anxiety-provoking situations.

The "Feeling of Knowing" Phenomenon

Another issue concerning recall failure is the significance of reports by individuals that they are sure that they will be able to recall certain information even though they are unable to do so at the moment. Such "blocking" is not uncommon in psychotherapy. It has also been studied in the laboratory, where it is referred to as the *feeling of knowing* or *tip-of-the-tongue* phenomenon (Brown & McNeill, 1966; Freedman & Landauer, 1966; Gruneberg & Monks, 1974; Gruneberg, Monks, & Sykes, 1977; Koriat & Lieblich, 1974).

The experimental literature in this area suggests, first, that this phenomenon can be produced even under circumstances in which the nonrecalled information is unlikely to be emotionally upsetting to the individual. For instance, a feeling of knowing can be produced by providing subjects with the dictionary definitions of somewhat uncommon words (sampan, ambergris) and then asking the subjects for the words being defined (Brown & McNeill, 1966; Koriat & Lieblich, 1974). This tip-of-the-tongue research suggests that a therapist ought to be cautious about inferring that a memory is emotionally significant simply because a patient reports "blocking on it."

The experimental work on the feeling-of-knowing phenomenon also provides data bearing on whether an individual who has such a feeling is actually more likely to have the pertinent information stored in memory than an individual who cannot retrieve the information and reports no feeling of knowing. In this body of experimental work (e.g., Freedman & Landauer, 1966; Gruneberg & Monks, 1974; Gruneberg, Monks, & Sykes, 1977; Koriat & Lieblich, 1974), there are two basic procedures used to demonstrate that subjects actually have stored in memory the information for which they have reported a feeling of knowing. One procedure involves giving subjects a retrieval cue (e.g., the first letter of the word being defined) and seeing whether the subject can then retrieve the information. The other is to present the subject with the correct information along with several alternatives (e.g., to present the word being defined along with several other words), to see if the subject can recognize the correct item of information. A subject who can retrieve the information when given a cue, or who can recognize it when presented with it, is presumed to have had that information stored in memory all along.

These studies have generally found that subjects who report having a feeling of knowing are more likely to have the information stored in memory than are subjects who cannot recall the information and have no such feeling of knowing. But the latter group of subjects, it has been found, do perform better on cued recall or recognition tasks than would be expected by chance. This suggests that at times these subjects too may have the information in memory even though they cannot immediately retrieve it.

Thus, a therapist should probably spend somewhat more time trying to help a patient recall information (e.g., about a past event) when that patient reports a feeling of knowing than when such a feeling is lacking. However, it is worth trying to assist the patient in recalling the information even when a feeling of knowing is absent.

DISTORTIONS IN RECALL

The research just reviewed provides evidence that a variety of factors can interfere with the recall of information that has been stored in memory. There is also information-processing research bearing on the factors that contribute to distortions in recall. By distortions in recall we mean any fault in memory that causes a person to report an event in a way other than in the way it was originally experienced and stored in memory. In practice, however, a clinician may have difficulty distinguishing between distortions in memory, which are by definition unintentional, and intentional misreports of what is remembered. It is worth attending here to Mark Twain, who in his autobiography asserted,

> For many years I believed that I remembered helping my grandfather drink his whisky toddy when I was six weeks old, but I do not tell about that any more now. . . . When I was younger I could remember anything, whether it had happened or not. (Neider, 1959, p. 3)

As with the discussion of recall failure, our discussion of distortion in recall will focus on the nonmotivational factors that can affect memory. Our argument is that psychotherapists ought at least to consider such factors in trying to account for the distortions in recall they encounter in their work. Since, as we suggest, it may be difficult in a specific instance to know whether a reported memory has been distorted, it is important that clinicians have a good understanding and awareness of the problem. Knowledge of when distortions are likely, how they come about, and their characteristics may enable the astute clinician to avoid erroneous assumptions and facilitate therapy.

Distortion of Material So That It Makes More Sense

One source of recall distortion is the process by which new information is integrated with information already in memory (i.e., with preexisting memory schemata). The consequence of such distortion in recall is that the information, as recalled, makes more sense than the original information did; the recalled information conflicts less with other information in memory than was the case with the original information.

Bartlett (1932) found evidence for distortions of this sort in his classic study of memory for prose. Bartlett presented his subjects with an Indian folktale entitled "War of the Ghosts." The subjects were asked to recall the tale a number of times over a period of months. The tale did not make sense in places, at least to someone from a non-Indian culture. Bartlett found that when the tale was recalled, it was frequently distorted in such a way that it made more sense. For instance, in the story, a young man is invited to accompany a group of warriors to attack a neighboring village; the warriors turn out to be ghosts. Some subjects, in recalling the story, described these warriors as ordinary human beings, distorting the story so that it made more sense within their own cultural framework.

Bartlett's findings are not unequivocally interpretable. Gauld and Stephenson (1967) have provided evidence that Bartlett's subjects may not have produced actual distortions in recall. Instead, these subjects may have been unable to recall certain parts of the story; in attempting to provide the experimenter with the material he was requesting, these subjects may have tried to make inferences about the nonrecalled aspects of the story based on their general knowledge. Gauld and Stephenson failed to find a substantial number of recall distortions when they told their subjects to report only those aspects of the original story that they were sure they recalled. These findings have import for clinical practice nonetheless, since they suggest that the instructions that clinicians give in trying to produce memories may influence the results obtained. Instructions of the kind often associated with clinical practice—not to censor thoughts, to say whatever comes to mind—could result in an overly lax threshold for determining what is "remembered."

A study by Coleman (1962) provides evidence for distortions in recall that do not seem to represent inferences made to fill gaps in memory. Coleman began by presenting the following sequence of words to one subject:

> about was good-looking way and treating made of that a him the quiet youngster nice he manners a them girls wild go with

Each of the words was presented on a separate card. After the subject had read the sequence, the cards were shuffled and the subject had to try to arrange them in their original order. Then the subject passed the deck, arranged in this way, to a second subject; this subject had to go through the deck and then, after the cards had been shuffled, arrange them in the order in which he or she had received them. After the original word sequence had been recalled in this fashion by 16 subjects, the word order had become:

> he was a youngster nice quiet with manners good-looking and a way of treating them that made the girls go wild about him

It thus appears that as each subject attempted to recall the sequence of words, he or she tended to distort the sequence so that it made more sense.

Spiro (1977) provides a similar example. He had his subjects read a story involving an engaged couple, Bob and Margie. Bob reveals to Margie that he does not want to have children. Margie, it turns out, very much wants children. A few minutes after they read this story, the subjects were informed either that Bob and Margie had married and were happy, or that they had broken their engagement. When subjects were asked to recall the story, those subjects who had been told that Bob and Margie had gotten married tended to distort the story in such a way that the couple's decision to marry made more sense (e.g., "They underwent counseling to correct the major discrepancy"). Subjects told that Bob and Margie had broken up did not produce such distortion.

It seems likely that recall distortions that render the original material more "sensible" would occur in psychotherapy. This might occur, for instance, when a patient reports experiencing an unpleasant affect each time he thinks of a baby-sitter who often took care of him during his childhood although there is no "rational" reason for the affective response. Even though the patient's affective response is associated primarily with a memory of a particular blue dress worn by the baby-sitter, an inquiry into the response by the therapist might produce a distorted version of the memory, one that makes more sense to the patient, for example, that the baby-sitter was always scowling and was often unkind. Or, another example of this type of recall distortion, a patient may experience anxiety every time he has the thought that success in an important enterprise is imminent. It is possible that in this situation the patient might recall the thought that preceded the feelings of anxiety as having been about failure rather than success; a failure–anxiety relationship would make more sense for most people than would a success–anxiety relationship.

One possible implication for treatment is that accurate recall of the thought that preceded distress in such circumstances might be possible only after the patient had been provided with a conceptual framework that made the actual sequence of events seem sensible. Thus, the therapist, in the present example, might have to discuss with the patient the reasons why some people fear success before the patient would be able to recall the success-related thoughts that preceded distress.

Inferential Elaboration

Another potential source of distortion in recall is a process of elaboration that appears to occur during the initial encoding of the material. The elaboration of the original material involves inferences about the objects or events being described. These inferences are based on the person's knowledge of the world, and it is possible that they occur without conscious effort on the person's part. This process of elaboration is illustrated by the following story (Reder, cited in Anderson & Reder, 1979).

> Alice went to Jimmy's house for lunch. Jimmy's mom served them tunafish sandwiches. Alice liked her sandwich very much and had almost finished it when, all of a sudden, her dentures fell out of her mouth.

The reader's surprise at finding that Alice wears dentures is likely to spring from an inference made about her age. Because a male with a boy's name invited Alice in for lunch, and because the male's "mom" served the lunch, the reader is likely to infer that Alice is a child. There are several studies that suggest that when people make these sorts of inferences, they sometimes mistakenly recall that the inferred material was actually presented.

Bransford, Barclay, and Franks (1972) presented their subjects with a number of sentences and afterward tested subjects' recognition of these sentences. Bransford *et al.* were interested in whether subjects tended to mistakenly recognize sentences resembling the original ones in certain ways. For instance, one of the sentences originally presented to some subjects was "Three turtles rested on a floating log, and a fish swam beneath them." Subjects presented with this sentence tended to mistakenly identify the sentence "Three turtles rested on a floating log, and a fish swam beneath it" as having been presented. This sentence was not falsely recognized by subjects who had originally been presented with the sentence "Three turtles rested beside a floating log, and a fish swam beneath them." Thus, it appears that the subjects who made the false recognition had inferred that since the turtles were on a log, a fish swimming under the turtles must have swum under the log as well. Later, these subjects could not distinguish their own elaboration from the material originally presented.

Sulin and Dooling (1974) provide another example of the way recall can be distorted through elaborations on the original material. These investigators presented their subjects with passages about famous persons and then tested for recognition of the material in the passages. For some subjects, the actual name of the famous person was used in the passage; for the other subjects, this name was replaced by a fictitious name. The text of one of these passages was:

> Helen Keller's need for professional help. Helen Keller was a problem child from birth. She was wild, stubborn, and violent. By the time Helen turned eight, she was still unmanageable. Her parents were very concerned about her mental health. There was no good institution for her problem in her state. Her parents finally decided to take some action. They hired a private teacher for Helen.

For some of the subjects, the recognition test included the sentence "She was deaf, dumb, and blind"; this was, of course, true of Helen Keller. It was found that subjects who had been presented with this passage showed a much greater tendency to identify this sentence as having been included in the passage than was the case for subjects presented with the identical passage but with the name "Carol Harris" replacing Helen Keller. Thus, it appears that subjects' recall of the passage was distorted by relevant information that they had acquired from other sources besides the passage itself.

In a follow-up study, Dooling and Christiaansen (1977) gave subjects the Carol Harris passage but told them, just before the recall test, that the passage had actually been about Helen Keller. Dooling and Christiaansen found that subjects tested a week after the passage had been presented tended to falsely recognize the "deaf, dumb, and blind" sentence. This was not true for subjects tested 2 days after the presentation of the passages. This finding suggests that if sufficient time has elapsed between the presentation of information and the attempt to recall it, additional information acquired at the time of recall can result in distorted recall of the original information.

The studies reviewed in this section provide evidence that recall can be influenced by information stored in memory but not contained in the to-be-recalled material itself. It seems likely that this type of recall distortion occurs in psychotherapy. For instance, patients might read about other people who have problems that seem similar to their own. Under these circumstances, it is possible that the patients' recall of their lives will be influenced by this new information. A patient who reads in a newspaper or book about the "alcoholic mother syndrome" might "remember" traits his mother possessed that are consistent with the stereotype he has now acquired. It would therefore be important for the therapist of patients who have been reading about their own psychological difficulties to be particularly careful in assessing the accuracy of memories that neatly dovetail with the scheme the patients have acquired through reading. Therapists themselves may often give clues to the memories they expect based on their own schemata. When the expectations of therapists match those acquired by patients, distortions in memory may be particularly likely.

It seems likely that therapists as well as patients will be susceptible to this kind of recall distortion. For example, if a patient has recently made a suicide attempt, the therapist's awareness of this attempt could distort his recall of the therapy hours following the attempt; the therapist might mistakenly recall material of the sort that would be expected from someone who had recently attempted suicide ("I feel hopeless"). Similarly, a therapist who learns that a patient has just committed suicide may now recall statements that the patient made in recent therapy hours that were indications that he was about to do so. If these "memories" are distortions based on the therapist's preconceived ideas about the way a suicidal patient acts, then they may serve to confirm in the therapist's mind beliefs that are actually false. If therapists were aware of the circumstances in which this sort of recall distortion could occur, they might try to minimize such distortions by assessing the accuracy of their memories with especial care (e.g., by actively trying to recall activities of the patient that seem inconsistent with the ones the therapist recalls most easily).

Biasing of Recall through Verbal Labeling

The verbal label that persons apply to a stimulus can also have a distorting effect on their later recollection of that stimulus. Two early studies demonstrated this by testing students' recall of ambiguous visual forms (Carmichael, Hogan, & Walter, 1932; Hanawalt & Demarest, 1939). Carmichael *et al.* presented their subjects with a set of ambiguous line drawings. As each drawing was presented, the experimenters supplied a verbal label for it. An example of the drawings used in this study is shown in Figure 1A. This particular drawing might be labeled by the experimenter as either "eyeglasses" or "dumbbells." It was found that when subjects had to reproduce the drawing from memory, their reproductions were distorted so that they looked more like the object referred to by the label (see Figures 1B and 1C).

Hanawalt and Demarest (1939) presented their subjects with the same set of ambiguous figures that Carmichael *et al.* (1932) used. However, in this study a verbal label was not provided at the time the figures were presented. Instead, a label was provided at the time of recall (e.g., "Draw the figure that resembled eyeglasses"). These labels biased subjects' reproduction of the figures just as the labels in the Carmichael *et al.* study did.

A more recent study by Loftus and Palmer (1974) provides another illustration of the effects of labeling on recall. Loftus and Palmer showed their subjects films of automobile accidents. After each film the subjects were asked questions about the accident portrayed. For some subjects, included among these questions was "About how fast were the cars going when they hit each other?" Other subjects were asked "About

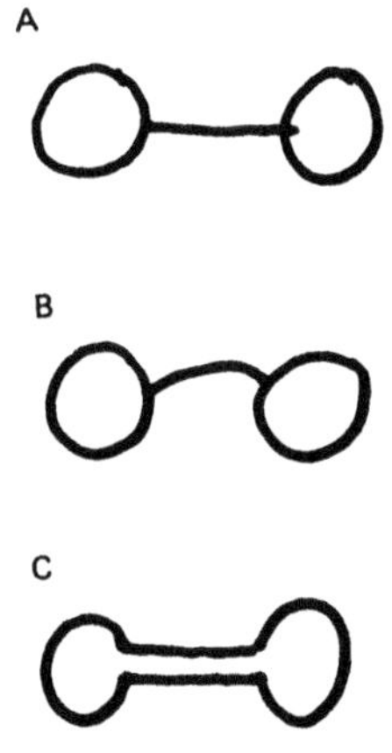

FIGURE 1. An ambiguous visual form and two distorted reproductions; (A) original ambiguous form; (B) example of reproduction of original form when it has been labelled "eyeglasses"; and (C) example of reproduction of original form when it has been labelled "dumbbell." Adapted from *Journal of Experimental Psychology*, 1932, *15*, 73–86.

how fast were the cars going when they smashed into each other?" The latter group of subjects gave significantly higher estimates of the cars' speeds than did the former group.

In clinical work, there would seem to be numerous opportunities for verbal labeling to produce distortions in recall. Here, the therapist is the potential source of the labels. The therapist could provide such labels at the time the stimulus originally appeared. A therapist might, for instance, label a set of feelings reported by a client as "hostile," with the consequence that in later discussion of those feelings the client would remember them as hostile even though that was not an entirely appropriate summation of the affect involved.

Alternatively, the therapist could provide a distorting verbal label long after the patient actually encountered the stimulus. The question "How enraged does your spouse become when your children misbehave?" might elicit a description of more intense anger than would the question "How irritated does your spouse become when your children misbehave?"

Therapists who are aware of the possible distorting effects of verbal labels should be better able to assess the veridicality of their patients' responses. Such therapists might, for example, ask the same question more than once during the course of treatment, varying the critical word or phrase (angry/irritated) to see whether the label affects the patient's responses to any significant degree. If the patient's responses are affected, it may be necessary to question the patient in more detail about the events of interest, in order to be more certain of their nature ("What exactly does your spouse do when your children misbehave?").

Biasing of Recall through Misleading Recall Cues

In another type of memory distortion, the individual "remembers" an object that was not actually present in the scene being recalled. Loftus has examined how the wording of the questions used to elicit recall can produce such distortions (Loftus, 1975; Loftus, Miller, & Burns, 1978).

In one of Loftus's studies (Loftus *et al.*, 1978), subjects were shown a series of 30 slides depicting the course of an automobile–pedestrian accident. The slides showed a red Datsun traveling along a street, then turning right at an intersection. The automobile then hit a pedestrian who was crossing at the crosswalk. There was a sign on the corner where the automobile turned right. This sign was visible on only one of the slides. Half the subjects saw a slide showing a stop sign; the other half saw a slide showing a yield sign. After viewing all 30 slides, the subjects filled out a questionnaire. For half the subjects, this questionnaire included the question "Did another car pass the red Datsun while it was stopped at the stop sign?" For the other half of the subjects, the same question was asked but with "yield sign" replacing "stop sign." The experiment was designed so that for half the subjects the question included the sign they had actually seen, and for the other half the sign in the question was not the one actually presented.

Twenty minutes after filling out the questionnaire, subjects were presented with pairs of slides and asked to choose which slide of each pair had previously been presented. The critical pair of slides was the one showing the red Datsun next to either a stop sign or a yield sign. The subjects who had previously been exposed to a misleading question (i.e., one that referred to the wrong type of sign) were more likely than the other subjects to identify the wrong slide as the one they saw earlier.

It is possible that the group differences found here were not because of the effects of the misleading question. Rather, it could be that the question referring to the correct sign facilitated the performance of subjects who received it. To rule out this possibility, Loftus *et al.* (1978, Experiment 3) repeated their earlier experiment, adding a control condition in which the critical question did not contain a reference to either sign. It was found that the misleading question increased the frequency of false recognitions above that of the control group.

Another possible explanation for the Loftus *et al.* finding is that the subjects given the misleading question did not actually recall seeing the wrong sign. Instead, when these subjects received the misleading question, they might have recognized that it referred to the wrong sign, but decided to "go along" with the experimenter; in this view, the false recognitions of the sign the subjects made were owing to the "demand characteristics" of the experiment.

Loftus *et al.* (1978, Experiment 2) attempted to rule out this possible explanation for their findings. They repeated their original experiment but this time included a "debriefing questionnaire" at the very end of the experiment. This questionnaire informed subjects about the critical question and asked them whether they had actually seen a different sign than the one referred to in the critical question. Of the subjects who received the misleading question *and* identified the wrong slide, only 12% reported that they had actually seen a sign different from the one referred to in the question. It is possible, of course, that the subjects in filling out the debriefing questionnaire were still trying to comply with the experimenter's wishes by denying what they had actually seen. This does not seem especially likely, however. Thus, it is probable that the misleading question did in fact cause a distortion in the subjects' recollection of the critical slide.

A study by Loftus (1975) provides another example of the sort of recall distortions that can be produced by including misleading information in the questions used to elicit recall. Subjects were shown a videotape of an automobile accident. The subjects then answered questions about the accident. For half the subjects, one of the questions was: "How fast was the white sports car going when it passed the barn while traveling along the country road?" For the other subjects, the analogous question was "How fast was the white sports car going while traveling along the country road?" In actuality, no barn appeared in the videotape.

A week after they saw the videotape, the subjects returned and answered a new set of questions about the accident. Among these was "Did you see a barn?" Subjects previously exposed to the question containing the misleading reference to a barn were significantly more likely to mistakenly recall seeing it.

Recall distortions of the sort Loftus has studied could plausibly occur in the clinic as well as in the laboratory. A therapist might come to believe that a particular patient becomes anxious (or depressed or angry) only in the presence of certain people or things. The therapist's questions about the patient's feelings might then assume that these people or things are present whenever the patient experiences distress. Such questions could potentially lead the patient to mistakenly recall the presence of these people or things even at times when they were not in fact present. This would then obscure the fact that there were other sources of the patient's feelings besides the ones that had already been determined. For example, at a certain point in treatment a therapist might reach the conclusion that a boy suffered anxiety attacks when he perceived that his parents were about to quarrel (e.g., because the boy feared the breakup of the marriage). Once the therapist had come to this conclusion, the therapist's questions to the parents might incorporate the assumption that both were present whenever the boy had an anxiety attack (e.g., "What did the two of you do when the anxiety attack began?"). Questions of this sort could cause the parents to "recall" that both had been present at the time of a particular anxiety attack, when this was not in fact the case. The parents might then fail to recall that anxiety attacks sometimes occurred when the boy was alone with one of the parents. It might be that the boy experiences anxiety in these single-parent situations for different reasons than in the two-parent situation; the therapist-induced recall distortions could serve to obscure this fact.

If the therapeutic situation does allow for recall distortions of this type, it would be very important for therapists to consider carefully whether the questions they ask a patient incorporate unwarranted assumptions about the presence of people or objects. Otherwise, therapists might be in danger of generating the very memories that they then take as the confirmation of their favored hypotheses.

Distortions in Recall of Frequency Information

The distortions in recall discussed up to this point have all involved the retrieval from memory of information presented at one particular point in time. There are also situations in which it is necessary to recall the frequency with which a certain type of event occurred in the past, when it is impossible to recall every single occurrence of that event. There is evidence that distortions can occur in the recall of such frequency information.

Tversky and Kahneman (1973) have proposed that distortions in frequency judgments are a function of the strategies we typically employ in making these judgments:

> Lifelong experience has taught us that instances of large classes are recalled better and faster than instances of less frequent classes; that likely occurrences are easier to imagine than unlikely ones, and that associative connections are strengthened when two events frequently co-occur. Thus, a person could estimate the numerosity of a class, the likelihood of an event, or the frequency of co-occurrences by assessing the ease with which the relevant mental operations of retrieval, construction, or association can be carried out. (p. 208)

Tversky and Kahneman report a number of studies that provide support for their views. In one of these studies (1973, Experiment 3) they asked subjects to estimate the relative frequencies of certain classes of words. For example, subjects were asked whether there are more words in the English language that begin with the letter K than there are that have K as their third letter. The great majority of subjects indicated that words beginning with K are more frequent than words with K as their third letter. In fact, the opposite is true. Tversky and Kahneman explain this and similar findings by proposing that subjects estimate the frequency of a particular class of words by attempting to gauge how easy it is to retrieve the members of that class, that is, how "available" the class members are. The easier it is to retrieve the words in a particular class, the more common the subjects think the class is. Since it is easier to search through memory for words that have a certain letter in the first than the third position, the former class of words tends to be seen as more common than the later, even when this is not the case.

In another study, Tversky and Kahneman (1973, Experiment 8) presented each subject with a list of names of well-known persons of both sexes; some names (Richard Nixon, Elizabeth Taylor) were better known than others (William Fulbright, Lana Turner). Each list contained 39 names of individuals, 19 of one sex and 20 of the other. The lists were constructed so that when a list had a majority of names from one sex, the names from the opposite sex were more famous. Subjects were asked to try to recall whether the names in the list with which they were presented were mostly of men or of women. The investigators hypothesized that since the more famous names would be easier to recall, subjects would tend to believe that these names had occurred most frequently. Thus, the sex associated with the most famous names ought to be judged to be the most frequent, even though this was not true for any list. This was found to be the case.

Tversky and Kahneman point out that the *illusory correlation* phenomenon studied by Chapman and Chapman (Chapman, 1967; Chapman & Chapman, 1967, 1969, 1975) represents another instance of a bias in frequency judgments. Chapman and Chapman use this term to describe their finding that subjects make systematic errors in judging how frequently stimuli have appeared together. These judgments tend to be biased in such a way that stimuli that are strongly associated with each other are believed to have co-occurred more frequently than they actually have. In Tversky and Kahneman's (1973) view, such distortions in recall occur because subjects do not make their judgments of the frequency of co-occurrence of a pair of stimuli by trying to retrieve individual instances of the occurrence of each member of the pair; instead, subjects base their judgment of the frequence of co-occurrence of the stimuli on the basis of the strength of the associative bond between the stimuli.

In the earliest investigations of the illusory correlation phenomenon Chapman (1967) presented his subjects with a series of word pairs. There were 12 different word pairs in each series, with each pair repeated a number of times. Each word pair occurred equally often. The 12 word pairs were constructed by taking one set of 4 words and one set of 3 words, and pairing every word in the first set with every word in the second set. The words were chosen so that of the 12 word pairs constructed in this way, 2 of the pairs consisted of words that were strongly associated (e.g., lion–tiger) whereas the other 10 pairs consisted of words that were weakly associated (clock–butter). The subjects' task was to observe the words as they were presented so that afterward they could report how often each word had been paired with every other word.

Chapman found that subjects tended to report that the 2 pairs of strongly associated words had occurred more often than the other 10 pairs, even though all pairs had occurred with equal frequency. This is a finding, then, of an illusory correlation.

Chapman and Chapman (1967, 1969) demonstrated in two studies that such illusory correlations could distort the frequency judgments of psychodiagnosticians. In one study (Chapman & Chapman, 1969), clinicians were asked what types of Rorschach responses were, in their experience, the most frequent correlates of male homosexuality. The Rorschach "signs" that the clinicians reported to be associated with male homosexuality turned out to be ones that in empirical investigation had not turned out to be correlated with it. At the same time, the clinicians failed to report the Rorschach signs that in fact had been empirically determined to be correlates of homosexuality.

Chapman and Chapman (1969) then sought to determine whether associative strength could account for the illusory correlations that the clinicians reported. They had college undergraduates rate each Rorschach sign reported by the clinicians, along with the two empirically established signs, for their associations with male homosexuality. It was found that the invalid signs did in fact have a much higher associative relationship with male homosexuality than did the overlooked valid signs (one of the invalid signs was "feminine clothing"; one of the valid signs was "a response on Rorschach Card V of a human or humanized animal").

Chapman and Chapman (1969) then tried to determine whether naïve observers (college undergraduates) would "discover" the same invalid correlations between certain Rorschach signs and male homosexuality that experienced clinicians had reported. They prepared a set of 30 Rorschach cards. Each card contained one Rorschach response; the area of the inkblot used in the response was circled, and the accompanying verbalization was typed on the card. In addition, on a corner of the card were typed two statements about the emotional problems of the patient who purportedly made the Rorschach response. The Rorschach responses used in this part of the experiment consisted of the invalid signs of male homosexuality reported by the clinicians, the two empirically validated signs, and two "filler" categories of responses (geographic features and food). There were four possible statements about the patient's emotional problems. The statement concerning homosexuality was "He has sexual feelings toward other men." When the statement about homosexuality was paired equally often with each of the categories of Rorschach responses (valid, invalid, and filler), subjects discovered the same illusory correlations that the clinicians had; that is, the subjects reported that the invalid signs had more frequently been paired with the homosexuality statement than the valid signs had been.

In another experimental condition, Chapman and Chapman (1969) constructed the stimulus materials so that the valid Rorschach signs actually appeared more often with the statement about homosexuality than did the invalid signs. Even under these circumstances subjects reported that the invalid signs had been more highly correlated with homosexuality than the valid signs had been; such illusory correlations therefore appear to be difficult to abolish.

As noted earlier, Tversky and Kahneman (1973) attempt to provide a theoretical explanation of the illusory correlation findings by hypothesizing that people judge the frequency of co-occurrence of a particular pair of stimuli by determining the strength of the association between the stimuli. Such a strategy makes more sense because the more frequently two stimuli have co-occurred, the stronger their associative bond is likely to be. The problem with this strategy, however, is that other factors besides frequency of co-occurrence can increase the strength of the association between stimuli. From this perspective, in Chapman's work it is the semantic relationship between certain stimuli that, by increasing their associative strength, causes judgments of their frequency of co-occurrence to be too high.

The Chapman and Chapman (1969) study demonstrates quite clearly the clinical relevance of experimental research on distortions in the recall of frequency information. It is pertinent here that Chapman and Chapman (1969) were concerned with associations that existed for both clinicians and nonclinicians. It seems likely that in clinicians of a particular theoretical orientation certain associations will be strengthened because of their theoretical importance. For instance, a psychoanalytically oriented clinician who, in the course in training, had given a great deal of thought to the causal relationship between difficulties in bowel training and obsessive personality traits would undoubtedly have established a strong associative connection between these phenomena. The clinician would then be vulnerable to overestimating the frequency with which they have co-occurred in patients. This, in turn, could cause the clinician erroneously to conclude that clinical experience had provided confirmation of the theory that had been taught.

Similarly, a cognitive therapist who had been taught that experiences of helplessness were associated with depression would likely begin clinical work with a strongly established association between learned helplessness and depression. This could lead to an overestimation of the frequency of co-occurrence of these particular phenomena and the mistaken conclusion that clinical experience had confirmed an *a priori* theory.

In the same way, a therapist who was taught that schizophrenia results from certain types of pathological parental relationships would be likely to have strongly internalized this association. As in the previous examples, such a strong *a priori* association could lead to the mistaken belief that clinical experience had confirmed theory; again, the clinician would be likely to overestimate the frequency of co-occurrence of the phenomena that, in theory, ought to co-occur.

Tversky and Kahneman's (1973) work on the relationship between "availability" and frequency estimation also has implications for the clinician. It is likely that a clinician who has a particular theoretical orientation will pay special attention to phenomena that the theory suggests are important. As a result, these phenomena will likely be easier to recall later, that is, will be especially "available" in Tversky and Kahneman's (1973) terminology. And, in turn, this ease of recall may lead the clinician to overestimate the actual frequency of occurrence of the phenomena. So, for instance, clinicians who place particular importance on the occurrence of sexual feelings in a child for the parent of the opposite sex will be likely to attend very carefully to reports of such feelings by their patients. As a consequence, such reports will be especially easy to retrieve from memory and their frequency may be overestimated.

In the same way, a clinician who has been taught that all dying patients go through certain stages of psychological adjustment would be likely to pay particular attention when a patient reports feelings thought to be characteristic of one of these stages. This additional attention would be likely to facilitate the later recall of such reports. As a consequence, their frequency might be overestimated.

A clinician of any theoretical orientation may be in danger of overestimating the frequency of occurrence of phenomena consistent with the theory acquired during training. Possibly, a clinician could reduce the likelihood of such overestimations by paying particular attention to phenomena that are inconsistent with the theoretical beliefs the clinician holds. The clinician might be careful to note instances in which one phenomenon occurs in the absence of another phenomenon which ought, in theory, to accompany it. And the clinician might make a special effort to remember instances in which a patient never reported the occurrence of a phenomenon that, in theory, ought to have occurred. It may be that the kind of distortion of frequency information discussed here accounts for the bewilderment clinicians sometimes feel when they discover that others have observed only infrequently phenomena that they themselves see "practically every day" in the consulting room.

CONCLUSION

The research we have reviewed provides evidence that both failures in recall and distortions in recall can be produced by other factors besides the emotional significance of the memories involved. We are not proposing that the findings of this body of research are automatically generalizable to the context of psychotherapy. However, we do argue that the implications of this work for clinical practice are sufficiently important that clinicians ought to be aware of it and ought at least to consider its implications when they are making inferences about the causes of the recall difficulties that their own patients exhibit.

We believe that this research, along with our own, makes a strong case for the relevance of empirical research on basic psychological processes for clinical practice. No amount of empirical work will render obsolete clinical experience and judgment when it comes to making the innumerable decisions that a therapist must make during even a single therapy hour. However, the most effective clinician is likely to be the one who exercises this clinical judgment within the context of a firm grasp of the relevant research findings. The gap that so often exists between the researcher in the laboratory and the clinician out on the front lines can and ought to be bridged.

Obviously, clinicians cannot keep abreast of the staggering quantity and variety of basic psychological research that may be relevant to clinical practice. Nor can they always make the connections between research and practice. Rather, we think that research clinicians able to bridge the gap must continue to be trained and supported. They need outlets for their work in books and journals. But they also need an audience of practicing clinicians trained to pay attention to evidence, to care about the scientific data base on which they are persuaded practice must rest.

REFERENCES

Anderson, J. R., & Bower, G. H. Recognition and retrieval processes in free recall. *Psychological Review*, 1972, *79*, 97–123.

Anderson, J. R., & Reder, L. M. An elaborative processing explanation of depth of processing. In L. S. Cermak & F. I. M. Craik (Eds.), *Levels of processing in human memory*. Hillsdale, N.J.: Lawrence Erlbaum, 1979.

Baddeley, A. D. *The psychology of memory*. New York: Basic Books, 1976.

Barclay, J. R., Bransford, J. D., Franks, J. J., McCannell, N. S., & Nitsch, K. Comprehension and semantic flexibility. *Journal of Verbal Learning and Verbal Behavior*, 1974, *13*, 471–481.

Bartlett, F. C. *Remembering*. Cambridge: Cambridge University Press, 1932.

Beck, A. T. *Cognitive therapy and the emotional disorders*. New York: International Universities Press, 1976.

Bransford, J. D., Barclay, J. R., & Franks, J. J. Sentence memory: A constructive versus interpretive approach. *Cognitive Psychology*, 1972, *3*, 193–209.

Brown, J. Reciprocal facilitation and impairment of free recall. *Psychonomic Science*, 1968, *10*, 41–42.

Brown, R. W., & McNeill, D. The "tip-of-the-tongue" phenomenon. *Journal of Verbal Learning and Verbal Behavior*, 1966, *5*, 325–337.

Carmichael, L., Hogan, H. P., & Walter, A. A. An experimental study of the effect of language on the reproduction of visually perceived form. *Journal of Experimental Psychology*, 1932, *15*, 73–86.

Chapman, L. J. Illusory correlation in observational report. *Journal of Verbal Learning and Verbal Behavior*, 1967, *6*, 151–155.

Chapman, L. J., & Chapman, J. P. Genesis of popular but erroneous psychodiagnostic observations. *Journal of Abnormal Psychology*, 1967, *72*, 193–204.

Chapman, L. J., & Chapman, J. P. Illusory correlation as an obstacle to the use of valid psychodiagnostic signs. *Journal of Abnormal Psychology*, 1969, *74*, 271–280.

Chapman, L. J., & Chapman, J. P. The basis of illusory correlation. *Journal of Abnormal Psychology*, 1975, *84*, 574–575.

Coleman, E. B. Sequential interference demonstrated by serial reconstructions. *Journal of Experimental Psychology*, 1962, *64*, 46–51.

Conrad, R. The developmental role of vocalizing in short-term memory. *Journal of Verbal Learning and Verbal Behavior*, 1972, *11*, 521–533.

Crowder, R. G. *Principles of learning and memory*. Hillsdale, N.J.: Lawrence Erlbaum, 1976.

Dooling, D. J., & Christiaansen, R. E. Episodic and semantic aspects of memory for prose. *Journal of Experimental Psychology*, 1977, *3*, 428–436.

Dooling, D. J., & Lachman, R. Effects of comprehension on retention of prose. *Journal of Experimental Psychology*, 1971, *83*, 216–222.

Dooling, D. J., & Mullet, R. L. Locus of thematic effects in retention of prose. *Journal of Experimental Psychology*, 1973, *97*, 404–406.

Freedman, J. L., & Landauer, T. K. Retrieval of long-term memory: "Tip-of-the-tongue" phenomenon. *Psychonomic Science*, 1966, *4*, 309–310.

Gauld, A., & Stephenson, G. M. Some experiments relating to Bartlett's theory of remembering. *British Journal of Psychology*, 1967, *58*, 39–49.

Godden, D. R., & Baddeley, A. D. Context-dependent memory in two natural environments: On land and under water. *British Journal of Psychology*, 1975, *66*, 325–332.

Goldstein, A. P., Heller, K., & Sechrest, L. *Psychotherapy and the psychology of behavior change*. New York: Wiley, 1966.

Gomes-Schwartz, B., Hadley, S. W., & Strupp, H. H. Individual psychotherapy and behavior therapy. *Annual Review of Psychology*, 1978, *29*, 435–471.

Greenspoon, J., & Raynard, R. Stimulus conditions and retroactive inhibition. *Journal of Experimental Psychology*, 1957, *53*, 55–59.

Gruneberg, M. M., & Monks, J. Feeling of knowing and cued recall. *Acta Psychologica*, 1974, *38*, 257–265.

Gruneberg, M. M., Monks, J., & Sykes, R. N. Some methodological problems with feeling of knowing studies. *Acta Psychologica*, 1977, *41*, 365–371.

Hanawalt, N. G., & Demarest, I. H. The effect of verbal suggestion in the recall period upon the reproduction of visually perceived forms. *Journal of Experimental Psychology*, 1939, *25*, 159–174.

Hersen, M., Eisler, R. M., Smith, B., & Agras, W. A token reinforcement ward for young psychiatric patients. *American Journal of Psychiatry*, 1972, *129*, 228–233.

Kanfer, F. H. The many faces of self-control, or behavior modification changes its focus. In R. B. Stuart (Ed.), *Behavioral self-management: Strategies, techniques, and outcomes.* New York: Brunner/Mazel, 1977.

Koriat, A., & Lieblich, I. What does a person in a "TOT" state know that a person in a "don't know" state doesn't know? *Memory and Cognition*, 1974, *2*, 647–655.

Light, L. L., & Carter-Sobell, L. Effects of changed semantic context on recognition memory. *Journal of Verbal Learning and Verbal Behavior*, 1970, *9*, 1–11.

Loftus, E. F. Leading questions and eyewitness report. *Cognitive Psychology*, 1975, *7*, 560–572.

Loftus, E. F., & Palmer, J. C. Reconstruction of automobile destruction: An example of interaction between language and memory. *Journal of Verbal Learning and Verbal Behavior*, 1974, *13*, 585–589.

Loftus, E. F., Miller, D. G., & Burns, H. J., Sematic integration of verbal information into a visual memory. *Journal of Experimental Psychology*, 1978, *4*, 19–31.

Lovaas, O. I., & Newsom C. D. Behavior modification with psychotic children. In H. Leitenberg (Ed.), *Handbook of behavior modification and behavior therapy*. Englewood Cliffs, N.J.: Prentice-Hall, 1976.

Malpass, R. S., & Devine, P. G. Guided memory in eyewitness identification. *Journal of Applied Psychology*, 1981, *66*, 343–350.

Meichenbaum, D. Toward a cognitive theory of self-control. In G. E. Schwartz & D. Shapiro (Eds.), *Consciousness and self-regulation*. New York: Plenum Press, 1976.

Neider, C. (Ed.), *The autobiography of Mark Twain*. New York: Harper Bros., 1959.

Overton, D. A. State-dependent learning produced by alcohol and its relevance to alcoholism. In B. Kissin & H. Begleiter (Eds.), *The biology of alcoholism* (Vol. 2). New York: Plenum Press, 1972.

Posner, M. I. *Cognition: An Introduction*. Glenview, Ill.: Scott, Foresman, 1973.

Spiro, R. J. Remembering information from text: The "state of schema" approach. In R. C. Anderson, R. J. Spiro, & W. E. Montague (Eds.), *Schooling and the acquisition of knowledge*. Hillsdale, N.J.: Lawrence Erlbaum, 1977.

Sulin, R. A., & Dooling, D. J. Intrusion of a thematic idea in retention of prose. *Journal of Experimental Psychology*, 1974, *103*, 255–262.

Tversky, A., & Kahneman, D. Availability: A heuristic for judging frequency and probability. *Cognitive Psychology*, 1973, *5*, 207–232.

Wagstaff, G. F. Helping a witness remember—a project in forensic psychology. *Police Research Bulletin*, 1982, Spring (38), 56–58.

Wolpe, J. *The practice of behavior therapy*. New York: Pergamon Press, 1972.

3

Implications of Nonclinical Research for Mental Health Policy

CHARLES A. KIESLER

INTRODUCTION

This chapter presents a sampling of nonclinical research that is germane to national mental health policy. Consideration of this research demands taking a somewhat different perspective than the practicing clinician is accustomed to. I shall first define mental health policy and consider how research can be used in the development of national policy alternatives. I shall then consider how nonclinical research is developed and applied to these themes.

I define national mental health policy as

> the *de facto* or *de jure* aggregate of laws, practices, social structures, or actions occurring within our society, the intent of which is to improve the mental health of individuals or groups. The study of such policy includes the descriptive parameters of the aggregate, the comparative assessment of particular techniques, the evaluation of the system and its subparts, human resources available and needed, cost-benefit analyses of practices or actions, and the cause and effect relationship of one set of policies, such as mental health, to others, such as welfare and health, as well as the study of institutions or groups seeking to affect such policy. (Kiesler, 1980, p. 1066)

Work on this chapter was completed in May 1982.

CHARLES A. KIESLER • College of Humanities and Social Sciences, Carnegie-Mellon University, Pittsburgh, Pennsylvania 15213.

There are several things worth noting about this definition. First, national mental health policy is not a carefully planned and well-organized national plan. It is really an *aggregate* of laws and practices that are often not well articulated or related to one another. Indeed, the intents of various legislative practices often conflict. To consider national mental health policy, one must ask what is done in the name of mental health, by whom, with what effect, and at what cost. Some of the outcomes for mental health are unintended. For example, it is clear that our national *de jure* policy of deinstitutionalization and outpatient care is being undercut by a national *de facto* policy of inpatient care largely funded through Medicare, Medicaid, and private insurance companies.

The second thing to note about the policy definition is that I am explicitly advocating a top-down approach to policy analysis and research, rather than the more typical bottom-up approach that this topic has seen in the past (see Kiesler, 1982a). Loosely phrased, a bottom-up approach to mental health policy is to ask what we intend to do in the name of mental health, and review how that is progressing. In a top-down approach one asks what is done in the name of mental health, with what effect, and what future policy alternatives appear open to us. I shall give one example of the differences in outcome between these approaches.

The core aspect of our national *de jure* policy in mental health is deinstitutionalization and the development of outpatient care. Looked at from one perspective, these policies have been very successful. The National Institute of Mental Health (1976) reported a twelvefold increase in outpatient episodes over a 20-year period. Professional underpinnings for this effort appear to be very solid. The number of psychiatrists has increased dramatically over the last 30 years, and the number of practicing professional psychologists has increased as well. A variety of steps in improving professional care has developed concordantly. Professional service review organizations (PSRO) exist around the country; psychiatric diagnostic service manuals have progressed through several editions; the number of insurance companies insuring outpatient care for mental health services has increased rapidly; all 50 states have at some point adopted licensing or certification laws for psychologists; steady progress can be documented in the education of practicing psychiatrists, and so forth.

Today there are approximately 50,000 psychiatrists and licensed/certified psychologists who deliver mental health services. Elsewhere I have estimated that the total number of service hours available to treat mental disorders is on the order of 67 million hr per year (Kiesler, 1980). Taken from this perspective, the mental health service system is something we can be proud of. It demonstrates steady progress in the number of service providers, professional preparation and competence, and in actual number of service hours delivered. One can see in federal testimony in Congress over the past 10 years a steady press for funds to provide increases in the number of people who provide such services. This is the bottom-up perspective and it leads to a very positive view of current mental health policy.

What is the same situation viewed from the top down? Top-down analysis demands some indication of what the need for such services is and some inspection of the matchup between services needed and hours available to provide them. Epidemiological estimates of the prevalence of mental health problems in the United States vary quite dramatically. Two reports prepared for the President's Commission on Mental Health (Mechanic, 1978; Regier, Goldberg, & Taube, 1978) estimated that 15% of the U.S. population were in need of mental health services at any one time. Neither of these was really based on a national epidemiological survey, but used existing data to try to estimate the national figures. A more thorough epidemiological analysis by Dohrenwend, Dohrenwend, Gould, Link, Neugebaur, and Wunsch-Hitzig (1980) suggests that more on the order of 22% of the U.S. population have diagnosable mental disorders and another 13% have severe psychological distress that does not fit existing psychiatric categories. They refer to the latter group as the "demoralized."

Suppose we accept the more conservative estimate for purposes of discussion. Fifteen percent of the U.S. population equals approximately 35 million people. The top-down approach to national policy research would stress that we have 35 million people needing mental health services and only a potential of 67 million service hr (the actual service hours currently delivered by psychiatrists and psychologists is much smaller). Thus we have a national potential of only about 2 hr of service delivery for each person in need. This fact has consequences for the way we organize national service delivery systems and for consideration of methods of treatment. If the treatment of choice consisted of thrice-weekly psychotherapy, then the treatment strategy could only reach approximately 2% of the people in need in the country.[1]

Some options seem clear. Increasing the number of service providers at the rate we have been will not handle the problem. We would need 50 times as many doctoral level service providers as we now have in order to meet the service need with traditional methods. This is neither a pragmatic nor a politically feasible policy alternative. Consequently,

[1]This leaves aside the questions of potential interaction of treatment modality with diagnosis, demographic characteristics of the patient, and residence.

methods of extending service provision and maximizing service outcomes seem more promising methods of national policy formation. Short-term therapeutic interventions would be one such alternative; drug treatment seems promising from this perspective; service extenders in the form of paraprofessionals and volunteer workers seem necessary; a better matchup of specific services to patient characteristics and diagnosis would seem promising, to maximize the outcome of each service hour provided. All are potential avenues that one could investigate in order to develop a national mental health policy that would lessen the mismatch between service available and service needed.

The bottom-up approach is not inaccurate. It suggests that we have made steady progress in the quality of service provision, the number of service providers available, and the development of national networks such as community mental health centers in which to deliver the services. These are all things to be proud of nationally and continued progress should be encouraged. On the other hand, the top-down approach suggests that, however praiseworthy these attempts have been and are, they will never come close to matching the national need unless sharpened, refined, and expanded. The implications of the two approaches are vastly different. One is very positive and suggests continued good progress along the same lines we have taken. The other suggests that, although the progress is steady, the outcome of it represents such a small proportion of the national problem that a dramatic overhaul of national policy would be appropriate.

The top-down approach to policy review is not one practicing psychologists (and others) are accustomed to. One's perspective of national needs and progress is dramatically channeled by one's day-to-day activities, whether one is a practicing mental health service provider, a neurosurgeon, or a policeman. The mental health service provider has reason to feel increasing collective competence in dealing with the nation's mental health problems. It is difficult, however, to switch levels entirely, to take the broadest possible view of mental health policy, and thereby to realize the limitations of traditional service provision in national mental health policy.

Given this top-down approach to policy issues, how should we define the nonclinical research relevant to it? When asked about clinical research, colleagues usually mention the efficacy of such treatments as psychotherapy or specific brands of psychotherapy, organized systems of care, and drugs. When we speak of efficacy of these treatments, we are usually referring to the outcomes as the result of a specific technique, perhaps regarding a specific diagnostic or functional problem, perhaps even narrowly defined for some specific population (such as children).

If that is clinical research, then this chapter would have to be addressed to the effects of all other research on mental health policy—which would be an insurmountable task. We must narrow the task.

Clinical research is often difficult to generalize or translate into policy. Consequently, one important starting point is overall methods of summarizing, evaluating, and assessing clinical research. One question is, How do we summarize groups of diverse research findings in a manner that allows us to assess both the ease and cost of implementing them into policy? Two such general technique are meta-analysis and cost–benefit analysis. Each involves a group of research problems of considerate national import. Each deserves some preliminary discussion before more substantive research issues are discussed.

META-ANALYSIS

Meta-analysis (Smith, Glass, & Miller, 1980) is a general technique of assessing results across studies. Its major advance is that instead of merely registering a study or a particular effect in a study as either significant or not significant, it allows one to assess the degree of impact of a particular technique and average that impact across studies. Thus, Smith *et al.* come to the general conclusion that across approximately 500 studies of psychotherapy, the average person undergoing psychotherapy is sufficiently improved to be comparable to a person at the 80th percentile in the control group. With some methodological refinement, the estimate of this impact is even larger (Landman & Dawes, 1982). The technique emphasizes a unit of measurement (called the effect size) which allows one to add such effects across studies rather than attending only to the statistical significance within each study. Meta-analysis represents a significant methodological advance in assessing what we may conclude from previous research findings. Given the simple summarizing statistic one can further aggregate or disaggregate studies on the basis of other criteria. These might include specific methodological strengths and weaknesses, innovative independent variables, different providers, or type of dependent variable measured. Thus one could ask whether studies with random assignment have statistically more powerful results than those without. Alternatively and in a more limited sense, one might ask whether psychotherapy has a more powerful effect on self-esteem than on productivity.

The development of meta-analysis as a technique is only in the beginning stages, particularly in its application to such complicated phenomena as the delivery of mental health services. Nonetheless, it is at least potentially a tool of enormous impact. This tool, or one that accomplishes the same ends, is necessary in order to assess summarily the current knowledge base. Policy formation is, of course, a complicated and not always satisfactory method of planning for the future. Laymen often have great confidence in knowing whether mental health services are worthwhile or not. The senator at a public hearing is likely to ask, "What is the effect of psychotherapy?" The scientist/witness begins to respond, "Well, Senator, it is a more complicated question that that . . ." And the impatient senator interrupts to move onto a new line of questioning. Meta-analytic techniques allow us to inform the layman, including congressmen and state legislators, in a simple and straightforward way what research about these services has amounted to.[2]

COST–BENEFIT ANALYSIS

Meta-analysis, of course, only attempts to inform us about the relative outcomes of experimental treatments. The Congress and the public often wish to know more. Specifically, they often wish to know what they get for their money and whether it is worthwhile. The former is really a question dealing with cost–benefit; the latter a question dealing with cost–effectiveness. Although these terms are seen fairly often in the psychological literature, they are most typically used loosely and inaccurately. Cost–benefit and cost–effectiveness analyses are highly technical terms in economic research, but they cannot yet be applied very precisely to mental health services.[3] This is partly because of a weak national data base regarding mental health services and their outcomes. Furthermore, however, research in mental health services has not gathered the kind of information that typically goes into cost–benefit and cost–effectiveness analyses. Even providing a proper baseline for lack of action can be difficult. Consider housing policy in the United States. We might consider, as a policy alternative, building 100,000 new houses. We can calculate the costs of doing so, and make a pretty good try at allocating the costs to the several segments of society that will directly or indirectly pay for it. We can probably state fairly well what the outcomes of the new housing will be. A further important point is that we

[2]I do not suggest an uncritical acceptance of meta-analysis. See, for example, Bandura (1978) and Cook and Leviton (1980).

[3]See Saxe (1980) for an excellent discussion of the issues in the efficacy and cost effectiveness of psychotherapy.

can forecast fairly specifically what will happen if we do not build the houses.

All of these issues are more ambiguous in mental health services. Determining the proper comparative baseline is one problem. When assessing the comparative benefit of a mental health service, we need to state the outcome of not providing the service. With the current data base, it is difficult to say whether an unused or unfunded mental health service is likely to lead people to: become divorced; lose their jobs; increase their absenteeism; develop a deteriorating condition ultimately involving hospitalization; seek out expensive but inappropriate care; develop physical illness and consequent treatment; and so forth. Consequently, even the baseline for considering what will happen if nothing is done is very fuzzy in mental health policy.

Consider the problems of cost–benefit analysis when a service is provided. First, the cost of providing the service must be weighed against the cost of not providing the service, not merely calculated on the basis of how many dollars the service *per se* costs. The appropriate baseline must be established before a reasonable cost figure can be calculated. Second, with regard to both benefits and effectiveness, we must be sure we have cast a broad enough net to pick up potential consequences of social import. Studies of psychotherapy, for example, have disproportionately emphasized as dependent variables various personality measurements such as self-esteem, anxiety reduction, and the like. Although those and self-report measures are scientifically important in assessing the impact of psychotherapy, they are much less important to policy researchers/analysts discussing public alternatives. Public policy outcomes that would be of interest to policy analysts would include such things as employment, productivity, absenteeism, completing educational goals, maintaining family relationships, living independently, staying off welfare, paying income tax, maintaining property, and so forth. These outcomes are not only valued by society, they are of economic benefit to the larger social group. They can therefore be construed as analogous to other public policies and the decisions made about them. That is, one could conceivably compare cost–benefit analyses of mental health services with those of, say, housing policies.

I do not suggest that psychotherapy and other mental health services do not affect more typical public policy outcome variables. Indeed, there is evidence to suggest that they do. However, very few studies have included such variables and there is literally no national data base regarding them that one might use in order to begin calculating cost–benefit and cost–effectiveness ratios for alternative policy strategies.

Increasing our sophistication in economic analysis in general, and cost–effectiveness and cost–benefit analyses in particular, would represent important steps in developing a coherent and well-articulated approach to national mental health policy. Unfortunately our collective thinking in mental health care has not been much oriented in this direction, and the consequence has been inability to be heard or to be interpreted seriously in national policy circles.

National policies inevitably involve a national system and a national problem. The development of mental health policy has been inhibited because of: (1) scientific controversies about the impact of therapy and other services; and (2) lack of detailed description of the nation's mental health problems, their match with potential mental health services, and the public's cost–benefit of providing such services. These two aspects of policy review—meta-analysis and cost–benefit analysis—are critical nonclinical areas of research. They need substantial development to enhance the systematic use of clinical research in policy analysis and formation. Both techniques help assess and communicate effectively what we already know about mental health services and their outcomes. The rest of this chapter deals with a sampling of substantive research. The issues are important for mental health policy, in the early stages of investigation.

ENVIRONMENTAL AND BEHAVIORAL FACTORS IN MENTAL HEALTH

Most mental health services and the planning of mental health policy deal with individuals who have some specific mental health problem. Typically the person with the problem will come to an individual with expertise to treat it. In the context of this self-selected bias in orientation toward provision of services, we often neglect to consider issues affecting us all that more or less interfere with, or detract from, both health or mental health. Bandura (1978), Fuchs (1974), and others suggest that the country is reaching a limit on the degree to which its health and mental health can be affected at the margin by the addition of new services. That is, Fuchs, as a health economist, claims that the training of new physicians or the adding of new technical equipment to the national health system will have little consequence for the average quality of health in the United States. In that sense, huge investments in health or the education of more M.D.'s would not be cost-efficacious. Fuchs, Bandura, and others suggest that the major gains are to be made in the change of habits of individual citizens. Quality of health in the

United States would be significantly enhanced by changing habits dealing with smoking, eating and diet, exercise, alcohol intake, dental habits, sleeping patterns, and the like. There is no doubt that effective change of such national habits could dramatically reduce the health costs of cardiovascular disorders, cancer, and other problems. The technology to change such habits seems to exist. The Stanford Cardiovascular Project (e.g., Maccoby & Farquahr, 1975) has demonstrated that a combination of mass communication and personal intervention techniques can effectively reduce the cardiovascular risk in large populations. The techniques employed in the Stanford Project to induce change in dietary habits, exercise, drinking, and smoking would appear to have widespread ramifications.

There is a further consequence of such health-enhancing habit changes. It is fairly often noted in the literature that mental health problems can lead to health problems, but less attention is paid to the reverse sequence. However, serious health problems can have serious consequences in terms of mental health (cf. Broskowski, Marks, & Budman, 1981). The strong bilateral relationship between the two is typically unnoted. The point is that the sort of habit change we have described, seen by many to be potentially the most consequential for the nation's health, should have a clear impact on the nation's mental health as well.

In addition to individual habits there are other environmental factors that affect both the health and mental health of specific subparts of the population. For example, Cohen and his colleagues (e.g., Cohen, Evans, Krantz, & Stokes, 1980) have studied the impact of aircraft noise on cognitive performance and development in children. They found a persistent, sizable effect on cognitive performance in children whose schools were in the flight path of the Los Angeles airport. Following these children up over the course of several years, they found an interesting problem of apparent self-selection. That is, those children who were most affected in their school performance and social relations by the noise tended disproportionately to have moved from the area in later testings. One suspects that the effect on these children was more general than simply cognitive, and that the parents were sufficiently concerned to take the extreme and expensive step of moving to a different part of the city. Such variables as noise, air pollution, lead poisoning, and the perceived dangerousness of the neighborhood should all have effects on the mental health of the subpopulation having to deal with them. The actions needed to enhance the environment are less matters of service provision or habit change than they are legislative and regulatory mechanisms. One could easily conceive of cost–benefit research at the national policy level dealing with the implications of such legislation and regulation for enhanced health and mental health of the population.

SOCIAL NETWORKS AND VOLUNTEER GROUPS

There is some evidence that a social network can play a critical role in undercutting adverse stress in one's life, enhancing the effectiveness of deinstitutionalization, and perhaps even lowering the probability of institutionalization (Christmas, 1978). Research also shows the potential of volunteer groups to effect amelioration of mental health problems (e.g., Sainer, 1976). However, research in these two areas has not been significantly encouraged through national incentives. Programs are lacking to assist in building social networks and develop the more effective use of volunteer services in dealing with mental health problems. The cost is low and the potential is great.

INCENTIVES IN THE MARKETPLACE

Policy research tends to be disproportionately economic research. One approach to general policy research investigates incentives to change markets. There are several such incentives pertinent to the delivery of mental health services. One has to do with incentives and disincentives in insurance mechanisms and the relationship to effectiveness of mental health treatment. In Medicare, Medicaid, and some private insurance plans, a typical arrangement is to provide disincentives for outpatient care. That is, the insuring organization assumes that outpatient care could be very popular, and thereby tries to create barriers against its overuse. Consequently, one typically finds that a client will share in the cost of any outpatient care, but can receive inpatient care for free. (This arrangement typically applies to drug abuse and alcoholism in addition to other mental health services.) This is called a disincentive in economics, but its effect in mental health services is unintentionally negative.

One could document quite effectively the notion that we put people in mental hospitals who could be treated more effectively elsewhere (Kiesler, 1982b). A Task Panel to the President's Commission on Mental Health estimated that somewhere from 43% to 75% of the patients put in mental hospitals could be more effectively treated in outpatient settings (Freidman, 1978). Nationally, inpatient care is more expensive per patient, and currently accounts for 70% of the mental health dollar.

The total cost for inpatient care is greater but, because of disincentives for outpatient care, it is less expensive for the person undergoing treatment. Since inpatient care is better covered there is also an inescapable message to the patient that it must be more effective. That is, the patient reasons: Why should insurance pay more of Treatment A than Treatment B, if A were not more effective?

As a result, we now have a national system that is effectively biased toward inpatient care: for many patients a less effective form of care, and for all, a more expensive form. Future national planning should take these issues into consideration. We do need policy research to ascertain the proportion of cases affected by such policies, and the degree to which such disincentives affect behavior of both patients and practitioners.

A second form of incentives in the marketplace involves the behavior of the practitioner. Consider, for example, a private practitioner versus someone who is employed in an HMO. The income received by the private practitioner is determined by the number of hours of treatment; therefore one might expect a set of incentives for the practitioner to elongate treatment. For the person working in the HMO, on the other hand, salary is not affected by length of treatment or number of patients. One's salary is constant across those variations. Therefore, one might expect a practitioner in an HMO to be concerned with saving time and that such savings would be a major incentive in the details of everyday practice. On this basis alone, one might expect the length of treatment in private practice to be longer than that in an HMO, other things equal. One might expect referral practices to be different as well, and that such differences would have an impact on mental health care. If we think of our two actors as nonpsychiatric physicians, their willingness to refer their patients to psychiatric care might vary accordingly. The nonpsychiatric physician in an HMO, because a major incentive is the saving of time, should be willing to refer a patient to psychiatric care and sooner in the treatment process than a colleague engaged in private practice. Early referral to mental care produces a savings in physical care. This is the medical offset (or decreased utilization of medical resources following psychotherapeutic intervention) that has been reported in the literature (Jones & Vischi, 1980). Our incentive analysis implies that such offset should be greater in an HMO than in private practice.[4] Indeed, all of the studies of medical offset reported by Jones and Vischi were concerned with organized systems of care, and not private practice. Since

[4]Indeed, one might wonder whether there could be any offset for private practice. However, Schlesinger and Mumford (1982) do find such effect in the Federal Employees Health Benefit plan data. See also Mumford, Schlesinger, and Glass (1981) for further discussion of these issues.

the efficacy of mental health care should be related to how early in the development of a problem a service provider is contacted, such differences in referral practices should affect the quality of care and the cost-efficacy of care.

One might expect other differences in treatment received in private practice or organized care. Since time is presumably a major incentive for the person practicing in an organized system, one should see more atypical practices such as telephone contact, group therapy, the use of paraprofessionals, volunteer groups, and the like. Of course, whether such practices lead to an equally efficacious outcome is a topic for separate investigation.

For mental health care, differences in incentives of the providers, their outcomes in terms of referrals, type of care, and efficaciousness, have been largely uninvestigated. However, since policy analysis and development depends on deciding where to put one's dollars or which systems of care have the greatest impact, such research needs to be encouraged by both those in charge of research money and those in practice. One does note that many aspects of the differences in styles of practice with regard to nonpsychiatric physicians alone have been well researched. The above assumptions of time and money as major incentives for practitioners in different practice sites are based on Mechanic's (1979) analysis of the same problem for nonpsychiatric physicians. I have only extrapolated that analysis to mental health service providers.

A similar economic analysis of incentives for patients needs to be carried out. We know very little about how potential patients in a mental health care system see the alternatives for treatment, and the incentives attached to them. We should expect patients to have as a major incentive short-term care and some sense that they have regained control over their lives. These incentives also have implications for how former patients will see the service rendered. For example, Cummings and Follette (1976) followed up patients some time after completion of a therapeutic intervention. They found disproportionately that patients remembered the personal problem that they had; had forgotten the original symptomatology (e.g., severe headaches, inability to sleep); were confident the problem had been handled; and felt the therapy had little impact on their ability to cope with the problem. In short, the therapy and therapist were extremely effective but not given credit for the outcome. Why this paradox occurs deserves serious investigation. For example, most psychotherapists intentionally promote the feeling of the patients that they are in control of their lives and are personally managing the problem being presented. The notion is that perceived self-responsibility will contribute to the long-term mental health and social competence of the patient. The Cummings and Follette data suggest that this is true, but it has the unintended consequence that the patient feels the service was of little help. Such interactions may contribute to some of the problems that mental health service providers have at a national level in developing a public image of being capable and cost efficient.

MODES OF PRACTICE

We have raised questions of the cost–benefit and cost–effectiveness of psychotherapeutic practices and other mental health interventions. There has been little comparative research into modes of practice. At a national level, and considering where to put the public dollars, one can contemplate several possibilities: CMHCs; HMOs; private practice; as well as a variety of less formal methods of treatment such as halfway houses, day care centers, and the like. In addition, one could consider innovative forms of treatment, such as incentives to promote collaboration of general practitioners and mental health service providers, thus obtaining mental health expertise in problems very early. Very little evidence is associated with any of these alternatives. Indeed, there is little tough-minded research about the outcome effectiveness of any one of them let alone comparatively. It is not necessarily the case, or even likely, that one or two of the alternatives are effective and the rest are not. It is much more likely that they are more or less effective for different types of people, under different sorts of circumstances. The policy questions, however, still remain: What is the proper mix in terms of planning at a national level? One can change the mix through economic and other sorts of incentives, if policy-relevant data are clear. But the data needed to plan effectively are lacking.

THE EDUCATED PATIENT AND THE PERSPECTIVE OF THE PRACTITIONER

A recent survey by the American Psychiatric Association showed that the public has a fairly good idea of the potential differences between using a psychiatrist or a psychologist for mental health services. In particular, the public seemed aware that if drugs are needed in treatment one should go to a psychiatrist, but if not, a psychologist was a reasonable choice. Although one should not overinterpret this finding, it does suggest a broader set of issues. Presumably, potential patients have some idea what kinds of services are available, who provides them, and the like. They probably also have some idea of what their basic problem is, and what sorts of alternative services might be effective. The point is not that these "ideas" are accurate, but simply that they exist and are hard to escape. To the extent that these public attitudes or perceptions are uniform, they will lead to substantial distortions in the perspectives of practitioners and centers for delivery of services. Unless they extensively specialize, practitioners tend to assume they are seeing a broad array of psychological problems in their practice. To the extent that the public holds common perceptions of who does what and with what effect, the perspective of the practitioner is distorted. For example, I have often hear psychiatric friends say that most of their patients need some form of drug regimen. I have also often heard clinical psychologists say that they seldom see patients who need drugs for effective treatment. Of course, part of this difference may be because of how the two groups of providers interpret the scholarly literature on treatment effectiveness. Also, the difference may partly stem from the divergent perspectives of the university departments in which they have been educated. Yet such differences might also be an accurate reflection of actual patient characteristics. That is, such differences in perspective may exist because the patients are sorting themselves out in advance of seeking assistance, and doing so with some degree of accuracy. Thus, practitioners of particular backgrounds or particular modes of practice could have radically different perspectives, because they perceive (inaccurately) that the patients they see are a random sample of all patients.

There are other limitations to generalizing from one's practice. Bandura (1978) points out that theoretical perspectives are distorted because service providers feel that the patients they see are a reasonable sample of people with those problems. That is, mental health service providers do not see a random sample of depressives, neurotics, and alcoholics. Rather, they see the unsuccessful ones who are led by their lack of success to seek help. The ones who succeed in the arts, the sciences, and in public life are not seen. Therefore, one's theoretical understanding of the concept of depression is distorted by the fact that one has seen only the unsuccessful depressives, and not a reasonable sample of all depressives.

RELATIONSHIPS BETWEEN HEALTH AND MENTAL HEALTH

Distortions can easily occur in one's perception of physical and mental health problems. Indeed, one might predict that specialists in physical health problems would underestimate the extent to which their patients also have mental health problems, whereas specialists in mental health problems would make the opposite mistake. This is a function partly of what one expects to find, partly of what one focuses on in the interaction. Broskowski *et al.* (1981) extensively discuss the substantial overlap between the two sets of problems and document the fairly high conditional probability that if one has a problem of Type A, one will also have a problem of Type B. To some extent this is a difficulty for practitioners in both physical and mental health. However, to a different extent, it is an especially difficult issue for policy analysts. There are almost surely such problems of overlap being distorted in the practitioner's office, and they may consequently also be ignored in the office of the policy analyst. Psychologists are fond of pointing to the data that indicate the substantial proportion of people with mental health problems who are inappropriately treated by nonpsychiatric physicians (who are not well trained either to treat such problems or to recognize them). On the other hand, there are fewer data attempting to ascertain the degree to which people being treated by mental health service providers also have physical health problems and the degree to which those could be recognized.

SUMMARY AND CONCLUSIONS

Many of the major policy research issues in mental health do not involve narrowly defined clinical research. Major policy questions regard the epidemiology of mental problems, the inadequate national policy data base, techniques of assessing current knowledge for policy (e.g., cost–benefit analysis and meta-analysis), broad environmental influences on mental health, the impact of paraprofessional and volunteer groups and the influence of social networks, and economic and sociological analysis of varying modes and sites of treatment.

Further emphasis needs to be placed on a top-down approach to policy analysis and research. This approach effectively reveals the need for increased knowledge in order to translate clinical research into feasible national policy. The wise practitioner seeks to avoid the natural conclusion that one's clinical experience, however rich, can easily be implemented into cost-efficient national policy.

REFERENCES

Bandura, A. On paradigms and recycled ideologies. *Cognitive Therapy and Research*, 1978, 2, 79–103.

Broskowski, A., Marks, E., & Budman, S. H. *Linking health and mental health*. Beverly Hills, Calif.: Sage Publications, 1981.

Christmas, J. J. (Coordinator). *Report of the Task Panel of Community Support Systems*. Washington, D.C.: U.S. Government Printing Office, 1978. (President's Commission on Mental Health, Vol. 2)

Cohen, S., Evans, G. W., Krantz, D. S., & Stokes, D. Physiological, motivational, and cognitive effects of aircraft noises in children: Moving from the laboratory to the field. *American Psychologist*, 1980, *35*, 231–243.

Cook, T., & Leviton, L. Reviewing the literature: A comparison of traditional methods with meta-analysis. *Journal of Personality*, 1980, *48*, 449–472.

Cummings, N. A., & Follette, W. T. Brief psychotherapy and medical utilization. In H. Dorken & Associates (Eds.), *Professional psychologists today*. San Francisco: Jossey-Bass, 1976.

Dohrenwend, B. P., Dohrenwend, B. S., Gould, M. S., Link, B., Neugebaur, R., & Wunsch-Hitzig, R. *Mental illness in the United States: Epidemiological estimates*. New York: Praeger, 1980.

Friedman, P. R. *Report of the Task Panel on Legal and Ethical Issues*. Washington, D.C.: U.S. Government Printing Office, 1978. (President's Commission on Mental Health, Vol. 4)

Fuchs, B. R. *Who shall live? Health economics and social choice*. New York: Basic Books, 1974.

Jones, K., & Vischi, T. Impact of alcohol, drug abuse, and mental health treatment on medical care utilization: Review of the research literature. *Medical Care*, 1980, (Suppl. 17), 12.

Kiesler, C. A. Mental health policy as a field of inquiry for psychology. *American Psychologist*, 1980, *35*, 1066–1080.

Kiesler, C. A. Psychology and mental health policy. In M. Hersen, A. E. Kazdin, & A. S. Bellak (Eds.), *The clinical psychology handbook*. New York: Pergamon Press, 1982. (a)

Kiesler, C. A. Mental hospitals and alternative care: Non-institutionalization as potential public policy for mental patients. *American Psychologist*, 1982, *37*, 349–360. (b)

Landman, J. T., & Dawes, R. M. A reanalysis of the Smith and Glass meta-analysis of psychotherapy outcome, controlling for metholodogical and statistical factors. *American Psychologist*, 1982, *37*, 509–516.

Maccoby, N., & Farquahr, J. W. Communication for health: Unselling heart disease. *Journal of Communication*, 1975, *25*, 114–126.

Mechanic, D. Report of the Task Panel on the Nature and Scope of the Problem. In President's Commission on Mental Health (Vol. 2). Washington, D.C.: U.S. Government Printing Office, 1978.

Mechanic, D. Physicians. In H. E. Freeman, S. Levine, & L. G. Reeder (Eds.), *Handbook of medical sociology* (3rd ed.). Englewood Cliffs, N.J.: Prentice-Hall, 1979.

Mechanic, D. *Mental health and social policy*. Englewood Cliffs, N.J.: Prentice-Hall, 1980.

Mumford, E., Schlesinger, H. J. & Glass, G. V. Reducing medical costs through mental health treatment: Research problems and recommendations. In A. Browskowski, E. Marks, & S. H. Budman (Eds.), *Linking health and mental health*. Beverly Hills, Calif.: Sage Publications, 1981.

National Institute of Mental Health. *The financing, utilization, and quality of mental health care in the United States* (Draft report). Rockville, Md.: Author, 1976.

Regier, D. A., Goldberg, I. D., & Taube, C. A. The de facto U.S. mental health services system: A public health perspective. *Archives of General Psychiatry*, 1978, *35*, 685–693.

Sainer, J. S. The community cares: Older volunteers. *Social Policy*, 1976, *7*(3), 73–75.

Saxe, L. *The efficacy and cost-effectiveness of psychotherapy*. Washington, D.C.: U.S. Government Printing Offices, Office of Technology Assessment, Congress of the U.S., 1980. (GPO stock #052-003-00783-5)

Schlesinger, H. J., & Mumford, E. *Mental health services and mental care utilization in the fee for service system*. Paper presented at the Mental Health Policy Symposium Series, Carnegie-Mellon University, 1982.

Smith, M. L., Glass, G. V., & Miller, T. I. *The benefits of psychotherapy*. Baltimore, Md: Johns Hopkins University Press, 1980.

II

SOCIAL PSYCHOLOGY

Although it is generally considered a fact that pathological behavior patterns occur in a social context, it has only been recently that research on social processes has been extrapolated and applied to psychotherapeutic activities. In this section of the book, the clinical implications of research in some of the major areas in social psychology are explored. In Chapter 4, Bunker and Julian focus on the leadership literature, placing the therapist–client relationship within the framework of various leader–follower models. The therapist's influence effectiveness is analyzed from the perspective of research on leadership styles and situational factors. In addition, the complexities of therapist–client interactions are explored from social exchange and social power theories.

The area of nonverbal communication is covered by Davis in Chapter 5. She argues that clinicians face certain difficulties integrating movement research with their activities. However, by focusing on specific symbolic activity and movement styles of patients, the practitioner enriches his or her perceptions of defenses, conflicts, and any psychotherapeutic change. Two case histories are presented to illustrate the correlation between shifting body positions and rapport. Davis proposes that the training of therapists and the practice of psychotherapy would benefit greatly by heightened kinesthetic self-awareness.

In Chapter 6, Worchel and Worchel focus on aggressive behavior in therapy in light of theories and research developed outside of clinical psychology. Why, they ask, are certain techniques effective in encouraging and discouraging aggression and what can research offer by way of guidance in the development of new methods? Certain key issues, particularly catharsis and displacement, are considered within the context of frustration–aggression and social learning theories. After reviewing the various antecedents and consequences that increase and decrease aggressive activity, Worchel and Worchel offer guidelines for treating patients who are either unable to express aggression or unable to inhibit their own aggressiveness.

As a topic, attitude change has obvious relevance to the change efforts of clinicians. Harvey, Weary, Maddux, Jordan, and Galvin assay this relationship in Chapter 7. After exploring the major and minor theoretical models of attitude change, they use one of these models, cognitive dissonance and the related research on effort justification, to understand therapeutic change as dissonance reduction. In the penultimate section of the chapter, the authors discuss some key treatment variables such as the credibility and trustworthiness of the therapist as well as the power of self-perception, reactance, and perceived control of patients.

Stereotyped attitudes in psychotherapy, in particular those that concern the patient's gender, are the topic of concern for Shafran in Chapter 8. After citing research supporting the continued existence of stereotypes based on gender, she turns to the issue of gender stereotyping and psychotherapy. The data, she argues, fail to demonstrate that gender bias interferes with the psychotherapy process. In the penultimate section of the chapter, social psychological data and concepts are integrated with case material demonstrating how gender stereotyping may influence the expectations, preferences, and assumptions of patients and therapists. Shafran notes, however, the danger in not looking beyond these stereotypes to the personal meaning of the situation for both participants.

Research in interpersonal attraction has traditionally focused on such variables as proximity, attractiveness, similarity, and reciprocity. In Chapter 9, Sprecher and Hatfield relate this literature to three stages of relationships—how they begin, grow, and how and why they end. Throughout the chapter, careful attention is paid to male–female differences in relationships and the relevance of research findings to couples therapy. The authors examine relationship initiation, the roles of mutual disclosures and equity in a growing relationship, the man's need for power as a frequent cause of breakups, and the different reactions of men and women to the termination of a relationship. The subjects of loneliness and gender differences in love and intimacy are also examined thoroughly.

The experience of injustice depends on whether the individual is the harmdoer, the observer, or the victim. The needs to believe in a just and in a controllable world have been linked to the notion of injustice. Research on these aspects of injustice is discussed in the second section of Steil and Slochower's chapter. The next part of Chapter 10 explores injustice in the context of psychoanalytic therapy, emphasizing the effect of the patient's history as the harmdoer or victim on the therapist's countertransference reactions. Recommendations are made for helping patients integrate their habitual roles into their life experience.

4

The Leadership Literature and Its Relevance to Psychotherapy

BARBARA BENEDICT BUNKER AND JAMES W. JULIAN

INTRODUCTION

When leadership is mentioned, it summons up images of great men and women, of power and its uses, of risk-taking initiatives, and imaginative forward movement. Yet the essence of leadership is that it is a relationship. For every leader there must be those who are willing to follow. Critical in understanding leadership is understanding the influence relationship that must develop between the people who are the leaders and those who are the followers. Not surprisingly, interest in the study of leadership has roots in the beginnings of recorded history; the empirical study of leadership has its origins in the beginnings of the social sciences (Allport, 1954). The concern of the present chapter is whether this long tradition of empirical study of leadership can contribute to our understanding of another very special influence relationship: psychotherapy.

Dynamic psychotherapy has at its core a relationship between two persons: the therapist and the client. This dyad represents an instance of what Georg Simmel has described: "The simplest sociological

BARBARA BENEDICT BUNKER AND JAMES W. JULIAN • Department of Psychology, State University of New York at Buffalo, Amherst, New York 14226.

formation, methodologically speaking, remains that which operates between two elements. It [the dyad] contains the scheme, germ, and material of innumerable more complex forms" (Simmel, 1950, p. 122). Thus, the therapist–client relationship is a very particular case of an already special class of relational events. It is an intimate social relationship, which is in some fashion subject to the dynamics that govern two-person relationships generally.

In this chapter we shall suggest ways in which the empirical study of leader–follower relationships has implications for the therapist–client relationship. To do so requires a glance at some of the recurrent findings and generalizations from the study of leadership broadly, and an extension of these findings to the more intimate relationship of the client and therapist. The therapist–client relationship is a peculiarly structured intimate association in which there is an obvious ordination of the participants. Clients, after all, come to therapists subordinate, that is, in need of help. By implication, at least, they acknowledge that the therapist has a certain power, in this case an expertise, to help them. As such, the therapist–client relationship is like others in our society where there is a distinct power differential between the members—for example, doctor–patient, teacher–student, leader–follower. Although the differentiation of power and status between the therapist and client points clearly to possible similarities in dynamics to leader–follower relationships, it is important to keep in mind some of the ways in which the therapy setting is distinctive. One of the most subtle distinctions of the therapeutic relationship is its task. The closest familiar analog is that between student and teacher. The teacher often has expertise and knowledge of both the *content* to be learned and the *process* of knowledge acquisition. The teacher aims to influence the student so as to facilitate the student's learning of the content, and, importantly, the skills and "techniques" for learning similar content in the future. Therapy can be thought of in this way as a special teaching relationship in which the process of *learning how to learn* from experience is also the special content of the therapeutic task. Hence, in therapeutic relationships, the emphasis is far greater on the process of learning.

In contrast to the therapeutic relationship, there is no necessary coincidence between process and content of the task in the leader–follower relationship. Indeed, in the extreme case, the leader may wish to influence the follower to contribute expertise to some task, precisely because neither the leader nor others share that expertise. In this instance leadership is much more a matter of conceptualizing the overall task, organizing, and directing the followers to facilitate their contribution. In this sense, then, the particular quality of the therapeutic task may limit any generalizations that we might make about the therapist–client relationship, as viewed from the study of leadership.

In the first part of this chapter we briefly discuss several broad topics in the literature on leadership. Our order of presentation is also a rough chronicle of their emergence in the history of leadership research. Each section, in addition to a brief description of the approach and findings, also includes a discussion of how some of the insights garnered from the study of leader–follower relationships are relevant to the practice of psychotherapy. Finally, we consider two ideas in the leadership literature, power and charisma, that have been topics of discussion throughout the history of leadership research rather than emergent in a particular time period. Since all social relationships are in some essential way governed by certain basic dynamics, we believe that the study of leadership also has relevance to the special, therapy influence relationship.

THE EMPIRICAL STUDY OF LEADERSHIP

THE STUDY OF LEADER ATTRIBUTES

One of the dominant themes of leadership research over its history has been the study of leader attributes. The aim of this work was to distinguish leaders from nonleaders on any of a host of possible individual differences, for example, height, appearance, intelligence, self-confidence, and many more. The assumption was that there must be "something" about people who lead that differentiates them from those who are led. The consistent verdict, however, from this voluminous body of research has been issued a number of times (cf. Mann, 1959; Stogdill, 1948): Leaders and led do not differ to any considerable degree. Two possible implications from this generalization may have bearing on therapist–client relationships. First, these findings affirm that therapists and their clients are cut from the same cloth. As a generalization this is certainly not apt to generate much controversy. It may be the basis for some of the bad humor about difficulties telling the inmates apart from the staff. It also reinforces the potential value of therapist self-understanding as a basis for insight and empathy in the therapeutic encounter. As in the leader–follower arena, gross dissimilarity may be a disadvantage to the therapist.

A second implication is that personal attributes of the therapist do not forecast therapeutic effectiveness. The findings from the study of leadership suggest that it is not crucial to be exceptionally bright, extroverted, articulate, achievement oriented, or whatever to be able to perform adequately as a leader. This issue may be irrelevant with the exception of those characteristics toward which selection to training institutions is biased. Chief among these is the current practice that only students who demonstrate exceptional verbal and mathematical skills, as indexed by high test scores, together with a record of their successful application, are admitted to candidacy for professional degrees. We do not mean to imply that we are opposed to the scientist–practitioner model, or to training bright, articulate therapists. Rather, we are suggesting that leadership research implies that the emphasis on these exceptional talents, along with others, possibly does not ensure "influence potential." The research literature says that no simple set of attributes is likely to generalize to effective influence across a wide variety of clients. Rather, effective influence seems to depend on the complex interaction of personal, interpersonal, and situational characteristics.

In spite of the long tradition of research devoted to examining leader attributes, it is only recently that attention has turned to gender as a potentially significant variable. We shall not explore here why a variable of such obvious importance has been largely ignored historically. More to the point, the recent research on gender in leadership shows a pattern of conflicting results from which it is difficult to generalize (Hollander & Yoder, 1980). It is safe to say that whether a man or a woman is performing the leader role makes a difference in terms of the responsiveness of those with whom they work (Bartol, 1974; Day & Stogdill, 1972). The dynamics that may account for these differences in influence will require much further research.

This generalization is perhaps important, however, in and of itself. By implication, it raises the issue of sex differences among therapists as a variable that may have an important impact on the therapy process. To be sure, this consideration is readily acknowledged by dynamically oriented therapists. Here we propose, however, that it is not necessary to invoke analytic concepts of personality development, or the significance of the transference in the therapy process, to argue that the sex of the therapist is an important parameter of the therapeutic situation. Unless we are willing to accept that the therapist–client relationship is immune from the dynamics of the more general influence process, then all therapists should be alert to the ways in which gender may affect the course of therapy. This seems especially important for therapists whose usual orientation does not represent therapist attributes as potentially important features of therapeutic work.

THE STUDY OF LEADER STYLE

As disillusionment with the trait approach gradually grew, another approach, an interest in leadership style, began to take root. This approach was still person centered, that is, the leader was the focus of study. This time, however, it was not fixed qualities that were of interest, but rather the behavior of the leader. What does a leader do? How is it done? These were the questions of interest. Notice the implicit assumption: Flexibility of leader behavior is possible because actions are under the leader's control. Leaders have choice and can adjust their behavior. Thus, notions of leader development and of training in leadership skills emerged from the leadership style approach.

In 1939, a decade before the interest in style began to emerge fully, Lewin, Lippitt, and White published a classic study in which they varied the style of the leader and studied the effect of democratic, authoritarian, and laissez-faire leadership on groups from a boys club. Despite the seminal nature of this study in showing significant effects of style, little follow-up research was immediately forthcoming. Only in the late 1940s, when the trait investigations had run their course, did a systematic program of research on leadership style develop at Ohio State under Stogdill and his colleagues (Shartle, Stogdill, & Campbell, 1949; Stogdill & Shartle, 1948). A decision was made to study the behavior of real leaders in their natural settings. The method was to have subordinates rate their bosses, describing what they did and did not do. The Ohio State studies produced evidence for nine behavioral dimensions that collapsed into two main factors: *consideration*, the interpersonal interaction patterns with subordinates that demonstrated friendliness, helping behavior, and a positive and supportive relationship; and *initiation of structure*, the "task"-oriented behaviors such as setting procedures, performance standards, and making role expectations explicit.

This post-World War II emphasis on the behaviors of leaders influenced the field as a whole. At the University of Michigan, Kahn and Katz (1953) examined the behavior of industrial supervisors using interviews with subordinates. They also identified two factors, this time labeled *production oriented* and *employee oriented*. The

> production-oriented supervisor emphasized planning, direction, and goal achievement. The employee-oriented supervisor was typified by rapport with subordinates, an open and accepting style, and a concern for the feelings and problems of subordinates. (Chemers, 1983)

Similarly, at Harvard, Bales and Slater (1955) developed an interaction coding system by observing undergraduates in "leaderless" groups.

In these groups two types of leaders emerged. The *task leader* paid attention to the group's goals and to moving the group in that direction. The *socioemotional leader* paid attention to the interpersonal relationships that were developing and attempted to help.

Also during this period, Fiedler (1967) began at Illinois his extensive research on leader orientation, based on his "esteem for least preferred co-worker" scale (LPC). Although variously interpreted, "the most enduring and widely used interpretation of LPC is as a measure of relative task vs. interpersonal orientation" of the leader (Chemers, 1983). Thus, we can see substantial agreement from a variety of different investigative traditions about two factors of leader style. These are clearly central in any understanding of leadership, and, we presume, in any understanding of influence relationships generally.

These two factors—consideration, and initiating structure—have been applied in the leader training literature and research. Blake and Mouton (1964) propose that leaders may be either high or low on each of these two dimensions, and that the most effective leaders are both task and person oriented. Another application, by Fleishman and Harris (1962), found that foremen low in consideration had higher grievance rates. They also found that high-consideration leaders were permitted by their followers to initiate more structure than those who were low.

Implications for Therapeutic Style

Are these two factors emphasized in the theory and practice of therapy? The Rogerians' "unconditional positive regard" is certainly similar to consideration. In Rogerian therapy this factor is seen as crucial to the therapeutic process (Rogers, 1951, 1954). Given the research on empathy and related factors in the therapy effectiveness literature, it appears that there has been both discussion and research on the consideration factor for many years. From the perspective of the leadership literature, it may be that the high-consideration therapist is effective not simply because of the regard shown for clients, but also because structuring is accepted with less resistance (Fleishman & Harris, 1962).

Various theories of therapy emphasize the initiating of structure. Behavioral therapies, for example, could certainly be described as high in the initiation of structure. Patients proceed through a very structured series of steps in their treatment. The therapist initiates and directs this learning process. Some interpretations of the leadership literature suggest that the person high in both factors is the most effective. Is the reinforcement process used in this therapeutic form the same as high consideration? It appears not, since rewards can be seen as separate from the therapist's personal affect toward the client. Has enough attention been paid to the impact of this factor in therapy done under this perspective? Is it possible that the leadership literature is correct, and that the most effective therapists are high on both factors?

This question brings us to dynamic therapy, which at first appears to be low in structuring. Yet a distinction may emerge between the parameters established at the outset and the subsequent therapeutic process. Initially, as the patient contracts with a therapist, the limits that are set to organize the relationship are highly structured—time, place, fee, role, verbal interaction, and so on. This structure remains in place during therapy, though it may not be discussed unless a violation has occurred. We might refer to this as the *surface structure*. It is clear and easily perceived by both parties. Once therapy has begun, structure is less obvious, at least to the patient. We might refer to this as the *deep structure*. The transference process, for example, is such a structure. The structure of the situation creates transference. The therapist works on its analysis with the patient. Thus the process is created via initiating structure, but the working-out of that process is not "directed" by the therapist. Nevertheless, it may be useful to think of that relationship as having a deep structure that would bear more articulation. For example, there are role prescriptions for both patient and therapist that have boundaries, and that provide structure. If they are absent or inconsistent, difficulties are apt to develop.

Theoretically, we suggest that it is the initiating-structure factor that has the most potential for understanding therapist behavior in a variety of therapeutic situations. The importance of the consideration factor has been well articulated in therapeutic literature (e.g., Rogers, 1951, 1954). However, too often it has stood alone rather than being seen as a natural companion to initiating structure. It is the relationship between consideration and structure that needs further elaboration.

SITUATION IN THE STUDY OF LEADERSHIP

Leadership research acknowledged from its beginning that the personality of the leader was not the only important set of variables. As Gibb (1954) and others have commented: "Leadership is always relative to the situation." Implicit in our earlier discussion of leader behavior is the recognition that leaders need to adapt their particular style to meet the functional demands of a specific situation. In the 1950s research began to focus more on the demands that different types of situations made on leaders, and less on the leader as a person. Situational factors that have been studied at various times are: task demands, the degree to which the task is structured, size of the group, history and resources of the group, the communication structure imposed by the task or the organization, and the source of the leader's legitimate authority, whether by appointment from "above" or as emergent, based on election by group members. Thus, the major thrust in the situational emphasis was to examine the leadership context in order to discover how the *functions* of the leader were being performed (Hollander, 1985).

More recently, the most systematic attention has been paid to situational variables as a feature of the "contingency models" of leadership (Evans, 1970; Fiedler, 1978; House & Mitchell, 1974; Vroom & Yetton, 1973). Fiedler's model qualifies the effects of the leader's task-oriented or relationship-oriented style in terms of the clarity and specificity of the task, the leader's formal authority or power to reward and punish followers, and the amount of loyalty and support given to the leader by group members. Relationship-oriented leaders, for example, have been found to be most effective in situations of moderate task complexity.

More recently, other investigators have begun to examine further the ways in which different situational features shape the relationship between leader characteristics and effectiveness (Calder, 1976; Herold, 1977; Pfeffer, 1977). Even more recently, perhaps as an outgrowth of general increasing awareness of ecological variables, the overall situational or organizational context has been proposed as important in understanding the leader influence process (Osborn & Hunt, 1975).

Although it is perhaps commonplace to acknowledge the significance of situational factors in leadership, as yet there is still no accepted description of which variables are the ones worthy of concern. We do not have an adequate taxonomy of task or of situational dimensions. What is clear is that when situational differences are examined in concert with, for example, the leader's style, they have been found to be important.

Situations in Clinical Practice

It may be fruitful, within the domain of the therapist–client relationship, to consider the effect of the situation. Certain broad features of the overall situation are likely to be obvious and consensually present for both client and therapist, such as: whether the client is in therapy voluntarily; whether the client is being seen individually or as part of a group; whether the client is an inpatient in a hospital setting or is seen as an outpatient; and whether money is being exchanged by the client and the therapist. This last factor is dealt with more below, but for now we suggest that settings vary in terms of the salience of the exchange of money. For example, there are some agency situations in which no money is exchanged by either the client or the therapist (contingent on seeing that client), whereas in other settings the exchange of money is a very important aspect of the therapist–client relationship.

Even more difficult to discern than the above situational factors are those potential variables that are analogous to variations in the so-called "task" structure. For example, what is the likely impact on the course of therapy of the "task" of dealing with a snake phobia, in contrast to the task of uncovering the basis of a chronic depression? Such differences in "situation" have had important implications for determining the effective style of leader behavior. Perhaps it has been the implicit recognition of these situational variables that has led to the development of different approaches to treatment.

Another potential situational variable is the expected duration of the course of therapy. To the extent that there is a shared understanding of how long the client and therapist will be meeting together, this contextual variable may qualify all of the ensuing interaction. For example, therapeutic activity is presumably much different in a long-term analysis, taking perhaps years of meeting several times a week, than in short-term, time-limited therapy where six 1-hr sessions are scheduled (Bellak & Small, 1978).

What we propose is that the study of leadership points to the imperative of considering situational variables in understanding the leader influence process. Although it is presumably impossible to transpose particular situational concerns in leadership to the therapeutic relationship, it may be useful to take the contribution of "situations" into account. As in applied leadership, both therapist and client already do take account of situations. Examining them directly may result in greater therapeutic leverage from making them explicit.

SOCIAL EXCHANGE APPROACHES TO THE STUDY OF LEADERSHIP

Behaviorism, which influenced general psychology, influenced the study of leadership as well. Basically, the behavioristic approach derives from the research and models of operant conditioning. The essence of this approach to social psychology is that behaviorists assume that individuals who maintain a social relationship do so because they are engaging in mutually reinforcing behavior (cf. Blau, 1964; Homans, 1961). That is, each person in the relationship must be emitting behavior that influences the other person by providing the other with satisfactions. A key concern of this approach was the process of exchange among group members, specifically the elements or "currency" of influence (Homans, 1961).

This social exchange perspective emphasized that the leaders were not only giving such things as structuring and consideration, but were also receiving something of great importance to themselves. Similarly, this approach required recognition that the followers were not only receiving directives and praise, for example, but were themselves influencing the leader in important ways. Thus, a major contribution of social exchange thinking was the emphasis on *mutuality* in the influence process.

Although mutuality of influence is a central idea in the literature, most research has focused on what leaders give and what followers receive in terms of consideration and structure. Thus, predominant attention has been given to one side of the social exchange relationship. What has received much less exposure has been what leaders receive and followers give to the maintenance of the relationship. That is, we still know comparatively little about the ways in which followers influence and shape leader behavior.

Hollander's (1985) recent review of this literature suggests that followers do indeed influence the pattern of leader behavior. Investigators have focused particularly on the influence of successful or competent task performance by group members. As Hollander points out, however, relatively few of these research designs in fact look at the process of mutual influence.

Findings from another research tradition bear on this same issue. In laboratory studies of emergent leadership, investigators have been able to alter the pattern of behavior in discussions by giving private feedback from the experimenter that ostensibly reflected endorsement or approval (Bavelas, Hastorf, Gross, & Kite, 1965). By implication, this finding indicates that leader behavior may well be altered as a function of the responsiveness of others to the leader's directives.

Although these notions of leader–follower relationships as an exchange are appealing, it is also clear that at this point the actual process of exchange remains largely unexplored. The exchange paradigm, however, has stimulated interesting distinctions that are relevant to the therapist–client exchange.

Equity is a concept that has received considerable attention in social psychological investigations in recent years (Walster, Walster, & Berscheid, 1978). Equity models build on the notion of social exchange by describing a criterion for evaluating the exchange. The underlying assumption is that a relationship that endures implies a mutually profitable exchange. A profitable exchange is one in which the net gain from rewards minus costs is positive for both persons; both parties to the exchange consider that they are receiving a fair return on their "investments." This line of thinking implies that there will be dissatisfaction whenever the exchange becomes unbalanced. It is perhaps obvious but nonetheless important that profit and cost relate to the values that individuals place on the elements of "currency" in the relationship. Hence, the issue of equity leads quite directly to a consideration of the determinants of these values. In the social psychological literature, variables that have received scrutiny are those that describe the context that individuals use to evaluate what they get from a relationship. In short, can they get it cheaper elsewhere? Thus, the value that is placed on the contributions that one's partner gives to the relationship is in part a function of their relative scarcity or abundance. An individual may endure high costs in a relationship because no ready alternative source is seen to be available. Similarly, the contributions of the other may be accorded greater value than one's own because the other's expertise, for example, is seen as precious.

To cast the leader–follower relationship into a social exchange framework is to see it as one manifestation of the more general paradigm. The distinctive flavor of this particular application is in terms of the task context. Leadership is an exchange in a setting where there is a mutual understanding that there is a job to be done. The leader provides competence at getting the task accomplished, and the followers "yield" to these initiatives. This one-dimensional characterization of the exchange, however, is oversimplified. Task accomplishment is clearly not the only currency of reward. As we have seen, followers respond to "consideration" from the leader, and there is tentative evidence that leaders similarly respond to the esteem and approval of followers (Barrow, 1976; Herold, 1977). Thus, even within this special context where there is a relatively explicit identification of a joint task, leaders and followers are influencing each other in complicated ways. They are exchanging emotional rewards and costs, as well as working on the task.

SOCIAL EXCHANGE APPROACHES TO THERAPY

A consideration of the therapist–client relationship as a mutual exchange reveals many parallels to features we have identified from the study of leadership. Once again the question to be posed is: What is it that is being exchanged? Surely therapists offer their expertise for a price. Part of that expertise is their ability to influence their clients, to lead them often in subtle ways to redefine their problems, to think about themselves and the nature of solutions from a different perspective. As in the case of leadership, therapists not only structure the process of therapy, they also play an active role in defining the common goal or task. Thus, some of what clients get from their exchange with therapists is the sense that they can make progress toward accomplishing a goal. Therapists begin to define a problem in a particular way and at least to imply that there are steps that can be taken to reach a solution. Much that a therapist offers, then, is a way of structuring their joint task, both in terms of the way that the "problem" is conceptualized, and also in terms of the process that is to be followed in trying to reach a resolution. At a minimum, the latter structuring involves such features of a "contract" with the client as how often and under what circumstances they will meet. Of course, sometimes the process of therapy is much more explicitly specified.

The therapist, then, initiates a structure, much as do leaders generally. What do clients offer in return? Of course, under many circumstances, the therapist receives money. But it is unlikely that money is a contingent element of the exchange, since the usual practice would be for the therapist to set standard fees at the outset for all clients. No doubt some clients do manage their payments in ways that make this transaction an active part of the process of therapy. At these times therapists may be influenced by their maneuvers in paying their bills. These sorts of contingencies may also be apparent for other parameters of therapy, for example, in the management of time. Any departures from an agreed-upon schedule by either the client or therapist may be seen as attempts to shape or influence the other. This suggests that both time and money are "exchanged" and thus have potential influence on both the client and therapist. These two basic features are fairly obvious, however, and for that reason their effects are likely to be monitored closely.

Other denominations of exchange may be more difficult to notice. As we have indicated, task performance is an important dimension of an influence relationship. The leader initiates and the follower yields, with the expectation for both that some progress will be made toward their joint goal. In terms of the therapist and client, the client is influenced by the initiatives that the therapist takes and the therapist in turn by the client's responsiveness to these initiatives. The important implication is that the social exchange paradigm recognizes that the client is in a powerful role, capable of yielding to or resisting the therapist's suggestions.

Of special interest are those clients who are either totally resistant to the therapist's initiatives, or overly responsive to every nuance of therapist behavior. From a social exchange view, these relationships are ones in which the client does not emit behavior that is reinforcing to the therapist. Ideally under such circumstances, therapists merely alter behavior over time so as to elicit signs of their influence, that is, they elicit behavior that is seen as contingent on their effort. Under less than ideal circumstances, however, therapists' feelings of anger and frustration are aroused. These negative "countertransference" feelings may then become important in the overall client–therapist relationship. (See Langs, 1973; or Sandler, Dare, & Holder, 1973, for reviews of the importance of countertransference to the therapy process.) A growing recognition of the significance of countertransference feelings is quite consistent with the conceptualization of the therapy relationship as an exchange process.

That therapist and client alike have affective reactions to the task of therapy raises the issue of how such exchange of affect may influence the course of therapy. In dynamic therapies, the importance of transference reactions is emphasized. That the client has strong affective reactions to the therapist is seen as significant. These affective reactions are not, however, generally discussed as significant to the therapist. The social exchange approach would suggest that it is also necessary to examine the impact of clients' expressions of esteem and regard for the therapist. Just as therapists may have countertransference reactions to clients' resistance, they may also react to clients' affective expressions. Generally, though therapists are expected to take account of their affective reactions to client expressions of feeling, they are not supposed to show such reactions in behavior directly. In other words, though client expressions of esteem or anger may influence the task behavior of the therapist indirectly, presumably they are not directly "reinforcing" to the therapist. Thus, this dynamic model of therapy seems to suggest that the client–therapist relationship is not a mutual exchange process in the usual sense. Therapists shape clients' attitudes and behaviors, in part with their affective communications; but the reverse, it is hoped, does not hold. The achievement of this unusual, one-sided influence relationship is probably the most exacting demand of the therapist role.

POWER AND CHARISMA: TWO IMPORTANT CONCEPTS IN LEADERSHIP

Concepts of power and charisma have been a generic part of the social psychological literature. They have not, however, been discussed and studied at any one particular time; nor has their investigation been limited to a discussion of leadership. Power, for example, is dealt with as part of interpersonal, group, and organizational processes. Still, both concepts are important for understanding leadership dynamics.

POWER

Power has had many definitions, both in the analysis of leadership and of social relationships generally. Often it is used to describe legitimate authority, the power one has as a result of occupying a specific role. Between therapist and client there is such an acknowledged power difference. French and Raven (1959) distinguish six bases from which power is exercised: legitimate, reward, coercion, informational, expert, and referent.

Legitimate or *role* power is exercised by the therapist in establishing the parameters of the relationship, for example, agreement to work with the client, schedule, fee structure, and missed appointments. It is based on the perception that the therapist has the power to set these conditions. A more subtle form of legitimate power occurs early in the relationship when the therapist helps the patient discover how to use the time available. In some forms of therapy, for example, behavioristic, the therapist is clearly an expert who gives instruction based also on expertise in the method. In dynamic therapy the process is less direct and other forms of power may enter into the influence process. Especially in schools of thought that regard the therapeutic relationship as egalitarian, for example, Rogerian, it becomes important to explore power dynamics.

In addition to legitimate power, *reward* is an important source of influence for therapists as they shape interaction. *Information* may sometimes be a source of influence in a more didactic mode. *Coercion* in some of its more subtle forms may or may not be acceptable in the process, but the therapist surely can affect patients by various forms of withdrawal—of attention, or even of services.

The structure of the therapeutic relationship controls power differences in some interesting ways. Knowledge about others is often one form of power (Heider, 1958). In most relationships a reciprocal process occurs in which there is mutual disclosure (Altman & Taylor, 1973). That means that each participant in the relationship has information and potential power to harm or benefit the other. In relationships of differential power in our society, those low in power may withhold information that could be harmful to them. However, there are relationships where higher-power others, such as lawyers, doctors, or therapists, need information that could increase their power to render professional services. In these relationships professional codes of ethical behavior and confidentiality strictly protect the lower-power client from any misuse of whatever information is given.

How much power or influence do therapists really have? Within the contractual agreement they clearly have a considerable amount. However, if the relationship is questioned, if the client wishes to dissolve the contract, then, with the exception of total institutions, the therapist has no more power than the client. One possible area for negotiating power in the relationship is the fee-paying arrangement. Here therapists set conditions and clients can accede, negotiate, or propose.

The notion of *idiosyncrasy credits* also concerns the management of the power imbalance in a social exchange. Hollander (1958) proposes that over time in a relationship, followers accord "credits" to the leader who demonstrates competence and success at the task. Over time these credits allow the leader greater and greater latitude to take initiatives, to depart from the norms, to innovate without threatening the integrity of the relationship. It may be said that the accrual of such credits heightens the leader's power in the relationship.

Does it make sense to apply the notion of idiosyncrasy credits to the therapeutic relationship? Do therapists develop increased influence (power) with patients on the basis of their performance as leaders in structuring the relationship and teaching patients how to learn in it? Interestingly, this idea of gradually increasing power runs counter to many theories of the therapeutic relationship in dynamic therapy. Here the focus of power (or influence) differential comes early in the relationship, when transference processes are thought to give the therapist unusual influence over clients. As the transference is worked through, the relationship becomes more balanced and less differential is experienced. The client becomes healthier, stronger, and thus increases power by being able to challenge, evaluate, and judge.

Although the notion of idiosyncrasy credits predicts increasing latitude from followers on both task and nontask behaviors, the research on this concept has almost exclusively studied follower yielding on task-related issues. This research demonstrates increased follower willingness to yield with increased leader-demonstrated competence at the task (Hollander, Julian, & Perry, 1966).

This suggests that, in subtle ways, therapist influence may grow stronger rather than weaker over time. If that is so, then it probably occurs in some dimensions and not others. It may be that both increased patient strength that emerges from dealing with transference and increased therapist influence occur, but on different dimensions. Perhaps over time therapists do have increasing latitude to violate some of the more formal rules of the relationship without disrupting the therapeutic process, for example, being late, permitting interruptions, making personal statements. It also seems likely that therapists' influence over patients as they "learn-how-to-learn" in therapy may lead to a change in client values. For example, clients may value openness, the exhanced expression and exploration of feelings as they proceed in therapy. In this sense the client becomes more like the therapist: They share more values. Thus, if the therapist appeals to these values we might expect that over time there would be increased client responsiveness. Clients would then be more "influenceable" and the therapist more "powerful."

Of course, the therapist–client social exchange is also affected by the client's own supporting others. Clients who have alternative sources of support will likely be less susceptible to dependence on the therapist. Clients, however, have been known to act independent and healthy in the therapy hour and then fail to sustain these behaviors in the real world. The "transference cure," a well-known phenomenon, can result not only from the power relationship between therapist and client, but also from the client's external system.

Power and influence processes, then, are both central to understanding exchanges between therapist and client, and also so subtle that they demand continuous scrutiny.

CHARISMA

Although the leadership literature has shown only sporadic interest in charisma, it is a fascinating topic that relates closely to therapeutic influences.

Charisma is that magical attracting quality that energizes followers (Weber, 1947); and a useful distinction has been made between the cognitive orientation of the leader and that person's affective style (Katz & Kahn, 1978). In pursuing this distinction Katz and Kahn describe the requirements in each dimension for effective leadership. At the cognitive level effective leaders have "systemic perspective," that is, a sensitivity "to the requirements that the organization must meet in order to maintain a state of equilibrium with its environment" (p. 540). This perspective includes a perception of the needs and functioning of the internal system as well.

The affective style of leaders when it is positive and compelling has been labeled "charisma." This capacity enables the leader to mobilize more support and trust than a rational analysis of the situation might engender. Charismatic leaders are those who at the very least appear to know how to blend the use of content—ideals, goals, and mission statements—with the use of affect—impassioned messages, personal commitment expressed through "how" something is said as well as "what." Thus both are important characteristics. Charisma uninformed by a systemwide perspective may lead to exciting human relations, but the ship may founder for lack of direction. Goals and mission statements that appear credible but are unsupported by fervor may lack the following required for implementation.

What creates charisma? Why is the word "magical" frequently associated with it (Hollander, 1985; Katz & Kahn, 1978; Weber, 1947)? First, distance from the leader is important. The charismatic leader is not human, fallible, or well known. "Day-to-day intimacy destroys illusion" (Katz & Kahn, 1978, p. 546). The leader must be distant enough to permit the magic of illusion to work.

At the same time, for identification with the leader to occur, that person must be seen as enough like the followers to gain their trust and allegiance (Kelman, 1958).

A third force that creates charisma has to do with external conditions. In times of crisis people seek a leader who is charismatic, in whom they can put trust, and to whom they can give over responsibility for the overly complex decisions that confront them. They invest the leader with more than usual power and authority, and pledge their loyalty (Weber, 1947). In less turbulent times a leader may need to act courageously or imaginatively in order to attract this same trust and loyalty (Katz & Kahn, 1978).

When these conditions are present, there is potential for the emergence of a charismatic leader. Interestingly, explanations for leaders' extraordinary powers have focused on both their cognitive orientation and their affective style. Theorists such as Burns (1978) describe the leader's capacity to articulate followers' needs and wishes just before these become conscious and articulate as an important factor in the development of charisma.

Theorists emphasizing affective style may focus on leaders' ability to embody the resolution of the internal conflicts of their followers. Fromm, for example, points out that a leader's personality structure "will usually exhibit in a more extreme and clear-cut way that particular personality structure of those to whom his doctrines appeal" (1941, p. 65).

Charisma, like power, has two faces, one positive, the other negative (McClelland, 1975). Charismatic leaders do not always lead well.

The classic study of encounter group leaders (Lieberman, Yalom, & Miles, 1973), for example, showed that more charismatic leaders had more risk of casualty in their groups.

For the follower, the charismatic leader displays a set of personal qualities that creates a highly personal sense of "knowing" the leader. This idenitification with the leader occurs despite and as a part of the necessary distance between leader and follower. It is a distinctly one-way process, since the leader is disclosing to followers who may be total strangers. Often it is "crafted disclosure"; leaders think very carefully about what they will say and do and what symbols they will invoke.

THERAPISTS AND CHARISMA

There are some interesting parallels between the therapist–client relationship and the leader–follower relationship. Neither is a social exchange relationship between persons of equal status where we would expect disclosure to be controlled by the norm of reciprocity (Altman & Taylor, 1973). Both can be described as one-way relationships in which the distance between the two participants is a barrier to personal knowing of the other. In dynamic therapy, the therapist's role may also be described as one of crafted disclosure, that is, disclosure with an eye toward the primary task of therapy rather than of social exchange. This controlled role limits what can be known about the therapist and is structured to produce a reaction in the client. In the leadership literature this reaction is charisma; in therapy it is transference. Thus, in both, affective arousal is important.

In times of crisis, followers look to leaders to understand and resolve complex situations. In therapy, clients (often in crises themselves) look to their therapist to have something analogous to the systemic perspective: to understand them and their problem and its context as well as the whole process of therapy. It is easy for a confused, uncertain client to ascribe to the therapist great or even complete understanding. The therapist is all powerful, in charge of the process, and even controlling it. This identification with the therapist closely parallels what followers feel about charismatic leaders.

In each of these relationships there is distance, a role-created barrier to normal social exchange. In leadership this barrier heightens the leader's charisma, that global and undifferentiated sense of the leader's power. In therapy, distance heightens transference processes. The sense of distance is also heightened by the structuring of the relationship. The therapist is seen as very valuable and important; for example, only a limited amount of time is available and at a high fee. All of this creates energy and motivation in clients that can be used in the therapeutic process.

But what of the therapist's affective style? We have already said that the charismatic relationship is experienced by followers as highly personal even though the follower may know little about the leader. The therapist's affective style is traditionally empathic, caring, responsive; in other words, it too is very personal. The therapist behaves like someone who knows one well, with whom one is close, even though as client one does not have the same knowledge about the therapist.

The combined impact of the cognitive and affective style of the therapist creates transference, that quality of identification with the therapist that, at least at the beginning, energizes the relationship.

Here Kelman's (1958) notions about change processes are instructive. He proposes three levels of permanence of change. *Compliance* is change that occurs only under surveillance. *Identification* is change that occurs because of the personal emulation of a referent other, and disappears when the other is no longer an ideal. *Internalization* is change that occurs congruent with inner values and is therefore "permanent."

In therapy the transference process is thought of as providing energy to move the client toward work that will result in internalization. Thus identification is a station-stop on the way to the goal. If, however, the client changes only as a way of pleasing the therapist, we say that there is a "transference cure." The client is stuck in the identification stage and does not proceed to internalization, a more permanent and desirable state. Thus, the goal is a process in which the client articulates and internalizes the structure of the self and how that will be embodied in everyday life. Therapists may, like leaders, help by giving voice to what is coming to consciousness, what the client is ready to deal with, what the client wants and needs.

In the leadership literature, distinctions between compliance and identification are often discussed under the topic of how to motivate followers. The process of internalization is dealt with in the concept of shared leadership and decision making. When leaders involve followers in shaping decisions with them, the follower has an opportunity to express values and become committed to solutions. The decisions made in a participatory process will be more internalized than when the leader decides alone (Vroom & Yetton, 1973). Thus in both leadership and therapy, charisma is an enabling process but not the goal of the relationship.

CONCLUSION

Social psychology is in large part the study of social interaction. Leadership, like psychotherapy, is a special case of social interaction. Some, though not all, generic principles that govern all interaction also apply in leadership situations.

We have reviewed some of the principal notions that have been useful in understanding leadership, and discussed their relevance to another special case of interaction, psychotherapy. Such reflections are intended to be provocative rather than prescriptive. It often seems, however, that the perspective of another discipline can cast in bolder relief issues that remain dim when treated within an accustomed framework. The study of leadership provides a different lens through which to view psychotherapy.

ACKNOWLEDGMENTS

The authors thank Robert W. Rice and Edwin P. Hollander for their comments on an earlier version of this chapter.

REFERENCES

Allport, G. W. *The historical background of modern social psychology.* Cambridge, Mass.: Addison-Wesley, 1954.

Altman, I., & Taylor, D. A. *Social penetration: The development of interpersonal relationships.* New York: Holt, Rinehart & Winston, 1973.

Bales, R. F., & Slater, P. E. Role differentiation in small decision making groups. In T. Parsons & R. F. Bales (Eds.), *Family, socialization and interaction processes.* New York: Free Press, 1955.

Barrow, J. C. Worker performance and task complexity as causal determinants of leader behavior, style, and flexibility. *Journal of Applied Psychology,* 1976, *61,* 433–440.

Bartol, K. M. Male versus female leaders: The effect of leader need for dominance on follower satisfaction. *Academy of Management Journal,* 1974, *17,* 225–233.

Bavelas, A., Hastorf, A. H., Gross, A., & Kite, R. W. Experiments in the alteration of group structure. *Journal of Experimental Social Psychology,* 1965, *1,* 55–70.

Bellak, L., & Small, L. *Emergency psychotherapy and brief psychotherapy.* New York: Grune & Stratton, 1978.

Blake, R. R., & Mouton, J. S. *The managerial grid.* Houston: Gulf, 1964.

Blau, P. M. *Exchange and power in social life.* New York: Wiley, 1964.

Burns, J. M. *Leadership.* New York: Harper & Row, 1978.

Calder, B. J. An attribution theory of leadership. In B. M. Staw & G. R. Salancik (Eds.), *New directions in organizational behavior.* Chicago: St. Clair Press, 1976.

Chemers, M. M. The social, organizational, and cultural context of effective leadership. In B. Kellerman (Ed.), *Leadership: Multidisciplinary perspectives.* Englewood Cliffs, N.J.: Prentice-Hall, 1983.

Day, D. R., & Stogdill, R. M. Leader behavior of male and female supervisors: A comparative study. *Personnel Psychology*, 1972, *25*, 353–360.

Evans, M. G. The effects of supervisory behavior on the path-goal relationship. *Organizational Behavior and Human Performance*, 1970, *5*, 277–298.

Fiedler, F. E. *A theory of leadership effectiveness*. New York: McGraw-Hill, 1967.

Fiedler, F. E. The contingency model and the dynamics of the leadership process. In L. Berkowitz (Ed.), *Advances in experimental social psychology*. New York: Academic Press, 1978.

Fleishman, E. A., & Harris, E. F. Patterns of leadership related to employee grievances and turnover. *Personnel Psychology*, 1962, *15*, 43–56.

French, J. R. P., & Raven, B. The bases of social power. In D. Cartwright (Ed.), *Studies in social power*. Ann Arbor: University of Michigan, 1959.

Fromm, E. *Escape from freedom*. New York: Rinehart, 1941.

Gibb, C. A. Leadership. In G. Lindzey (Ed.), *Handbook of social psychology* (Vol. 2). Cambridge, Mass.: Addison-Wesley, 1954.

Heider, F. *The psychology of interpersonal relationships*. New York: Wiley, 1958.

Herold, D. M. Two-way influence processes in leader-follower dyads. *Academy of Management Journal*, 1977, *20*, 224–237.

Hollander, E. P. Conformity, status, and idiosyncrasy credit. *Psychological Review*, 1958, *65*, 117–127.

Hollander, E. P. Leadership and power. In G. Lindzey & E. Aronson (Eds.), *The handbook of social psychology* (3rd ed). Reading, Mass.: Addison-Wesley, 1985.

Hollander, E. P., & Yoder, J. Some issues in comparing women and men as leaders. *Basic and Applied Social Psychology*, 1980, *1*, 267–280.

Hollander, E. P., Julian, J. W., & Perry, F. A. *Leader style, competence, and source of authority as determinants of actual and perceived influence* (Tech. Rep. No. 5, ONR Contract 4679). Buffalo: State University of New York at Buffalo, 1966.

Homans, G. C. *Social behavior: Its elementary forms*. New York: Harcourt, Brace, 1961.

House, R. J., & Mitchell, T. R. Path-goal theory of leadership. *Contemporary Business*, 1974, *3*, 81–98.

Kahn, R. L., & Katz, D. Leadership practices in relation to productivity and morale. In D. Cartwright & A. Zander (Eds.), *Group dynamics*. New York: Harper & Row, 1953.

Katz, D., & Kahn, R. L. *The social psychology of organizations* (2nd ed). New York: Wiley, 1978.

Kelman, H. C. Compliance, identification and internalization: Three processes of attitude change. *Journal of Conflict Resolution*, 1958, *2*, 51–60.

Langs, R. *The technique of psychoanalytic psychotherapy*. New York: Jason Aronson, 1973.

Lewin, K., Lippitt, R., & White, R. K. Patterns of aggressive behavior in experimentally created "social climates." *Journal of Social Psychology*, 1939, *10*, 271–299.

Lieberman, M. A., Yalom, I. D., & Miles, M. B. *Encounter groups: First facts*. New York: Basic Books, 1973.

Mann, R. D. A review of the relationships between personality and performance in small groups. *Psychological Bulletin*, 1959, *56*, 241–270.

McClelland, D. *Power: The inner experience*. New York: Irvington, 1975.

Osborn, R. N., & Hunt, J. G. An adaptive-reactive theory of leadership: The role of macro variables in leadership research. In J. G. Hunt & L. L. Larson (Eds.), *Leadership frontiers*. Kent, Ohio: Kent State University Press, 1975.

Pfeffer, J. The ambiguity of leadership. In M. W. McCall, Jr., & M. M. Lombardo (Eds.), *Leadership: Where else can we go?* Durham, N.C.: Duke University Press, 1977.

Rogers, C. L. *Client-centered therapy*. Boston: Houghton Mifflin, 1951.

Rogers, C. L. *Psychotherapy and personality change*. Chicago: University of Chicago Press, 1954.

Sandler, J., Dare, C., & Holder, A. *The patient and the analyst*. New York: International Universities Press, 1973.

Shartle, C. L., Stogdill, R. M., & Campbell, D. T. *Studies in naval leadership*. Columbus: Ohio State University, Personnel Research Board, 1949.

Simmel, G. *The sociology of Georg Simmel* (K. H. Woolf, Ed.). London: Free Press, Collier-Macmillan Ltd., 1950.

Stogdill, R. M. Personal factors associated with leadership: A survey of the literature. *Journal of Psychotherapy*, 1948, *25*, 35–71.

Stogdill, R. M., & Shartle, C. L. Methods for determining patterns of leadership behavior in relation to organization structure and objectives. *Journal of Applied Psychology*, 1948, *32*, 286–291.

Vroom, V. H., & Yetton, P. W. *Leadership and decision making*. Pittsburgh: University of Pittsburgh Press, 1973.

Walster, E., Walster, G. W., & Berscheid, E. *Equity: Theory and research*. Boston: Allyn & Bacon, 1978.

Weber, M. *The theory of social and economic organization* (T. Parsons, Ed.; A. M. Henderson & T. Parsons, Trans.). New York: Free Press, 1947.

5

Nonverbal Behavior Research and Psychotherapy

MARTHA DAVIS

A HISTORICAL PERSPECTIVE

> *In spite of . . . difficulties in conscious analysis, we respond to gestures with an extreme alertness and, one might almost say, in accordance with an elaborate code that is written nowhere, known by none, and understood by all.*
>
> (Sapir, 1968, p. 556)

> *He that has eyes to see and ears to hear may convince himself that no mortal can keep a secret. If his lips are silent, he chatters with his finger tips, betrayal oozes out of him at every pore.*
>
> (Freud, 1905/1959, p. 94)

There have been many assertions of the importance of body movement, facial expression, voice tone, and other aspects of nonverbal behavior in face-to-face interaction. And, if one includes research from anthropology, communication, and ethology as well as psychology, there is now an extensive and fascinating literature on bodily communication

---

MARTHA DAVIS • Department of Mental Health Sciences, Hahnemann University, Philadelphia, Pennsylvania 19102. The author's recent research has been sponsored by Hahnemann University in Philadelphia, aided until 1982 by grants administered through the Institute for Nonverbal Communication Research, New York.

that is of potential value to the clinician (M. Davis, 1972; M. Davis & Skupien, 1982).

There are now some excellent anthologies and research overviews of nonverbal communication (F. Davis, 1973; Harper, Wiens, & Matarazzo, 1978; Knapp, 1978; LaFrance & Mayo, 1978; Weitz, 1979). And the clinician interested in nonverbal communication can also find many specialized texts such as on facial expression (Ekman, Friesen, & Ellsworth, 1972), skill in perceiving nonverbal behavior (Rosenthal, Hall, DiMatteo, Rogers, & Archer, 1979), ethological and quantitative assessment of face-to-face interaction (Duncan & Fiske, 1977; Kendon, Harris, & Key, 1975), periodicity in communicative behavior (M. Davis, 1982; Jaffee & Feldstein, 1970), nonverbal correlates of power relationships (Henley, 1977), gender and nonverbal behavior (Mayo & Henley, 1981), as well as books devoted to nonverbal communication in psychotherapy (Scheflen, 1973; Waxer, 1978).

From the 1930s until recently the network of nonverbal communication researchers has been intricately interrelated. The zeitgeist for the current "movement movement" was germinating as early as the 1930s when psychiatrist Harry Stack Sullivan and linguist Edward Sapir were close associates (Sullivan, 1939) and anthropologist Eliot D. Chapple was working with psychiatrist Erich Lindemann on the measurement of therapy interaction rhythms (Chapple & Lindemann, 1942). It can be seen growing at the anthropology department of Columbia University where Franz Boas, who made some of the earliest attempts to study cultural movement styles from film, would directly influence Gregory Bateson and Margaret Mead's film study of the Balinese (Bateson & Mead, 1942) and David Efron's classic analysis of Jewish and Italian gestures (Efron, 1942/1972). It shows in the influence of Sapir and linguists Trager and Smith on anthropologist Ray L. Birdwhistell, and in the impact of communications in the mid-1950s among Frieda Fromm-Reichmann, Birdwhistell, and several important linguists attending the Palo Alto Institute for the Advanced Study in the Behavioral Sciences—from which Bateson, Birdwhistell, and later Albert E. Scheflen would be spurred to examine communication in psychotherapy (Birdwhistell, 1970). By the mid-1960s Charny (1966), Condon (1968), Lomax (1968), Byers (1972), and many other would confer intensely with Birdwhistell and begin various lines of kinesic research, while Chapple was directly influencing Feldstein, Jaffee, and a growing number of psycholinguistic researchers. And by the 1960s Bartenieff (Bartenieff & Lewis, 1980) would be bringing Rudolph Laban's work on movement analysis and notation to bear on her studies of psychiatric patients with this author (Bartenieff

& Davis, 1975), examination of infant movement patterns with Kestenberg (Kestenberg, 1975), and cross-cultural studies of movement with Lomax and Paulay (Lomax, 1968).

At a larger level the movement movement appears one manifestation of the trend encompassing communication systems research, interpersonal theories of psychiatry, and efforts to diffuse distinctons of mind–body, right and left brain hemisphere functions, and behavioral form–content. And the current respect for visible behavior is surely a powerful expression of the impact of film and television.

A therapist surveying this field may be struck by certain clinical potentials that are very close to being realized. First, the field is a remarkable source for operationalizing key psychological concepts that have largely eluded empirical assessment, such as bonding, regulation of intimacy, interpersonal trust, empathy, rapport, repression of affect, and splitting. Many of the research examples to be discussed are in effect studies of the visible manifestations of these processes. For example, it is conceivable that, if very complex movement observation assessments such as those done by Kestenberg and her associates can be standardized, movement rhythms and muscle tension analyses may empirically confirm Freud's theory of psychosexual stages. Kestenberg (1975) has found "tension flow" rhythms in infants and young children that correspond to specific forms of drive discharge, "oral," "anal," "phallic," and "genital" motor rhythms that appear in a stage-specific way in the first 4 years of life, and rhythmic predominances that reflect specific fixations. I hope to show with later examples that very diverse intrapsychic and interpersonal processes can be observed in movement depending on what details of it the researcher studies.

Another general impression from this field has been strongly asserted by Kiesler:

> The most crucial place to search for relationship is in the nonverbal behavior of the interactants. [T]he total available methodology of assessment for paralanguage, kinesics, proxemics, touch, etc. is centrally relevant for assessment of client and therapist relationship factors. (1979a, p. 303)

Early research regarded bodily expression as a royal road to the unconscious with affective states and emotions expressed from "within" the individual "outward" to the perceiver. The crude and often contradictory beginnings of facial recognition research are now supplanted by highly sophisticated, empirically validated research on the face and emotion (Ekman & Friesen, 1975). But the last 20 years of nonverbal communication research have been most notable for discoveries of how subtly and complexly kinesics and proxemics convey "relationship messages" as well. Not coincidentally, Kiesler (1979) has written a position paper stressing the importance of nonverbal behavior in psychotherapy within an interpersonal, Sullivanian framework.

Finally, the clinician surveying this literature may get a disconcerting feeling that there is something too elusive about the subject. How can it promise so many insights and yet be so neglected in therapy practice and training; how can it excite and fascinate at one level yet seem so trivial in the real world; how can someone like Reich present such rich analysis of character from the way his patient walks and sits whereas one's own patients appear motorically indistinct, unremarkable, bland? In actual therapy practice what *is* its relevance?

Despite protestations in the literature that the goal is not to someday read body language like a book with the aid of a dictionary of precise definitions, there is a concerted effort perceptible throughout the research to *delineate* specific motions and positions and then to determine, or at least try to approximate, the most context-free, commonly shared meaning(s) of them. This requires that the action, posture, movement, or vocal characteristic have a perceptible "discreteness" and specificity. Positions, instrumental actions, the signs of American Sign Language, and certain hand gestures (such as what Ekman and Friesen have called "emblems") have this property of discreteness and specificity.

By now it is possible to construct a glossary of specific meanings from the literature on facial expression of emotion (Ekman & Friesen, 1975), postures and attitude expression (Mehrabian, 1972), emblems (Kendon, 1980; Morris, Collett, Marsh, & O'Shaughnessy, 1979), and, of course, formal sign language (Stokoe, 1972). In this literature discrete movements and positions with their respective meaning(s) become the figure; and the "display rules," "blends" (Ekman & Friesen, 1975), individual and cultural variations, that is, the qualifications, become very much the ground.

Certain features of nonverbal behavior that are highly discrete and specific, such as facial expressions of happy, sad, disgust, anger, fear, and surprise, have highly specific interpretations both in their "pure" forms and as identifiable elements of facial blends. Specific emotions can also be decoded from types of voice tone (Ostwald, 1963; Scherer, 1979), although not as yet with the weight of the empirical evidence produced for facial recognition.

Once the observer attends to more diffuse, continous, difficult-to-demarcate activity such as fidgeting, postural shifts, degrees of tension, and so on, the interpretations become more general. For example, Waxer

(1978) has demonstrated that observers can distinguish anxious and depressed patients from soundless videotapes, and he can tease out of their reports what clusters of movements appear to contribute to their impressions. But the nonverbal referents are not nearly so specific and particularized as those for discrete facial expressions or even voice quality. The glossary becomes disappointingly general (e.g., depressed patients display longer averted gaze and lowered head positions).

Although there have been advances in decoding facial, vocal, and postural cues of potential value to the clinician, a great deal of the clinically valuable research is not and cannot be so concrete. The interpretation of the meaning of specific body movements is greatly influenced by the knowledge of the context (Birdwhistell, 1970; Scheflen, 1973). There are also notable differences in the *ability* to decode (perceive and interpret) nonverbal behavior. Rosenthal *et al.* (1979) have offered evidence that individuals will be more or less accurate in interpreting alternative meanings of the expressions of a person presented in soundless film segments or content-obscured sound tapes. With the PONS (Profile of Nonverbal Sensitivity) test they can generate individual skill profiles indicating what aspects of nonverbal behavior one more readily perceives than others. To further complicate efforts to specify what a movement means, there is the enormous complexity referred to in the literature as "channel discrepancy"—the numerous ways movements may anticipate, contradict, amplify, or slant speech content; the discrepancies between what one says and how one says it—in short, the complexity of deciphering ongoing behavior from diverse "channels."

The research on specific meanings of discrete expression is based primarily on the reliable judgments of detached observers or listeners who are not in the situations photographed or taped. There is a vast difference between the perceptions of one looking at a photo or videotape and the experience of the interactants. And this difference—"detached decoding" versus "participant observation"—is crucial to the clinician.

There is now a sizable experimental literature on observer judgments of nonverbal behavior in simulated interviews and psychotherapy sessions (Waxer, 1978). Counseling psychologists in particular have shown a strong interest in nonverbal communication and have contributed considerably to this literature (Gladstein, 1974). It is directly related to earlier facial recognition research in that experimental subjects, often college students, are asked to judge photographs, films, or videotapes in terms of what affect(s) or attitude(s) are expressed in the facial expression, visual behavior, and positions of the interactants. This literature would seem a good place to start for a clinician interested in nonverbal behavior in psychotherapy; yet for a number of reasons it can be a misleading and very limited source for the clinician.

First, one would not necessarily be learning about nonverbal behavior in psychotherapy. One would be reading about *recognition* and *interpretation* of contrived behaviors presumed related to psychotherapy. Frequently, the behaviors to be judged are so contrived as to generate either clinically trivial or artificial findings. For example, the "therapist" is to sit forward or back at controlled points in a simulated interview to examine perceptions of rapport with the client.

Besides problems inherent in experimental manipulations and use of naïve judges, there is another drawback to the decoding research that is crucial to the clinician. It cannot safely be assumed that the client or therapist will interpret kinesic, proxemic, and paralinguistic patterns as raters do because there is a profound difference between impressions by one *separate from* the interaction and impressions as a participant *in* the interaction (Bakken, 1978). In fact, we know very little about how the perception of nonverbal behavior varies by context. There is, however, a great deal of indirect evidence that perceptions while in the interaction are dramatically different from those of a naïve observer or a researcher viewing a videotape.

For this decoding research to be of clinical value the research must honor at least some of the clinician's needs—namely, that the behavior examined is relevant and important to therapy; that there are some data offered about the patterns of its frequency and occurrence in therapy; that the analysis of the behavior is not so gross and simplistic that it adds nothing to what the clinician would intuitively grasp anyway; that the researcher is consistently clear as to *who* is rating *what* with *what terms* and for what ends, and does not split too readily into unwarranted generalizations to therapy.

Because nonverbal behavior is ubiquitous and visible and has such important psychological valence, because it is in a sense at the nexus of social perception, affective expression, and "presentation" and experience of self, it affords a way of empirically examining relationships between perception of "the other" during interaction versus detached "objective" observation versus "subjective" experience of one's behavior and manner. For the therapist this could elucidate such important clinical concerns as transference–countertransference reactions or parataxic distortions.

For example, in the research on posture mirroring there are clues as to what new insights could be generated from study of these multiple perspectives. "Objective" onlookers offer a "group consensus" that mirroring reflects heightened rapport (Trout & Rosenfeld, 1980). Charny (1966) found that the client's verbal behavior was more "positive" and "bound to the therapeutic situation" during periods of mirrored positions. Scheflen (1973) has outlined some interesting alternations in and out of mirrored positions that relate to shifts in alliances during a family therapy session. LaFrance (1982) presents evidence that posture mirroring precedes the interactants' experience of rapport and may be a mechanism for establishing it. But it would also help to learn about the patient's and therapist's experience during periods of mirroring, both when aware and when not conscious of it. In other words, it should be possible to examine "subjective" and "objective" reactions in relation to naturalistic occurrences of psychologically cogent nonverbal behaviors such as posture mirroring and to learn much more than that they "mean" rapport.

MOVEMENT AND PERSONALITY

> I shall consider here, particularly, ways of thinking and perceiving, ways of experiencing emotion, modes of subjective experience in general, and modes of activity that are associated with various pathologies . . . it is clear that there are many interesting aspects of style that cannot even be touched on here—for example, body-movement styles. (Shapiro, 1965, pp. 1–2)

Shapiro's passing reference to movement style is an appropriate footnote to the scattered literature on movement and personality. Beyond Reich's classic essay on movement and character development (1949), it is not woven into the mainstream of literature on personality and psychodiagnosis. Before the 1930s, texts on differential diagnosis of dementia praecox, mania, and depression gave ample space to descriptions of motor symptoms (M. Davis, 1972). Within early analytic writing, patterns of inadvertent actions and styles of body movement were readily considered part of the "psychopathology" of everyday life, and it has been argued (Murphy, 1976) that Freud did not take issue with Ferenczi's or Reich's formulations on the analytic importance of body motion, but with the introduction of active techniques of intervention or in Reich's case with the political direction of his work. In experimental psychology, Allport and Vernon (1933) completed an ambitious study confirming with various mechanical quantitative measures that individuals have distinctive motor styles that persist over time and task and that appear cogently related to personality. Yet after the 1930s the subject is addressed by a very few isolated writers, most of them psychoanalysts reporting clinical observations.

Individual differences in activity type from infancy have long been recognized (Fries & Lewi, 1938), and the persistence through childhood of motor manifestations of temperament can be found in the developmental research (Escalona & Heider, 1959). Mittelmann (1958) postulated a motor phase of development around age 2 and presented numerous clinical examples of the crucial importance of movement in ego development. Reich (1949), Lowen (1958/1971), and more recently Kestenberg (1966, 1975) have formulated elaborate theories as to the role of movement in the early evolution of psychological defenses, drive discharge, and psychosexual phases in early childhood. And recent microanalytic film research on mother–infant interaction documents the crucial role of motion in bonding (Beebe & Stern, 1977). As Pine (1977, p. 77) has said, "this mutual preverbal cueing and its consequent shaping effects is the earliest glue of the object tie."

Yet experimental research examining relationships between personality and movement, posture, breathing patterns, and other bodily expressions is very sparse (Allport & Vernon, 1933; Christiansen, 1972; Takala, 1975). In their 355-page synthesis of nonverbal communication research, Harper *et al.* (1978) devote only 2 pages to personality variables and body movement and 5 pages to individual differences in proxemics. And the research they cite is on very specialized topics (e.g., while they are interacting with a male interviewer, daughters of widows display more "uncomfortable" and less "interactive" nonverbal behavior than daughters of divorced parents, Hetherington, 1972).

The notable paucity of research on movement and personality does not seem to stem from poor evidence of this subject's importance. No doubt there are many problems in systematically examining such a complex subject, yet complexity has not dissuaded other personality researchers. There is the fact that the current renaissance in body motion research has developed primarily in relation to interpersonal and social psychological topics. Personality is not at all a popular research topic in nonverbal communication research, although Rosenthal *et al.* (1979) have extensively pursued individual differences in nonverbal sensitivity and decoding skill. Whatever the reasons for the paucity of research on movement and personality, it would seem important to clinicians interested in nonverbal behavior. This section will conclude with some clinical and research examples demonstrating the clinical potential of this subject.

Descriptions of movement correlates of personality can be divided into two types—those that focus on *specific* symbolic actions (Deutsch, 1952, 1966; Mahl, 1968, 1977, 1978), and those that focus on analysis of *movement style* including chronic features of posture, facial expression, and breathing (Christiansen, 1972; M. Davis, 1970; Lowen, 1958/1971; North, 1971; Reich, 1949). Deutsch (1966), for example, describes specific actions and positions clients assumed on the couch and their relationship to identifications, resistances, and intrapsychic conflicts. Mahl (1968, 1977) describes how specific actions may precede verbalizations of repressed themes, and he draws on extensive clinical observations of this anticipatory phenomenon to elaborate a theory in which kinesthetic experience or awareness of the action plays a major role in making unconscious material conscious. For example, Mahl (1978) describes the transference significance of a client's proclivity for curling up on the couch. She assumed a curled-up position on her side during the third, fourth, and fifth interviews and then reported a dream that included an image of a "beautiful Indian girl" lying curled up on a couch whom she covers with a blanket. Intense recurrent curling up was accompanied by "unconflictfully" described wishes to regress and finally to her relating a story about how an illness of her mother when she was 3 years old meant she went suddenly from maternal overindulgence to strong pressures from her father to be self-sufficient. The patient never assumed a curled-up position again after describing this traumatic separation.

Mahl (1977) has asserted that what he calls the "A→B phenomenon" in which certain actions anticipate subsequent verbalizations is not a rare event, and he has even estimated that clear-cut examples occur about once every 3 or 4 analytic hr. He contends that such sequences stem from "the most important aspects of the person's life" and may occur during "significant phases of the analytic process." Within the enormous complexity that is psychotherapy, the ability to attend to what is "most" essential or urgent at any one time is an art. Behaviors that reflect key dynamics of the moment would be of interest to therapist and patient, and, if Mahl is correct, specific A→ B, action→verbalization sequences deserve particular attention as cues to what is most urgent. In general, there is the implication within the literature that movement idiosyncracies and striking symbolic actions often prove to reflect essential personality dynamics; that the therapist concerned with what to focus on among many details would not be distracted by trivia in attending to this nonverbal behavior.

A classic example of the value of attending to movement style in analysis is presented by Reich in his description of a man he described as an "aristocratic" character:

> The patient is good-looking, of medium height; his facial expression is reserved, serious, somewhat arrogant. What is striking is his measured, refined gait. It takes him quite some time to get from the door to the couch;

> plainly, he avoids—or covers up—any haste or excitation. His speech is measured, quiet and refined; occasionally, he interrupts this with an emphasized, abrupt "Yes," at the same time stretching both arms in front of him, and afterwards stroking his hand over his forehead. He lies on the couch in a composed manner, with his legs crossed. His dignified composure hardly ever changes at all, even with the discussion of narcissistically painful subjects. (1949, p. 180)

Reich, most of all, conveys the importance of such impressions, the "character armoring" function of movement and voice patterns, and their genesis in early experience. These are notions that now so permeate the theory and practice of psychotherapy that one can easily lose track of where they were first clearly articulated. In this example, Reich describes how his first impressions persisted until he "found himself" telling the patient that he was playing the role of an English lord. And the patient then related a very significant element of his family story, a fantasy, based on fragments of family lore, that he was in fact of aristocratic lineage in contrast to his father. Reich then describes the evaluation of his fantasy from 4 years old, its relation to oedipal themes, and its translation into his demeanor by adolescence. "His arrogant behavior, then, had the structure of a symptom: it served the purpose of warding off of a drive as well as its satisfaction" (in this case sadistic urges). There is now a considerable literature extending this part of Reich's work. Clinical case examples with fine attention to motion, posture, and breathing can be found in Murphy (1976), Lowen (1958/1971), Kestenberg (1966), Mahl (1968), and Keleman (1980).

In the 1940s Chapple and his associates reliably demonstrated that we can differentiate depression, mania, hysteria, paranoid states, and anxiety neuroses from the activity rates of patients and the patterns of interruption and silence during interviews. But efforts to systematize *movement* and *posture* analysis in order to more rigorously examine relationships between movement and personality are as yet too cumbersome and elaborate. Kestenberg (1975) has developed an observation method for obtaining "movement profiles" based on concepts from the movement analysis systems of choreographer Rudolf Laban and from her own investigations of "tension flow" patterns first visible in the diffuse activity of infants. She has conducted longitudinal studies of a few children that indicate that individual motion patterns visible in early infancy can be traced to adulthood. Kestenberg has trained a number of psychoanalysts, teachers, and dance therapists in her movement analysis method, and efforts are being made to test the reliability of the observations.

Several movement analysts influenced by Laban's work have developed individual profile methods (Bartenieff & Davis, 1975; North, 1971; Ramsden, 1973), but these also lack reliability. However, the author (M. Davis, 1970) developed a movement diagnostic inventory for psychiatric patients, and preliminary reliability data suggest it is a more feasible and manageable observation instrument than most Laban-based movement profiles. With this inventory it is possible to generate profiles of highly specific types of movement disorganization, constriction, and exaggeration that appear to correlate with differential diagnoses.

Psychodiagnosis from movement would seem a relatively minor clinical asset compared to the potential of such observations for assessing change and enriching perception of the ongoing process of therapy. For example, the author (M. Davis, 1977) completed an intensive film analysis of a family therapy session in which the movements of the designated patient corresponded dramatically to momentary changes in his mental status and manifestations of thought disorder. Gestures that involved very large spatial sweep of the arms (yet little or no visible movement in the torso) accompanied belligerent, paranoid outbursts. Long series of perseverative, "monotoned" gestures (one type of accent repeated in one plane, another dynamic limited to another direction) accompanied incoherent, rambling speech. And phases in which he was listless and his movement was integrated but without intensity paralleled depressed but coherent statements.

RELATIONSHIP MESSAGES

Earlier examples of symbolic actions and individual movement patterns may seem deceptively simple. We are used to looking at *what* individuals do, and we have a sense, however peripheral, of how distinctively they do it. Clinicians or researchers citing examples of dramatic actions and movement styles have distilled these observations from enormous complexity; they are so literary, so recognizable that they can be read as all that is important of the nonverbal channel. Entire books have been written on individual movement characteristics of patients with little report of what actually occurs in the sessions apart from the dramatic "productions" of the patient.

But 1-hr viewing a film or videotape of a therapy session without sound can impress one with a sense of how difficult it is to "distill out" individual characteristics and delineate symbolic actions, to perceive patterns in the interaction, and to determine what details to focus on.

One is confronted with the details of greetings and leave-takings, positions assumed through the session, listening behavior, eye contact patterns, the complexities of gesticulation, patterns of fidgeting, manipulation of objects, and a myriad inadvertent actions such as nose wipes, hair preens, and lip biting.

In time the observer may perceive relationships *between* the movements of patient and therapist. A clinician who is fortunate enough to have a videotape in which the camera is focused throughout the session on both the patient and therapist in full-body shots will probably still focus on one interactant at a time. From Darwin through Reich, the focus was on the individual "in vacuo." Anthropologists attending to body motion, however, shifted the focus to what occurs *between* speakers, to the interaction and "coactions" of two or more people in a face-to-face encounter.

Chapple (1949) appears to have been the first to put observers on one side of a one-way screen to record the *interaction* of two people in terms of the patterns of turn taking, pauses, interruptions, and monologues irrespective of speech content. Because we think of conversation in terms of speaking turns, it is easiest to attend to alternations such as A speaks, then B speaks, then A speaks, and so on, rather than to focus on the constant simultaneous actions of both. Describing interaction as alternating turns lends itself, however, to linear, causal renditions of the event in which A's actions are perceived to "cause" B's which in turn "trigger" A's and so on. Birdwhistell (1970) and Scheflen (1973) have eloquently argued against linear action–reaction descriptions of social behavior. In fact, Chapple did not succumb to this; he attended to a cogent aspect of social behavior, the patterned alternation of speaking and listening activity, the interaction rhythms perceptible in turn taking. Alternation of actions as in turn taking, and coactions as in the complex coordinations of positions that Scheflen describes (1973), are not contradictory perspectives. They complement each other, and both offer important insights for the clinician.

One of the finest examples of interaction research is the work of Daubenmire and her associates in the Ohio State University School of Nursing (Daubenmire & Searles, 1982). In the early 1970s they videotaped medical patients in surgery recovery rooms and in private hospital rooms. A patient could be taped all day for weeks interacting with various staff in this attempt to examine relationships between therapeutic interaction and favorable medical outcome. Daubenmire and her associates drew from diverse kinesics and communication research to evolve their observation methods, and their synthesis of behavioral observation and computer pattern assessment represents an increasingly occurring marriage in the seventies between ethological description and quantitative analysis. It is hard to appreciate the depth of this study because it is hard to imagine 25 columns of speech and motion details coded at 1-sec intervals for several weeks of videotape. In one aspect of the data analysis the researchers completed computer analyses of patterns of similarity and convergence of behaviors between patient and nurse. Daubenmire and Searles (1982) illustrate how this analysis elucidates therapeutic interactions with a description of a 20-min exchange between a nurse and a man dying of cancer:

> For the first 7 minutes RN3 is standing in the patient's exploratory space, a distance of 7 to 12 feet from the patient. Verbal exchange includes evidence of entry and approach behaviors and a discussion on mouth care to relieve some of Mr. B.'s discomfort. During minute 7, Mr. B. mentions he slept well and was able to go six to eight hours without much pain. At this point, RN3 moves into the patient's interpersonal space, a distance of 4 to 7 feet from the bed and sits down facing the patient. She also verbally recognizes Mr. B.'s comment and Mr. B. discloses his concern about taking morphine. . . . Patterns for eye contact, limb and body movement show an increase in similarity from minutes 7 to 10. The discussion of pain medication also continues until minute 10 when Mr. B. attempts to change the subject and RN3 directs him back to the discussion by a question about how he plans to "stretch out" the time between medications. . . .
>
> [After a brief interruption by another nurse] RN3 once more returns the conversation to the topic of Mr. B.'s pain medication. From minute 12 to 13, there is a measurable increase in similarity of eye contact, or increased mutual gazing. By minute 15, the S [similarity] value + 1 indicates continued mutual eye contact. Patterns for limb and body movement also show a shift to +1 similarity. This total similarity for all three variables lasts until minute 23. . . .
>
> During minute 23, the nurse moves toward Mr. B.and touches his arm. The conversation centers around how Mr. B. will specifically talk with his physician about nausea from chemotherapy and his concerns about spacing dosages of morphine. . . . From minute 26 to 29, Mr. B. begins to express his feelings about how difficult it has been for him to live during the past several years. . . . Nurse responds, "You and I can do something about the quality of your life and what you do with your life. You have that choice in determining what it is that you will do with your day." . . . During this conversation, Mr. B. continues to look away from the nurse. The break in eye contact, increased limb and body movements, and verbal content suggest Mr. B. was very uncomfortable talking about these problems. This shift in similarity during minute 23 is somewhat similar to the shift observed in minute 10. . . . In both patterns, there appears to be a period of self-disclosure which lasts for several minutes, the patient apparently becomes uncomfortable with the conversation, and similarity of eye contact and movement decreases. (1982, pp. 312–313. Reprinted by permission of Human Sciences Press)

Careful analysis of such therapeutic interactions reveals a remarkable capacity on the part of the therapist to tune in, accommodate to, pick up on, and participate in interaction sequences that seem above all devoted to building rapport. One cannot describe this as something the therapist *does*, and certainly he or she does not do it consciously. Nevertheless, I think that such examples of therapeutic intervention—such as those of good nursing from Daubenmire's study, or examples of psychotherapy sessions with very experienced psychotherapists such as Scheflen has described (1973) and more recently this author has been studying (M. Davis, 1974)—compel one to conclude that therapists establish rapport with patients at the nonverbal level by a process of attuning to the patient's communication style and toning down their own, evolving periods of posture mirroring, exquisitely pacing their comments and interventions, modulating their inquiry when the patient withdraws or shows distress, and sustaining concentrated focus on the patient while "softening" this direct gaze with head tilts and subtle variations in head nodding.

Scheflen (1973), for example, describes the intricate way two family therapists varied their positions in relation to the mother and daughter they were interviewing. He discovered that one therapist would unconsciously pace the intervention of the other with a series of pipe-smoking actions. At another level he discerned a rhythmic alternation of postural orientations—mother with daughter, then daughter with one therapist, then back again to mother with daughter—as it reflected a momentary "loosening" of the rigid mother–daughter attachment.

The following example of nonverbal interaction analysis (M. Davis, 1974) also illustrates therapist sensitivity and an exquisite ability to nonverbally tune in. It involves a very disturbed paranoid young man videotaped 10 months into his therapeutic work with the late Harley Shands.

The session begins with the patient agitatedly shifting his position until he bends over, elbows on knees, with his head down so his face cannot be seen. He does this just as the therapist sits down (synchronous or echoing position shifts are common in therapy). The therapist assumes two fairly typical therapist listening positions in the first 10½ min, but what is interesting is the correspondence between the type of position and the mental status of the patient. In the first half, the more open and oriented therapist position corresponds to a phase when the patient is most agitated and disorganized; when the therapist is sitting in a more aloof, contained position, the patient gradually collects himself. At 10½ min into this 22-min session the therapist gets up and leaves the room. (He had gone to check for the patient that no staff member was watching him beyond the one-way screen, although the patient did not object to the videotaping for research purposes.) This divides the session almost exactly in half. The patient again goes through an agitated phase as if unable to settle into a position while he waits for the therapist. This time, however, he sits up and places his hands on his lap when the therapist sits down in a position that is turned slightly away with arms tightly folded. The patient stays for a while at his most integrated in terms of movement and (as it turns out from later analysis of the speech) at his most verbally coherent.

At this point the therapist dramatically shifts toward the patient and sits with his right hand to his face. The patient momentarily appears unnerved and begins to laugh, then assumes the same arm position. He again becomes disorganized verbally and nonverbally, but throughout this phase their positions become more and more identical until the therapist seems to gradually adjust his hand until it is in exactly the same hand–finger configuration as the patient. It is as if the therapist has upped the ante—saying literally and figuratively, "You progressed to a coherent, appropriate way of relating to me when I turned away, can you when I move closer?" The therapist does not shift positions again until he signals the end of the session. The patient's position shifts now define the phases of the session as he gradually works toward the last phase in which he is most coherent and appropriate, one that is again characterized by sitting in a simple, hands-on-lap position. The graphs at Figure 1 also indicate subtle relationships between the movement patterns of patient and therapist, and perhaps a "deeper" level of attunement. Note, for example, how the therapist's movement is most spatially precise and complex when the patient's movement is most disorganized, and is less intense when the patient is more organized.

In no sense can these patterns be read simply as causal relationships. The dance of position shifts and the subtle fluctuations in movement organization, intensity, mobility, and spatial complexity shown in Figure 1 defy a simple action–reaction interpretation. Exquisite timing, mirroring, and careful modulation of intensity and spatial direction appear manifestations of the therapist's ability to empathize and create a context in which the patient could become more integrated and tolerate painful self-disclosures. As Freedman describes in the introduction to an excellent book on psychoanalysis and communication research (Freedman & Grand, 1977, p. 8), "the microscopic analysis of communication behavior may provide a more operational view of the integrative process."

Scheflen (1973) has argued that what is regarded as "crazy" behavior can be seen to fit coherently within the context of the interaction. He has described (private communication, 1967) splicing together different sections of a film of a patient to evoke different diagnostic impressions. Conceivably this could be done with film segments of the three different movement themes described earlier (M. Davis ; see p. 99).

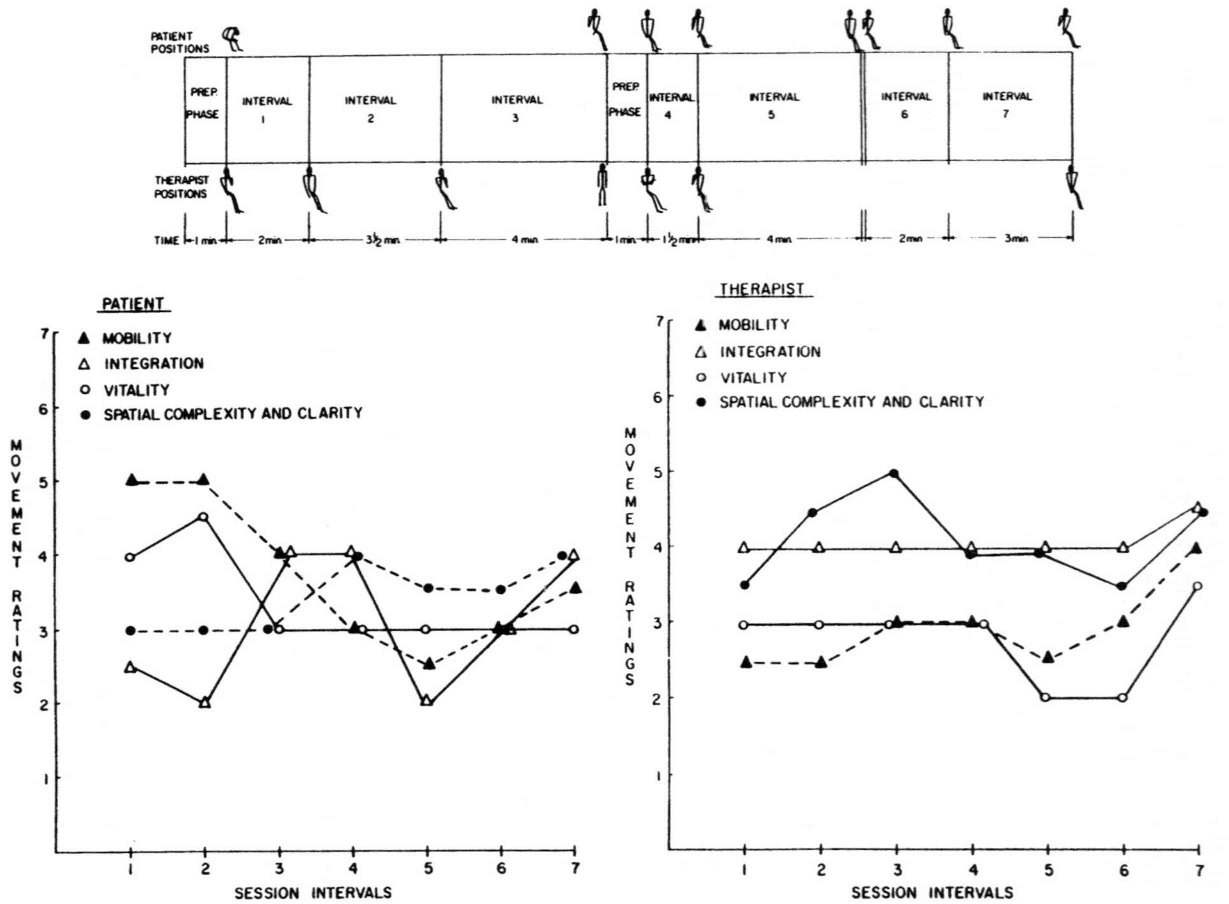

FIGURE 1. Alternations of body positions and movement ratings per interval. Movement ratings refer to the degree of body

Each one seen in isolation would evoke a different diagnostic impression: severe depression *or* chronic undifferentiated schizophrenia *or* paranoid schizophrenia. Perhaps this does not constitute evidence of the unreliability of the nosological system so much as it demonstrates the potential contribution of nonverbal communication research to behavioral assessment. When the camera (and the observer's orientation) "pans back" from the patient to allow focus on the interaction, the behavior takes on new, clinically important meanings. In the film analysis just cited, the patient's behavior meshed with a family constellation and therapist interventions in such dramatically visible ways that etiology, diagnosis, and prognosis could be sketched from nonverbal behaviors displayed in this one session.

Studies of miscommunication and disturbed behavior are, of course, as potentially valuable to the clinician as analyses of peak therapeutic interactions. Drawing on Scheflen's observation methods and theoretical orientation, McDermott and Gospodinoff (1979) have shown how "rational" and ordered the disorderly behavior of a child in a classroom appears after careful analysis of the behavioral contexts in which it occurs. LaFrance and Mayo (1976) have demonstrated that eye contact patterns of blacks during listening and speaking are almost the opposite of those for whites such that interracial pairs must suffer a serious interactional mismatching. Erickson (1979) has carefully analyzed interview behaviors and presented dramatic evidence of when and how the organization of eye contact, head nodding, and vocal inflection patterns serving smooth turn taking and flow of conversation can break down in interracial interviews.

The relatively rare film analyses of nonverbal interaction described here have as yet limited value for psychotherapy process research. Most are single cases and some took as long as 4 years of continuous film study to complete (Scheflen, 1973). Ways have to be developed to speed up the analyses as well as to make them reliable and replicable. The author is currently testing out a systematic and replicable nonverbal process coding method that would capture interaction patterns such as Scheflen, M. Davis (1975a), and others have discovered working on films or videotapes in isolation.

There is a considerable literature on nonverbal interaction process research if one includes more than studies strictly related to psychotherapy. And this area—systematic interaction process analysis of actual encounters—does promise to make a significant contribution to psychotherapy research. Into this group should be included microanalyses of mother–infant interaction, such as those of Beebe and Stern (1977), ethological studies of the organization of social encounters and conversational behavior (Kendon *et al.*, 1975), Gottman's marital interaction research (1979) and Horowitz's method for therapy process analysis (1978), as well as the more purely kinesic studies described earlier.

The early literature on movement and personality illustrated how it is impossible to tell the dancer from the dance. In the interaction analysis diagrammed in Figure 1 the patient is unmistakably the patient, and he would take his movement style with him out of the consulting room just as the therapist's movement style could be recognized elsewhere. But for the 22 min they interacted, their movements *also* reflected the shape of their relationship and their bodies changed form and tempo and articulation together. It would seem clinically very useful to understand how body movement reveals and is part of the person. It would also seem clinically very valuable to understand what a few minutes of nonverbal interaction can reveal about the therapeutic relationship and how it engenders—or does not promote—psychological integration and personal growth.

IMPLICATIONS FOR THERAPY TRAINING AND PRACTICE

However compelling the illustrations of the importance of nonverbal behavior in psychotherapy, the loop back to actual practice is far from completed; the best interaction analyses are single-case studies that cannot be readily generalized. And, as discussed earlier, much of the recognition or observer judgment studies are of dubious value to a clinician. Clinicians are not likely to arrange their chairs 30 in. apart because that is the distance college students found comfortable with an interviewer (Knight & Bair, 1976); touch clients on the back of the upper shoulder as they enter because in one study this correlated with deeper self-exploration (Pattison, 1973); smile and lean forward in more open positions to enhance rapport and trust, and so on. In fact, covert manipulations of kinesic and proxemic behaviors are seriously done to effect behavioral change according to literature on "neurolinguistic programming" and hypnotic techniques (Bandler & Grinder, 1979). But many therapists would object to such covert manipulations on ethical and clinical grounds.

Presumably, most therapists would not be comfortable with *active* and *explicit* intervention at the nonverbal level either. There is a rich literature of dramatic case examples that suggest how powerful "body therapies" such as bioenergetics or dance/movement therapy can be (Chaiklin, 1975; Lowen, 1958/1971; Murphy, 1976; Schutzenberger & Geffroy, 1979; Siegel, 1973). But no one seems to adopt such active interventions simply from reading about the importance of nonverbal behavior or the value of body therapy techniques. Conceivably, unless a "traditional" or dynamically oriented therapist trains with someone experienced in such interventions, he or she will not likely incorporate them into practice even if their efficacy is someday reliably demonstrated. Some psychoanalysts who believe in the value of the body therapies advocate a therapeutic collaboration in which the patient may work with a body-oriented therapist while also in traditional analysis or intensive psychotherapy (Braatoy, 1952; Kestenberg, 1973; Solnit, 1973).

Less controversial but equally untested is the value of videotape replay as an active intervention that can help patient and therapist focus on details of visible behavior that are out of awareness (Berger, 1978). It has been found useful as feedback to the patient (especially if done immediately after the session, Kagan, 1978), and as an electronic cotherapist who can help the therapist be objective about his or her own subjectivity (Whitaker, 1978). At the very least, videotapes of actual therapy sessions—that focus on both patient and therapist in full-body shots throughout the session—would seem valuable for training. But there is no guarantee that turning off the sound and carefully training one's eye to be more sensitive to nonverbal details will actually affect one's practice of therapy. It could atomize one's perception, distracting the therapist and disrupting his or her participant observation. The effectiveness of active interventions, videotape replay, and formal training in nonverbal observation for psychotherapy practice is still to be rigorously assessed.

Intuitively, though, we suspect that listening with the third ear is largely a matter of seeing more and even kinesthetically feeling more; that the third ear is in large measure eye and muscle. Jacobs (1973) describes cases where awareness of his own actions while listening to a client alerted him to important dynamics. For example, as a patient described an anxiety reaction, Jacobs found himself tugging slightly on his belt as if it were too tight. Reflecting on this uncharacteristic action led to an association to the patient's traumata surrounding abdominal surgery. Relating this interpretation to the client produced a flood of associations to these early experiences. Jacob's article is one of the few discussions of the value of consciously trying to heighten one's kinesthetic awareness and utilize it

> either in the service of providing clues to the meaning of the patient's communications, or in facilitating the recognition of previously undetected attitudes and feelings in the analyst himself. (1973, p. 92)

The loop back to practice is most likely made in one's training experience. Good training tapes based on actual sessions, extensive seminar discussion of the nonverbal level, and supervision with clinicians who have experience in consciously attending to and in some coherent way utilizing perceptions of nonverbal behavior are essential. Equally essential are many more relevant, sophisticated, and well-illustrated studies of psychotherapy from the nonverbal communication researcher.

ACKNOWLEDGMENTS

I am deeply grateful to Dr. Israel Zwerling of Hahnemann, who first encouraged me to pursue this research 20 years ago and whose support has been crucial to the development of my work over these years. Dr. Zwerling's support—in terms of both providing dynamic settings in which to work and ensuring the freedom to pursue unorthodox lines of research—has been pivotal to the careers of a remarkable number of authors cited in this chapter. In addition, I wish to gratefully acknowledge the support of the Board of Directors of the Institute for Nonverbal Communication Research, and of the late Harley Shands, M.D., of the Roosevelt Hospital where the psychotherapy research project was initiated. My thanks also to Rachel Hott and the members of the observation coding seminar at INCR who helped me clarify my recording methods.

REFERENCES

Allport, G. W., & Vernon, P. E. *Studies in expressive movement.* New York: Macmillan, 1933.

Bakken, D. G. Behavioral adjustment in nonverbal immediacy: A methodological note. *Personality and Social Psychology Bulletin,* 1978, *4,* 300–303.

Bandler, R., & Grinder, J. *Frogs into princes: Neurolinguistic programming.* Moab, Utah: Real People Press, 1979.

Bartenieff, I., & Davis, M. Effort-shape analysis of movement. In M. Davis (Ed.), *Research approaches to movement and personality.* New York: Arno Press, 1975.

Bartenieff, I., & Lewis, D. *Body movement: Coping with the environment.* New York: Gordon & Breach, 1980.

Bateson, G., & Mead, M. *Balinese character: A photographic analysis.* New York: New York Academy of Sciences, 1942.

Beebe, B., & Stern, D. N. Engagement-disengagement and early object experiences. In N. Freedman & S. Grand (Eds.), *Communicative structures and psychic structures.* New York: Plenum Press, 1977.

Berger, M. M. *Videotape techniques in psychiatric training and treatment* (2nd ed). New York: Brunner/Mazel, 1978.

Birdwhistell, R. L. *Kinesics and context.* Philadelphia: University of Pennylsvania Press, 1970.

Braatoy, T. Psychology vs. anatomy in the treatment of "arm neuroses" with physiotherapy. *Journal of Nervous and Mental Diseases,* 1952, *115,* 215–245.

Byers, P. *From biological rhythm to cultural pattern: A study of minimal units.* Unpublished doctoral dissertation, Columbia University, New York, 1972.

Chaiklin, S. Dance therapy. In D. X. Freedman & J. F. Dryud (Eds.), *American handbook of psychiatry.* New York: Basic Books, 1975.

Chapple, E. D. The interaction chronograph: Its evolution and present application. *Personnel*, 1949, *25*, 295–307.

Chapple, E. D., & Lindemann, E. Clinical implications of measurements of interaction rates in psychiatric interviews. *Applied Anthropology*, 1942, *3*, 1–11.

Chapple, E. D., Chapple, M. F., & Repp, J. A. Behavioral definitions of personality and temperament characteristics. *Human Organization*, 1955, *13*, 34–39.

Charny, E. J. Psychosomatic manifestations of rapport in psychotherapy. *Psychosomatic Medicine*, 1966, *28*, 305–315.

Christiansen, B. *Thus speaks the body: Attempts towards a personology from the point of view of respiration and postures*. New York: Arno Press, 1972.

Condon, W. S. Linguistic-kinesic research and dance therapy. *American Dance Therapy Association Proceedings*, Third Annual Conference, 1968, 21–44.

Darwin, C. *The expression of the emotions in man and animals*. Chicago: University of Chicago Press, 1965. (Originally published, 1872)

Daubenmire, M., & Searles, S. A dyadic model for the study of convergence in nurse-patient interactions. In M. Davis (Ed.), *Interaction rhythms: Periodicity in communicative behavior*. New York: Human Sciences Press, 1982.

Davis, F. *Inside intuition: What we know about nonverbal communication*. New York: McGraw-Hill, 1973.

Davis, M. Movement characteristics of hospitalized psychiatric patients. *Proceedings of the Fifth Annual Conference of the American Dance Therapy Association*, 1970, 25–45.

Davis, M. *Understanding body movement: An annotated bibliography*. New York: Arno Press, 1972.

Davis, M. *Analysis of movement styles and interaction in psychotherapy*. Paper presented at the First International Symposium on Nonverbal Aspects and Techniques of Psychotherapy, University of British Columbia, Vancouver, June 1974.

Davis, M. *Focusing beyond words: Nonverbal dimensions of therapeutic interactions* [Videotape]. Philadelphia: Communications in Medicine Department, Hahnemann Medical College, 1975. (a)

Davis, M. (Ed.). Psychoanalytic perspectives of movement. In *Body movement: Perspectives in research*. New York: Arno Press, 1975. (b)

Davis, M. *Towards understanding the intrinsic in body movement*. New York: Arno Press, 1975. (c)

Davis, M. *Methods of perceiving patterns in small group behavior*. New York: DNB Publications, 1977.

Davis, M. (Ed.), *Interaction rhythms: Periodicity in communicative behavior*. New York: Human Sciences Press, 1982.

Davis, M., & Skupien, J. (Eds.). *Nonverbal communication and body movement: An annotated bibliography 1971–1981*. Bloomington: Indiana University Press, 1982.

Deutsch, F. Analytic posturology. *Psychoanalytic Quarterly*, 1952, *21*, 196–214.

Deutsch, F. Some principles of correlating verbal and non-verbal communication. In L. A. Gottschalk & A. H. Auerbach (Eds.), *Methods of research in psychotherapy*. New York: Appleton-Century-Crofts, 1966.

Duncan, S. Jr., & Fiske, D. W. *Face-to-face interactions: Research, methods, and theory*. Hillsdale, N.J.: Lawrence Erlbaum, 1977.

Efron, D. *Gesture, race and culture*. The Hague: Mouton, 1972. (Originally published, 1942)

Ekman, P., & Friesen, W. V. *Unmasking the face: A guide to recognizing emotions from facial expressions*. Englewood Cliffs, N.J.: Prentice-Hall, 1975.

Ekman, P., Friesen, W. V., & Ellsworth, P. *Emotion in the human face*. New York: Pergamon Press, 1972.

Erickson, F. Talking down: Some cultural sources of miscommunication of interracial interviews. In A. Wolfgang (Ed.), *Nonverbal behavior: Applications and cultural implications*. New York: Academic Press, 1979.

Escalona, S., & Heider, G. M. *Prediction and outcome: A study in child development*. New York: Basic Books, 1959.

Freedman, N., & Grand, S. (Eds.). *Communicative structures and psychic structures*. New York: Plenum Press, 1977.

Freud, S. Fragment of an analysis of a case of hysteria. In S. Freud, *Collected papers* (Vol. 3). New York: Basic Books, 1959. (Originally published, 1905)

Fries, M., & Lewi, B. Interrelated factors in development: A study of pregnancy, labor, delivery, lying-in period and childhood. *American Journal of Orthopsychiatry*, 1938, *8*, 726–752.

Gladstein, G. A. Nonverbal communication and counseling psychotherapy: A review. *Counseling Psychologist*, 1974, *4*, 34–52.

Gottman, J. M. *Marital interaction: Experimental investigations*. New York: Academic Press, 1979.

Harper, R. G., Wiens, A. N., & Matarazzo, J. D. *Nonverbal communication: The state of the art*. New York: Wiley, 1978.

Henley, N. *Body politics: Power, sex and nonverbal communication*. Englewood Cliffs, N.J.: Prentice-Hall, 1977.

Hetherington, E. M. Effects of father absence on personality development in adolescent daughters. *Developmental Psychology*, 1972, *7*, 313–326.

Horowitz, M. J. *States of mind: Analysis of change in psychotherapy*. New York: Plenum Press, 1978.

Jacobs, T. J. Posture, gesture, and movement in the analyst: Cues to interpretation and countertransference. *Journal of the American Psychoanalytic Association*, 1973, *21*, 77–92.

Jaffee, J., & Feldstein, S. *Rhythms of dialogue*. New York: Academic Press, 1970.

Kagan, N. Interpersonal process recall: Media in clinical and human interaction supervision. In M. M. Berger (Ed.) *Videotape techniques in psychiatric training and treatment*. New York: Brunner/Mazel, 1978.

Keleman, S. A case study of excitement and personality. *Journal of Biological Experience*, 1980, *2*, 3–17.

Kendon, A. (Ed.). *Nonverbal communication, gesture and interaction*. The Hague: Mouton, 1980.

Kendon, A., Harris, R. M., & Key, M. R. (Eds.). *Organization of behavior in face-to-face interaction*. The Hague: Mouton, 1975.

Kestenberg, J. S. Rhythm and organization in obsessive-compulsive development. *International Journal of Psycho-Analysis*, 1966, *47*, 151–159.

Kestenberg, J. C. Comments on movement therapy as a psychotherapeutic tool. *Journal of the American Psychoanalytic Association*, 1973, *21*, 347–350.

Kestenberg, J. S. *Children and parents: Psychoanalytic studies in development*. New York: Jason Aronson, 1975.

Kiesler, D. J. An interpersonal communication analysis of relationship in psychotherapy. *Psychiatry*, 1979, *42*, 299–311.

Knapp, M. L. *Nonverbal communication in human interaction* (2nd ed.). New York: Holt, Rinehart & Winston, 1978.

Knight, P. H., & Bair, C. K. Degree of client comfort as a function of dyadic interaction distance. *Journal of Counseling Psychology*, 1976, *23*, 13–16.

Labov, W., & Fanshel, D. *Therapeutic discourse: Psychotherapy as discourse*. New York: Academic Press, 1977.

LaFrance, M. Posture sharing and rapport. In M. Davis (Ed.), *Interaction rhythms: Periodicity in communicative behavior*. New York: Human Sciences Press, 1982.

LaFrance, M., & Mayo, C. Racial differences in gaze behavior during conversation: Two systematic observational studies. *Journal of Personality and Social Psychology*, 1976, *33*, 547–552.

LaFrance, M., & Mayo, C. *Moving bodies: Nonverbal communication in social relationships*. Monterey, Calif.: Brooks/Cole, 1978.

Lomax, A. *Folk song style and culture*. Washington, D.C.: American Association for the Advancement of Science, 1968.

Lowen, A. *The language of the body*. New York: Macmillan, 1971. (Originally published, 1958)

Mahl, G. F. Gestures and body movements in interviews. In J. M. Shlien (Ed.), *Research in psychotherapy*. Washington, D.C.: American Psychological Association, 1968.

Mahl, G. F. Body movement, ideation and verbalization during psychoanalysis. In N. Freedman & S. Grand (Eds.), *Communicative structures and psychic structures*. New York: Plenum Press, 1977.

Mahl, G. F. *The embodiment of interpersonal relationships: Notes and observations*. Paper presented at the Second National Conference on Body Language, New York, 1978.

Mayo, C., & Henley, N. (Eds.). *Gender and nonverbal behavior*. New York: Springer-Verlag, 1981.

McDermott, R. P., & Gospodinoff, K. Social contexts for ethnic borders and school failure. In A. Wolfgang (Ed.), *Nonverbal behavior: Applications and cultural implications*. New York: Academic Press, 1979.

Mehrabian, A. *Nonverbal communication*. Chicago: Aldine/Atherton, 1972.

Mittelmann, B. Psychodynamics of motility. *International Journal of Psycho-Analysis*, 1958, *39*, 196–199.

Morris, D., Collett, P., Marsh, P., & O'Shaughnessy, M. *Gestures*. New York: Stein & Day, 1979.

Murphy, J. *Body movement in psychotherapy: Theories and techniques of working with body phenomena in psychotherapy*. Unpublished manuscript, 1976.

North, M. *Personality assessment through movement*. London: Macdonald & Evans, 1971.

Ostwald, P. F. *Soundmaking: The acoustic communication of emotion*. Springfield, Ill.: Charles C Thomas, 1963.

Pattison, J. E. Effects of touch on self-exploration and the therapeutic relationship. *Journal of Consulting and Clinical Psychology*, 1973, *40*, 170–175.

Pine, F. Discussion of papers. In N. Freedman & S. Grand (Eds.), *Communicative structures and psychic structures*. New York: Plenum Press, 1977.

Ramsden, P. *Top team planning: A study of the power of individual motivation in management*. New York: Wiley, 1973.

Reich, W. *Character-analysis* (T. Wolfe, trans.) New York: Orgone Institute Press, 1949.

Rosenthal, R., Hall, J. A., DiMatteo, M. R., Rogers, P. L., & Archer, D. *Sensitivity to nonverbal communication: The PONS test*. Baltimore: Johns Hopkins University Press, 1979.

Sapir, E. Selected writings of Edward Sapir. In D. G. Mandelbaum (Ed.), *Language, culture and personality*. Berkeley: University of California Press, 1968.

Scheflen, A. E. *Communicational structure: Analysis of a psychotherapy transaction*. Bloomington: Indiana University Press, 1973.

Scherer, K. R. Acoustic concomitants of emotional dimensions: Judging affect from synthesized tone sequences. In S. Weitz (Ed.), *Nonverbal communication: Readings with commentary*. New York: Oxford University Press, 1979.

Schutzenberger, A. A., & Geffroy, Y. The body and the group: The new body therapies. In S. Weitz (Ed.), *Nonverbal communication: Readings with commentary* (2nd ed.). New York: Oxford University Press, 1979.

Shapiro, D. *Neurotic styles*. New York: Basic Books, 1965.

Siegel, E. V. Movement therapy as a psychotherapeutic tool. *Journal of the American Psychoanalytic Association*, 1973, *21*, 333–343.

Solnit, A. J. Commentary on movement therapy as a psychotherapeutic tool. *Journal of the American Psychoanalytic Association*, 1973, *21*, 344–346.

Stokoe, W. C., Jr. *Semiotics and human sign language*. The Hague: Mouton, 1972.

Sullivan, H. S. Obituary of Edward Sapir 1884–1939. *Psychiatry*, 1939, *2*, 159.

Takala, M. Studies of psychomotor personality tests, I. Reprinted in M. Davis (Ed.), *Research approaches to movement and personality*. New York: Arno Press, 1975.

Trout, D. L., & Rosenfeld, H. M. The effect of postural lean and body congruence on the judgment of psychotherapeutic rapport. *Journal of Nonverbal Behavior*, 1980, *4*, 176–190.

Waxer, P. H. *Nonverbal aspects of psychotherapy*. New York: Praeger, 1978.

Waxer, P. H. Therapist training in nonverbal behavior: Toward a curriculum. In A. Wolfgang (Ed.), *Nonverbal behavior: Applications and cultural implications*. New York: Academic Press, 1979.

Weitz, S. (Ed.). *Nonverbal communication: Readings with commentary* (2nd ed.). New York: Oxford University Press, 1979.

Whitaker, C. A. The use of videotape in family therapy with special relation to the therapeutic impasse. In M. M. Berger (Ed.), *Videotape techniques in psychiatric training and treatment*. New York: Brunner/Mazel, 1978.

6

Aggression and Psychodynamic Therapy
From Theory to Practice

STEPHEN WORCHEL AND JASON WORCHEL

Man's inhumanity to man has perplexed psychologists for decades. Through human ingenuity the instruments of aggression have been refined to such precision that there exists the real possibility that humans will succeed in accomplishing the feat that has eluded disease, predators, and natural disasters: They will succeed in destroying the human race. Freud (1930) expresses a similar concern:

> The fateful question for the human race seems to me to be whether and to what extent their cultural development will succeed in mastering the disturbance of their communal life by the human instinct of aggression and self-destruction. . . . Men have gained control over the forces to nature of such an extent that with their help they would have no difficulty in exterminating one another to the last man. They know this and hence comes a large part of their unrest, their unhappiness, and their mood of anxiety. (p. 92)

Freud and some other investigators (McDougall, 1908) of his time argued that the human animal came prewired with a set of behavioral tendencies. Among these tendencies or instincts was aggression. According to this view, aggression is propelled by forces internal to the individual. This reasoning suggests that attempts to eliminate aggression

---

STEPHEN WORCHEL • Department of Psychology, Texas A & M University, College Station, Texas 77843. JASON WORCHEL • Virginia Institute of Short-term Dynamic Psychotherapy, Charlottesville, Virginia 22901. This chapter was prepared with the support of a grant from NIMH (5B01MH33468-02).

are bound to end in dismal failure. Rather, the forces of the aggression instinct can, at best, be guided into nonhurtful activities such as sports involving controlled violence or constructive substitute behaviors where the individual can energetically pursue goals.

Given that the urge to aggress was ignited by internal forces, Freud developed a plan for therapy that was aimed at giving the individual an arena in which to express these internal forces. Psychoanalytic therapy was designed to encourage patients to purge themselves of destructive impulses whose repression was responsible for anxiety.

Although there have been numerous hybrids of psychodynamic therapy and criticisms of Freud's instinct theory fill the literature, one aspect of Freud's work remains unimpeachable. His method of therapy was closely tied to his theory of behavior. The reason for using each technique could be directly traced to a theoretical superstructure. Given this close kinship between method and theory, Freud was able to use observations obtained in therapy to modify his theory. And modifications of theory were generally accompanied by refinements in therapeutic techniques.

The practice of psychodynamic theory today presents a somewhat different picture. Practitioners have borrowed, modified, and improvised in their efforts to find the approach that seems best. Although still sinking roots deep into the psychodynamic tradition, the range of psychodynamic therapies runs the gamut from group therapy to short-term therapy. In many cases, the decision to adopt a particular *modus operandi* has been made on the basis of what seems to work; the therapist finds that a technique produces desired results and so adopts this approach. Although there is much to recommend this form of evolution of therapy, one may lament the loss of the Freudian elegance that relied on a close marriage of practice and theory.

However, before mourning the passing of a golden age, it is instructive to examine the practice of psychodynamic therapy in light of theories that have been developed outside of clinical psychology. As we shall see, many of these theories offer new directions for the development of psychodynamic therapy (Sheras & Worchel, 1979). In other words, though the relationship between therapy and theory may not be as direct as one may desire, there is still a kinship.

One arena in which this closeted relationship exists is that of aggressive behavior. As we pointed out, human aggression occupied a central position in Freud's theories and psychoanalytic therapy. Although clinicians have continued to develop techniques aimed at dealing with aggression and the related emotion, anger, the advances in theories of aggression have not, for the most part, been spawned in clinical literature. Rather, social psychologists have taken up the gauntlet to explain the causes of human aggression. The research and the theories of these social psychologists have opened up new vistas in our understanding of violence. Unfortunately, there has been little effort in directly applying these theories in such settings as therapy.

Like two rivers that run parallel courses but rarely intersect, clinicians and social psychologists have applied their skills to dealing with and understanding aggression. This historical separation of the fields has, we feel, hindered the development of both areas. The observations of the clinician could provide vital data for the refinement of social psychological experimentation and theories on aggression. Similarly, the theories developed by social psychologists could offer important insights into why various therapy practices are effective and could serve to guide the integration of new techniques into psychodynamic therapy.

We shall here confine ourselves to the latter of these topics, namely, the contribution that social psychological theories of aggression can make to a better understanding of psychodynamic therapy aimed at dealing with aggressive behavior. We shall first briefly summarize the leading theories and research on aggression, and then weave the social psychological literature into the therapy session.

THE SOCIAL PSYCHOLOGY OF AGGRESSION

FRUSTRATION–AGGRESSION THEORY

A major charge against the instinct theories of aggression was that they failed to give adequate attention to situational factors. People do not aggress in random fashion. Rather, they are likely to respond with violence to certain situations. Drawing heavily on Freud's hydraulic view of human behavior, Dollard, Doob, Miller, Mowrer, and Sears (1939) argued that "aggression is always the consequence of frustration" and that "frustration always leads to the instigation to some form of aggression" (p. 1). The greater the degree or number of frustrations, the greater would be the instigation to aggress. The hypothesis that frustration leads to aggression has received a great deal of support. For example, Harris (1974) had a confederate cut into lines of people waiting in theaters, grocery stores, and movies. The confederate cut into the line either in front of the second person or the twelfth. Observations showed that the second person reacted with more aggression (verbal abuse) than the twelfth person; in other words, the greater the frustration, the greater the aggression.

The frustration–aggression theory was important because it located the instigation for aggression outside the individual. However, for our present purposes, other aspects of the theory are more important. The investigators drew on Freud's hydraulic model of personality when they argued that once instigated, the drive to aggress would press for expression. Catharsis, or a reduction in the instigation to aggress, would follow aggressive behavior. Thus, allowing the individual to engage in aggressive activity after frustration would reduce tension and, subsequently, the instigation to aggress. Without this catharsis experience, the drive to aggress would increase as additional frustrations were experienced until the cumulative tension would eventually compel aggressive action.

The catharsis hypothesis has been the center of a great deal of controversy. Some investigators (Hokanson, Burgess, & Cohen, 1963) have found reduction in physiological arousal (systolic blood pressure and heart rate) when angered subjects are allowed to aggress against their tormentor. Similarly, Bramel, Taub, and Blum (1968) found that subjects reported "feeling good" after aggressing. However, Geen and Quanty (1977) review numerous studies that found aggression often led individuals to experience guilt, remorse, or fear. These feelings were most likely to occur when the target of aggression was of high status, the aggressor felt that the aggression was inappropriately intense or foolish, and there existed a strong possibility of counteraggression by the target. Thus, although there are cases when a little aggression can feel good, there are also many situations in which it causes the individual to feel uncomfortable. An interesting point is that in both cases, the likelihood of future aggression is reduced. However, in the first case the reduction is the result of lowered tension and catharsis; in the latter case, the result of increased arousal, inhibition, and guilt.

Another important hypothesis of the frustration–aggression theory centers on the Freudian concept of displacement. According to the theory, displacement can occur in two ways: displacement in form, whereby the individual may substitute verbal aggression for the more preferred, but riskier, physical aggression, and displacement of target, in which the individual may either fear or be unable to attack the thwarter and will instead aggress against some object or person that is related to the original thwarter. The displaced aggression hypothesis has been suggested as one root for prejudice (Allport, 1954; Hovland & Sears, 1940). The effect of displaced aggression on the individual's arousal or instigation to future aggression is largely an unexplored area. Hokanson and

Shelter (1961) found a slight decrease in physiological arousal when subjects were allowed to aggress against an experimental confederate after being angered by the experimenter. However, it has been pointed out (Baron, 1977) that guilt may be more likely to accompany displaced aggression than direct aggression; in the former case, the individual has attacked an innocent target.

SOME MODIFICATIONS OF THE FRUSTRATION–AGGRESSION THEORY

The frustration–aggression theory generated voluminous research and a host of criticism. Much of the early criticism centered on whether or not frustration always presupposed aggression. Miller (1941) responded to these criticisms by modifying the theory. He admitted that aggression may be only one of the behaviors instigated by frustration, although he maintained that aggression may still be the preferred response.

Some of the criticisms, however, were not so easily resolved. One centered on the type of frustration that is likely to motivate aggression. Pastore (1952) argued that people will not be instigated to aggress by reasonable and understandable frustrations; that only arbitrary and unexpected frustrations instigate aggression. In order to support his position, he had subjects read a number of short incidents involving both arbitrary and nonarbitrary frustrations, and report how they would respond to each of the cases. As predicted, only the arbitrary frustrations instigated significant amounts of aggression. Pastore's position received support from Kulik and Brown (1979), who found people in natural settings most likely to aggress against arbitrary thwartings. This is an important modification, because it suggests that the desire to aggress may be reduced by reinterpreting frustrations as reasonable and nonarbitrary.

However, Burnstein and Worchel (1962) argue that the research on type of frustration overlooked a critical point: that people may be motivated to aggress by nonarbitrary frustration, but that they inhibit aggression in this situation. Norms in our culture dictate that only children aggress and throw tantrums when frustrated by reasonable events. In a sense, Burnstein and Worchel suggest that nonarbitrary frustrations may be the most annoying, because the individual not only has the original drive thwarted but is placed in a normative straitjacket in responding to this frustration. To demonstrate their position, Burnstein and Worchel put subjects in a group whose task was to arrive at a solution to a problem. However, a confederate thwarted the group by asking members to repeat statements. In some cases, the confederate was obviously not paying attention and his thwarting was arbitrary. In other cases, however, the confederate was having trouble with a hearing aid and his failure to hear was understandable. The results indicated that although subjects were openly more aggressive against the arbitrary thwarter, they also expressed aggression against the nonarbitrary thwarter when such aggression was allowed in a private and nonidentifiable manner. Thus, both types of frustration instigated aggression, but repression was more likely in the nonarbitrary case. There is still disagreement about the effects of various types of frustration (Zillman & Cantor, 1976), and it would be premature to conclude that the instigation to aggress can be eliminated by pointing out the reason for frustrations.

Berkowitz (1965) suggests another modification to the frustration–aggression theory—namely, that aggression does not always follow frustration and that instead frustration leads only to the readiness to aggress (or anger). Aggression itself will not be exhibited unless appropriate aggression cues are present in the environment to draw out the aggression. In our culture, most people associate guns and knives with aggression. However, some people may have learned to associate different names, places, or songs with aggression. Thus, aggression is most likely to be exhibited when the frustrated person is presented with an aggression cue.

Finally, the characteristics of the target have been found to influence the likelihood of aggression following anger. Zimbardo (1970) suggests that people feel inhibited to attack an individual with whom they may empathize. Empathy is most likely to occur when the target is a unique individual and clearly suffers from the attack. This empathy may well result in the attacker feeling guilt. Therefore, Zimbardo suggests that people would more readily attack a deindividuated or dehumanized target. A target can be dehumanized by removing its unique identity, masking its response to the attack, forcing it to remain anonymous, and removing other unique qualities from it. In support of Zimbardo's position, Worchel and Andreoli (1978) found that aggressors attempted to deindividuate the target of their attack by selectively "forgetting" information that identified the victim (e.g., victim's name, personal characteristics, and personal problems).

Taken together, the research suggests that frustration does lead to anger or a readiness to aggress. However, whether or not aggression will actually occur depends in large part on characteristics of the situation and the target. In some cases these characteristics will determine whether or not the individual will experience guilt or remorse if aggression is encouraged. As we shall see in our examination of therapy, the therapist must not only be concerned with facilitating expression, but must give careful attention to the nature of this expression. The above-mentioned research factors may influence both the likelihood of expressing aggression and the person's response to acting aggressively.

SOCIAL LEARNING AND AGGRESSION

Although frustration–aggression theory included external events as the root of aggression, it also relied heavily on internal processes to explain violence. The theory, like that of Freud, viewed behavior as only the tip of the iceberg; the bulk of activity occurs subcutaneously. Theorists such as Watson and Skinner eschewed this preoccupation with internal processes; they argued that behavior and external events are the iceberg, and that behavior could be predicted and explained by examining only the external antecedent conditions. Although not of the strict Skinnerian school, the social learning approach to aggression (Bandura, 1973; Bandura & Walters, 1963) also cast its attention on external events.

According to this position, aggression, like other behaviors, is learned through two mechanisms. The first is reinforcement. In their efforts to obtain goals, individuals may find that aggressive actions often pave the road to success. This reinforcement stamps in the behavior and the individual will resort to it in future situations. The second mechanism, and in many cases the most important, is modeling. Individual learn behaviors by observing others. Aggression is one behavior for which models are seldom lacking. Parents use aggression to discipline their children; aggression abounds in sports; the media are filled with models of aggression.

Bandura, Ross, and Ross (1961, 1963a, 1963b) conducted a series of studies in which children observed a model aggressively interacting with a Bobo doll. When later given the opportunity to play with a Bobo doll, the children reproduced many of the observed behaviors. The research further showed that imitation occurred with live or filmed, children or adult, and real or cartoon models.

Although individuals can learn to aggress by observing models, they will not express this aggression unless the situation offers rewards for such behavior. Bandura (1965) demonstrated this by having subjects watch a model who was punished for aggression. When subjects were placed in a similar situation, they did not aggress. However, when the experimenter specifically asked them to reproduce the model's behavior, the subjects were able to do so with uncanny accuracy.

The thrust of social learning theory has been on dealing with overly aggressive individuals. The theory argues against the catharsis mechanism of the frustration–aggression theory. According to Bandura (1977), aggression will beget more aggression. That is, encouraging aggression will serve to reinforce more aggressive behavior. In order to reduce aggression, it is necessary to punish aggressive behavior and provide a nonaggressive model to teach alternative behaviors. The punishment itself must not be aggressive; rather, it should include a withdrawal of love or reward.

Proponents of social learning theory (e.g., Baron, 1977; Eron, 1980) have marshaled an impressive store of data showing that the observation of aggressive models in the media does teach the audience to aggress. This evidence has also been used to argue against the notion that the vicarious experience of aggression can lead to catharsis. An important finding along these lines is that the viewing of televised aggression tends to reduce physiological responsiveness to the observation of subsequent aggression (Thomas, Horton, Lippincott, & Drabman, 1977). In other words, the observation leads to an insensitivity to it.

Social learning theory and frustration–aggression theory offer very different approaches to dealing with aggression. As we shall outline in the therapy section, these differences are most easily put into practice when dealing with different types of aggression-problem patients. However, even in the psychodynamic therapy session, implications from both theories are presently utilized.

AGGRESSION AND CONTROL

Recently, social psychologists working in a number of areas have begun to focus on the importance of control over one's personal environment and life. Seligman (1975) developed a theory of depression based on a loss of control. According to this approach, when people feel that their actions have no bearing on their outcomes, a state of learned helplessness develops. Responses to learned helplessness closely resemble the clinical state of depression.

Allen and Greenberger (1980a, 1980b) used the concept of control to explain some aggressive behavior. They were especially impressed with destructive behavior such as vandalism that seems to provide little reward for the individual. According to these investigators, aggression and other less direct forms of destructive behavior may be aimed at establishing control over one's environment. The ultimate in control is demonstrating the ability to destroy an object or individual. They argue that when people feel that they are losing control over their environment, they may turn to aggression as a means of restoring it.

In a number of studies, Allen and Greenberger found that giving individuals the opportunity to engage in simple destructive acts (destroying a tower of wooden blocks) enhanced the subjects' feelings of control. Furthermore, subjects who engaged in this destructive behavior reported feeling good. Although not presented as an alternative explanation for the catharsis hypothesis, it is interesting to view these data in that light. We might then expect that "catharsis" should result when the individual's aggression leads to increased feelings of control. Furthermore, this position suggests that catharsis can result from either direct or displaced aggression; the important ingredient is that the act enhance control.

This theoretical position not only offers an explanation for aggressive behavior, but has direct implications for how to deal with individuals suffering from aggression-related disorders. The restoration of feelings of control can be achieved through a number of routes without resorting to aggression. Psychodynamic therapy is designed to enhance the patient's feelings of control. This point in itself may harbor some important therapeutic effects.

EXCITATION-TRANSFER THEORY: A COGNITIVE APPROACH

Thus far we have concentrated on drive theory and learning theory approaches to aggression. Cognitive psychology has also been drawn upon to explain aggressive behavior. Zillman (1978) argues that the likelihood of aggression is determined not only by the situation that thwarts the individual, but also by the attribution the individual makes for his or her arousal. Zillman also suggests that the likelihood of aggression is determined by how angry individuals decide they are. This decision is influenced by the degree of arousal and by the interpretation given to the arousal.

To illustrate the point, Zillman (1971) had subjects watch either a violent, sexually arousing, or nonarousing film. The subjects were then insulted by a confederate and given the chance to aggress against him. Aggression was highest from subjects who had watched the violent film or the sexually arousing film. According to excitation-transfer theory, subjects in these conditions were aroused by the film and by the insult. However, they attributed their high state of arousal as being anger. Subjects who had watched the nonarousing film and were then insulted were not as aroused as the subjects in the other conditions and, hence, their anger was not as great. This excitation-transfer effect has been found with a wide range of arousing activities including humorous movies, physical exercise, and arousing noise.

At first blush, the cognitive approach that deals with the interpretation of present arousal would seem to have little to offer the psychodynamically oriented therapist. However, as we shall see, the process through which insight is gained in therapy is closely allied with the interpretation given for arousal experienced during the session. The psychodynamics of how present states of arousal become labeled as anger or some other feeling can yield important clues to past experience.

FROM THEORY TO THERAPY

This overview of the social psychological theories of aggression indicates that present approaches differ greatly from those views presented by Freud. Despite these differences, the social psychological theories offer some interesting interpretations of the techniques used in psychodynamic therapy to deal with aggression. Before applying these theories to practice, however, two points need to be discussed.

The first involves an important difference between research and therapy that necessitates caution when applying laboratory-obtained results to the practice of therapy. Whereas the researcher studying aggression has the liberty of defining the population to be studied and the hypotheses to be tested, the therapist has the extremely difficult task of initially determining whether or not the presenting symptoms have any connection with aggression. This appears simple when the presenting problems entail excessively destructive behaviors. However, most psychiatric patients will present with a host of complaints that on the surface have little if anything to do with aggression. For example, they may complain of dysphoric states of anxiety or depression, physical symptoms such as headaches or paralysis, sexual difficulties, writing blocks, or vague feelings of loneliness and worthlessness. Freudian and neo-Freudian psychodynamic theory provides an in-depth, comprehensive understanding of the psychic relationships between aggressive impulses and these apparently unrelated symptoms, but it may not indicate the most effective and efficient means of treatment. This has, to a large extent, contributed to the current variety of techniques employed by therapists who come under the rubric of psychodynamic psychotherapy.

A second issue concerns the field to which we apply the research results. In other words, what do we mean by psychodynamic psychotherapy? This continues to be a topic of impassioned controversy and confusion. Although there appears to be a wide spectrum of therapies that could be labeled psychodynamic psychotherapy, they basically can be divided into three groups: psychoanalysis proper, psychoanalytic psychotherapy, and psychodynamically oriented psychotherapy. All share both Freudian and neo-Freudian theories on the origins of psychopathology and personality development. Thus they subscribe to the concept of developmental psychosexual stages in which there emerges a dynamic interplay or conflict between instincts as wishes and defenses or controls against such impulses and learned fears or inhibitions necessary for socialization (Maddi, 1976). Acute symptoms and ultimately character structure are seen as maladaptive solutions to the conflict that are employed because they bring a measure of primary gain (reduction of distress through keeping instinctual forces "unconscious") and secondary gain (getting others to meet instinctual and learned needs). The conflict resolution strategies are called defenses (Alexander, 1952; Brenner, 1955; Fenichel, 1945; A. Freud, 1946) and become a major focus of intervention. Because each of these three therapy groups has differed in the importance given to intrapsychic processes as opposed to interpersonal, social, or cultural factors (Hine, 1972; Kardiner, Karush, & Ovesey, 1959; Rado, 1956, 1962) and as to which instinctual drives are important (Salzman, 1962; Thompson, 1950), they have developed differing goals, techniques, and approaches to transference (Paolino, 1981). Here we shall concentrate on how the actual techniques of psychodynamic therapies support or need to be modified based on the implications of the cited research on aggression.

Technique can be understood as a conscious attempt by a therapist to influence a patient to change in a direction to meet specified goals (Bibring, 1954). Psychoanalysis proper is the most rigid in its prescription of technique. Stone (1951) saw five major components of psychoanalytic technique: (1) contractual sessions of specific frequency, duration, and longevity; (2) use of free association as the most important means of communication; (3) specified financial agreement; (4) physical posture of the patient to be recumbent; and (5) interpretation as the primary curative intervention of the analyst. The techniques employed by both the psychoanalytic psychotherapist and the psychodynamically oriented psychotherapist are more flexible. In addition, other interventions intended to encourage abreaction, promote advice and suggestions, manipulate the external and social environment, and support defenses may be utilized. As Paolino emphasizes (1981), it is not so much the differences between these techniques that distinguish these therapies as it is their attitudes toward the techniques.

From its very inception, psychodynamic theory has been directed toward treating those individuals whose *inhibitions* against instinctual drives (e.g., aggression) are so excessive that alternate maladaptive behaviors (symptoms) emerge. Those who cannot tolerate the strong aggressive impulse without acting upon it are considered poor candidates for classical analysis (Kernberg, Burnstein, Coyne, Appelbaum, Horowitz, & Voth, 1972; Paolino, 1981). Thus Tardiff (1974) found that less than half of Boston-area psychiatrists treat violent patients. In dealing with acutely aggressive patients, the niceties of psychodynamic techniques give way to "using five people to do this. Four hold the patient's limbs and the fifth locks and unlocks doors. . . . Don't stand around trying to reason with him. Get out of range and get help. . . . We give them sedatives as soon as we can" (Turpin, 1982). When aggressive behaviors become an aspect of the personality of the patient, as in antisocial personality disorders, "the prognosis can be summed up as bleak. Here, as in some clinical situations, the doctor can only do his best and hope for the best" (Cloniger, 1978). Only to the extent that the aggressive behaviors are defensive against the emergence of other, more threatening, impulses, or arouse distress that may be experienced as guilty-fear (Rado, 1956) and shame, could the current techniques of psychodynamic therapists be useful. Perhaps the greatest contributions of the research on aggression are in its treatment implications for just those patients who are the least likely to benefit from current psychodynamic techniques.

As early as 1881, catharsis and abreaction were viewed as important curative elements in the treatment of neurotic disorders. Breuer labeled the technique he used to treat Anna O. the *cathartic method*. Freud found that both catharsis and abreaction produced symptom relief. "Only one factor marred Freud's satisfaction with the cathartic method: the necessity of using hypnosis to execute it" (Fancher, 1973, pp. 52–53). Ultimately Freud turned away from the cathartic technique and the subsequent pressure technique (use of cathartic technique but using pressure applied to the forehead instead of hypnosis) in favor of free association. Subsequently, the cathartic method began to diminish in importance relative to other techniques without its effectiveness ever being challenged.

Particularly for the overly inhibited individual, the cited research on catharsis demonstrates that given certain conditions true catharsis and abreaction can both reduce arousal and promote a sense of feeling good. These necessary conditions fall into two basic categories: those that actually promote the experience and release of the previous inhibited (aggressive) impulse, and the absence of conditions that would subsequently nullify the beneficial effects of such a release. The former category applies most when the impulse is intensely experienced and then expressed toward the original object of frustration. Obviously, if the person is available and willing to participate in the therapy situation, such as in family or couples therapy, the task of facilitating catharis is easier and theoretically more effective. When the original object is unavailable or uncooperative, other techniques must be employed. Moreno (1946) attempted to reproduce the unavailable object through realistic psychodrama, Perls (1969) by using the two-chair technique, and Erickson (see Zeig, 1980) through hypnotically induced imagery. The more classical psychodynamic therapist attempts to become the original object of frustration by utilizing a variety of techniques designed to enhance the transference relationship. These have included therapist anonymity, abstinence, frequent sessions, and remaining out of the patient's view. Bach's (1971) training manual outlines how to structure situations where couples and/or patient–therapist can fight both verbally and nonverbally. Afterward, the "catharsis factor" is measured based on "to what extent the fighter came out of the fight with a purged or cleansed feeling because of the release of aggressive tensions" (p. 169). More recently, Davanloo (1978) has shown through the use of videotapes of entire courses of short-term dynamic psychotherapy that by using his technique of systematically confronting the patient's defenses against the inhibited impulse, an intense transference relationship develops in which the inhibited impulse is readily experienced and expressed.[1]

The research indicates that once there has been a release of the pent-up impulse it is necessary to prevent any subsequent experience of guilt or remorse in order for there to be a true catharsis. This requires that patients not experience their behaviors as being inappropriately intense or foolish, that there be no threat of retaliation, and that their "target" not be seen as having high status.

The following illustrates the dangers inherent in attempting to deliberately create a cathartic event. A 50-year-old man with a severe passive-aggressive personality disorder and suffering from depression, stomach pains, and multiple phobias, had been in group therapy for 1½ years following an 8-year unsuccessful analysis. He had no prior history of aggressive behaviors excepting an incident when he was 10 years old in which he threw a stone at his brother resulting in minor injury. He had observed many fights utilizing Bataca bats within the group setting

[1]In "Case of the Man with an Impulse to Murder His Stepfather" Davanloo shows how by systematically confronting this man's defenses against his rage toward both his stepfather and mother there arises an intense transference relationship through which the impulse can be experienced and the defenses interpreted. This can then be related to both past and current situations in what is called a TCP interpretation (Davanloo, 1978, pp. 293–299).

but had been reluctant to participate in them. He was finally persuaded to express his hostile feelings toward one of the therapists by engaging in a "constructive fight" using the Bataca bats. After several exchanges, the patient dropped his bat and struck the therapist in the mouth with his fist. The group members and the other therapist immediately grabbed him until he was calm. Neither the group members nor the attacked therapist ever truly "forgave" the man, who dropped out of the group a year later with intense feelings of guilt, shame, and remorse.

Any cathartic technique that encourages physical aggression runs this kind of risk. Most psychodynamically oriented therapists therefore prefer to focus on verbal expressions of intense affect that are more likely to be viewed by both parties as appropriate responses. Furthermore, verbal assaults, though they can express the intensity of the affect, do not carry as great a fear of counterattack. The simple expression "I'm angry at you" does not represent the inner experience of the impulse but merely labels the affect. The therapist should persist until the experience of the impulse is communicated—"I'd like to strangle you with my bare hands." Clearly, a history of poor impulse control or evidence of psychotic processes would preclude intensifying the affect to levels that the patient cannot tolerate without a need to act out physically.

The catharsis research has little to offer a therapist treating patients who are already acting out their angry feelings in destructive and violent ways. At best, the therapist may wish to channel the aggression into less destructive avenues (sublimation) such as recreation. This, however, runs counter to the excitation-transfer data that suggest the increased arousal could lead to further aggression. A possible compromise may be the wilderness therapies for adolescents where the physical exercise may increase aggressive behaviors but the object of the aggression becomes an inanimate object such as a mountain to climb.

For highly inhibited individuals, the experience of their own aggressive impulses, especially when directed toward those for whom they feel affection, can be overwhelmingly frightening. The research on deindividuation suggests strategies for helping such individuals communicate their hostile feelings. The data show that when people can deindividuate their target, the expression of pent-up aggression is more likely. Psychoanalysts already facilitate the deindividuation process by actually remaining out of the patient's field of vision. Furthermore, the strict adherence to anonymity can also be seen as encouraging deindividuation. The research suggests that therapists should refrain from discussing their own personal backgrounds, opinions, attitudes, and belief systems. In addition, socializing with patients would be counterproductive. It is interesting to note that although the data on deindividuation would support current techniques, they would also suggest that as the transference relationship develops and patients move from viewing the therapist as an anonymous individual to one who represents a specific, highly individuated person in their past life, there should be an increasing resistance to the expression of aggressive feelings.

For those patients who are overtly aggressive and thus likely to have already been systematically deindividuating their victims, the research would suggest that the therapist should make every effort to focus such patients on the unique, individual qualities of their targets. This can be accomplished through imagery or by actually having the patient interview the victims. Too often, parents, teachers, or others attempt to hide their pain from the aggressive patient and respond by withdrawing. This may actually increase their becoming deindividuated and thus encourage further aggression. The research suggests that destructive behaviors, including suicide, may follow a period during which the patient has deindividuated *himself*. Nichtern (1981, p. 4), in discussing violent adolescents, notes that "the loss of self is a recurrent characteristic" of these individuals. The most severely violent adolescents appear to be depersonalized and may turn to violence in order to achieve greater object constancy. In concert with the aforementioned techniques, the therapist could intervene by bringing to focus those unique and enduring qualities the patients possesses that differentiate them from others.

The research on misattribution theory is most helpful to a therapist attempting to treat overtly aggressive patients. By bearing in mind that an "explanation" for the aggression may include the possibility that it stems from a heightened emotional and physical state created by other emotions, the therapist can intervene to reduce the environmental stimulation. One of the most common inpatient techniques for treating acutely violent patients is seclusion (Rada, 1981), thereby reducing external stimulation. For example, for the latency-age child who is attempting to repress sexual impulses, contact with erotic situations can produce great anxiety. In such a heightened state of arousal, the potential for misattribution is increased. Because the aggressive impulses may be less threatening than sexual impulses, they are more likely to be expressed. Clinicians should be alert to the possibility of covertly seductive parents or other adults when presented with aggression in these children. They can then intervene to reduce the erotic stimulation the child is experiencing rather than employ behavioral techniques designed to extinguish the aggressive behaviors.

In attempting to apply the research on aggression cues to the therapy situation, it must be remembered that not all violent acts are committed by psychiatrically ill patients. Simply recognizing that certain environmental or social conditions existed that facilitated an aggressive act is not of much value to the therapist who is searching for the dynamics of the "symptomatic" behavior. However, identifying specific aggression cues becomes important for therapists in two ways. First, they can intervene by eliminating the specific cues and thus potentially halt the behaviors. This is dramatically seen in treating paranoid patients who may become violent if they believe certain objects in the room contain microphones or transmitters, removing these objects will often calm these patients. Second, identifying the aggression cues can provide insight into the dynamics of the aggressive behaviors.

A woman came into the office complaining that her husband, for no apparent reason, would get enraged, beat her, and then leave to get drunk. By examining the most recent episode in detail, a different picture emerged. The woman said that just 2 days ago she had decided to give her husband a surprise. She prepared his favorite meal, lit candles, and wore a sexy negligee. He was most appreciative and affectionate until they entered the bedroom; at that point he became violent. Exploring this transition period in more depth revealed that the woman had casually mentioned that the man's brother had bought his wife a new mink coat. Following this remark, the man "for no reason" became violent. The aggression cue for the husband was the negative comparison with his brother. Further investigation revealed that all such past violent episodes followed similar comparisons with his brother. By eliminating aggression cues, the violence stopped. The therapy was then able to focus on the woman's need to make such comparisons and the man's unresolved sibling rivalry issues.

The therapist must also pay close attention to the aggression cues in the overly inhibited individual even though at first glance they may appear absent. In these individuals the aggression cue may arouse the desire to aggress, but its expression, and at times even its presence, remains inhibited. What emerges is other, more acceptable though symptomatic, behavior. By realizing this, the dynamic therapist can then explore the past unresolved conflicts that created the need for the repression.

A 38-year-old male librarian suddenly became extremely anxious while driving home from work one day. He pulled his car off the road, hyperventilated, and felt an impending sense of doom. This event lasted about 20 min and was followed by weeks of insomnia. By recognizing that there may have been an "aggression cue" in this otherwise passive individual, the therapist explored in detail what transpired just before the anxiety attack. After confronting the resistances to reliving the situation, what emerged was that he "remembered" the radio had been playing the Golden Oldies. One of these songs had been on the radio when, as a child, he witnessed his father beating his mother. After striking his father on the back of his head, he had suddenly become passive and his father had thrown him across the room. The event had been "forgotten" until he heard the song on the radio. The song can be construed as having cued a massive rage against his father, which then gave rise to the extreme anxiety as this rage was beginning to be experienced. The therapy then focused on the unresolved grief over the father's death, through which the rage at his father could be fully experienced.

Although the terminology is different, the research findings on the perceived loss of control as an impetus for violence are consistent with current psychodynamic concepts of aggression. People who find themselves dependent upon others are likely to experience a sense of vulnerability and profound loss of control over their destiny. When Harbin (1980) studied 30 battered-parent families, he found that "the one common trait our studies do reveal in these violence-prone children is that they exhibit enormous dependency needs" (p. 14). Rada (1981) also found that feelings of helplessness and forced dependency lead to violence. Blackman, Weiss, and Lamberti (1963) found that homicidal adolescents may be reacting against a perceived weak masculine identification as well as an intolerable state of forced dependency. McCarthy (1978) noted that aggression may represent a defensive response to lowered self-esteem, which can easily be seen as a perceived loss of control over self and environment. Turpin (1982) notes that

> prior to committing their crimes, dangerous patients often report that they experienced humiliation during either an actual event or their delusional perception of it. They also experienced a pervasive sense of powerlessness not unlike that expressed by suicidal patients.

When the aggressive behavior is so disruptive as to preclude more formal psychodynamic techniques, therapists can rapidly intervene to alter patients' environments so as to give them a greater sense of control. For violent bedridden patients, this might be as simple as assuring that the nurses answer their call button rather than ignoring it. For disruptive behaviors in the classroom, allowing the child to be the teacher for a period of time may provide a sense of control sufficient to halt the aggressive behaviors. Incarcerated criminals often become less overtly hostile when they are shown that they can exert some control over their destiny by learning how to use the legal system to their own benefit. Once the patient's increased sense of control has been successful in reducing the aggressive behaviors, then more traditional techniques can be employed to build ego functions so that the patient can become more independent and individuated, increase self-esteem, and strengthen sexual identification.

It is important to realize that aggressive behaviors can also be a disguised attempt to ask others to take control. Although this runs counter to the theory of loss of control leading to aggression, it is nevertheless seen in clinical practice. A young child may become frightened that his aggression has been too successful. The child may suddenly find himself in control of his family through his aggressive intimidation and become terrified because he suddenly realizes he does not have the skills or knowledge to sustain that position. He may then increase his aggression until he finds someone to put limits on him and thus return him to the relative safety of being a child. An example of this was a 6-year-old boy who had terrorized both of his elderly parents and frustrated his teachers at school to the point where he was about to be expelled. When traditional play therapy failed, a second therapist instructed the parents to bring in several of the child's favorite toys. He then told the boy that each week one of his toys was going to be destroyed right there in the office unless his misbehaving ceased. The boy lost three toys but spared a favorite train by having a "flight to health." The parents were then brought into the therapy situation and taught how to be more effective in placing limits. Even though psychoanalytic theory might suggest that castration anxiety was increased, and social learning theory that the boy actually learned to be more aggressive, the result was that the boy became well behaved and other developmental issues could then be addressed.

Several authors have pointed out the similarities between psychoanalytic and behavioral therapies (Birk, 1973; Dollard & Miller, 1950; Feather & Rhoads, 1972; Woody, 1971), but the contributions that social learning theory could make to psychoanalytic technique have largely been ignored. This is particularly unfortunate given the emergence of such comprehensive and innovative approaches as the multimodal therapy advocated by Lazarus (1973). Psychoanalytic technique has downgraded the importance of two ubiquitous and powerful elements of all therapies that are also the cornerstones of social learning theory: modeling and reinforcement. This is surprising given that Freud wrote extensively about the process of identification and introjection, which is, for all practical purposes, the same as modeling. Modeling has been shown to be effective in reducing aggressive behaviors even in autistic children (Lovaas, 1966; Lovaas, Koegel, Simmons, & Long, 1973; Nedelman & Salzacher, 1972). Furthermore, it has also been effective in the overly inhibited individual (Bandura, 1971). Interpretations become prized by the patient and become "tokens" of a reinforcement contingency paradigm. Few experiences are as pleasurable as the "aha" phenomenon following a correct, well-timed interpretation. Since the efficacy of modeling has been well established, the challenge for psychodynamic therapists is how to enhance their roles as models and dispensers of reinforcers within their current therapy styles.

Although it is impossible for therapists not to become models for their patients, psychoanalysts have tended to downgrade modeling because it conflicts with their greater goal of fostering a transference neurosis by becoming blank screens for their patients' projections. Since social learning theorists would argue that aggressive behaviors have been learned through modeling, it follows that the therapist must become an even more important model than those original models that demonstrated the aggressive behaviors. Otherwise, the likelihood of new learning is reduced. This will clearly be a difficult task if the original models have been parents, valued peers, or highly admired movie stars. Thus the work of Turner and Berkowitz (1972) is of great importance in that they show what conditions are necessary for models to become effective in changing behaviors. Their research shows that effective models must be perceived as having high status or being powerful. Simply being licensed and recognized as a socially sanctioned healer begins the process in which the therapist is perceived as possessing both power and status. These qualities are further enhanced by several of the "nonspecific" aspects of the therapist. Having "evaluation sessions" to see if the patient qualifies for therapy increases the perceived power of the therapist. Placing the patient on a waiting list further shows how highly esteemed the therapist is to other people. The high fees charged and expensive office furnishings also enhance an image of financial power and prestige. The personal attractiveness of both therapist and receptionist can project an image of power and status. Many of these variables can be seen as encouraging "identification" with the therapist (Halleck, 1978).

A possible modification of technique comes from the work of Mason (1970) and Aronson (1969) who have shown in both animal and human situations that interpersonal attraction is greater when an individual is finally able to gain the approval of another who was initially seen as rejecting. Thus, therapists who from the initial contact are profuse with verbal and nonverbal rewards of total acceptance are less likely to become objects of identification. Once identification has occurred, the therapist can then model the desired behaviors. For the aggressive child, he may simply hold him, thus modeling restraint. When a patient attempts to frustrate the therapist, he can observe how this new model reacts in nonaggressive ways. For the overly inhibited individual the therapist models a lack of fear toward approaching any "dangerous" impulse. He also models cognitive approaches of gaining personal insights.

The rather complete obfuscation of reinforcement contingency learning paradigms in psychoanalytic technique is ironic given that the process of working through requires first the development of a regressive transference neurosis that, by definition, is designed to foster a regressed individual who must in turn relate to a nonregressed analyst. The therapy relationship comes to recapitulate the original parent–child relationship. As Flemming has noted, "the object need in many adults reproduces in many ways the functional relationship between mother and child—the diatrophic feeling without which the analytic process meets with difficulty" (cited in Erle & Goldberg, 1979, p. 65). Perhaps at no other time in life are individuals more likely to be both subjected to and susceptible to contingency reinforcement learning than in the early childhood relationship with their parents. Since psychoanalytic techniques all encourage a recapitulation of this relationship, they harbor, perhaps more than any other therapy, the unique opportunity to utilize contingency reinforcement learning to help the patient bring about beneficial changes. Although most psychoanalytic psychotherapists maintain that their goal is to remove past obstacles that prevented normal development, no therapist can interact with a patient without employing reinforcement contingency learning. Thus, if it cannot be avoided, there is little sense in not utilizing it for the patient's benefit. Nor can this issue be dismissed by claiming that only patients know what is in their best interest, because all therapists possess beliefs about what constitutes a successful therapeutic outcome.

Once the therapist has become an important model for the patient, he is in a position to supply reinforcement contingencies. For most individuals the "tokens" are the pleasures of pleasing and being accepted by the therapist as well as their opposite—avoiding displeasing or angering the therapist. For those patients who need to provoke or anger the therapist in order to satisfy a particular intrapsychic conflict, the therapist may withhold reinforcement by not satisfying the "real" intrapsychic need until the patient understands the nature of his conflicts and acts in a more appropriate manner. For certain individuals and children, the source of pleasure is originally derived more from objects or food than from interpersonal relationships. Here the therapist can intervene by linking himself with the objects by becoming their dispenser. This is more graphically seen in the token economies (Fairweather, 1964), but also in individual therapies where toys, food, cigarettes, medications, and privileges become the tokens. In treating the overly inhibited individual, most psychodynamic therapists challenge resistances against the unconscious (aggressive) impulse within the transference relationship until the patient is able to express directly the inner experience of his rage. The therapist then accepts this communication and links it to both current and past situations where the impulse has been blocked from being expressed. The patient gains reinforcement both through having pleased his therapist as well as through gaining greater freedom from painful fears that have constricted him for many years. There is also pleasure from having mastered what once appeared to be an impossible task. In treating the actively violent patient, aversive responses are either implicitly or explicitly employed. In order not to model aggression, the patient is often simply held or placed in seclusion. Once calmed and able to verbalize his anger rather than act upon it (acceptable behavior), he is rewarded by regaining his freedom and staff approval. In individual therapy, "unacceptable" communications are usually responded to in an indirect fashion with silence on the part of therapist.

Early attempts at maintaining changes achieved through reinforcement contingency schedules were disappointing. This could have been a result of the therapist not appreciating the importance of the therapy relationship, possibly as a reaction against the psychoanalytic preoccupation with this relationship. Thus an important aspect of the reinforcement process was lost. This can be overcome by attending to those techniques that enhance identification with the therapist. Then the reinforcement contingencies become intrapsychic and thus always operative. Frequently, patients will report having had mental conversations with their therapist between sessions. It is through these "sessions" with the introjected therapist that the continued reinforcement that is necessary to maintain intrapsychic and behavioral changes takes place.

The scheduling of reinforcement is as important as the type of reinforcements. Social learning theory has demonstrated that positive, intermittent scheduling is most effective. This coincides with the emphasis in the psychoanalytic literature on the "timing" of an interpretation. Social learning theorists have demonstrated that waiting too long after the desired behavior before supplying the reinforcement will either lead to extinction of the changes or significantly reduce their likelihood of being permanently learned. As Brenner (1955) puts it, "a common reason for transference behavior that is unmanageable is undue delay in interpreting to a patient his transference wishes and conflicts."

CONCLUSION

We have attempted to suggest how theory and research on the social psychology of aggression can be applied to psychodynamic therapy. If the task has at times been difficult, it is not because the two areas deal with different topics or behaviors; the congruence is close. Rather, the two fields have adopted a rather myopic view of aggression, dealing only with those aspects that are deemed important to the respective field and even developing a closed jargon that can only be translated by people in the respective fields. The approach is reminiscent of an ancient Hindu fable (cited in Wheeler, Deci, Reis, & Zukerman, 1978):

> There were once five blind men who stood by the roadside every day and begged from the people who passed. They had often heard of elephants, but they had never seen one; for, being blind, how could they?
>
> It so happened one morning that an elephant was driven down the road where they stood. When they were told that the great beast was before them,they asked the driver to let him stop so they might see him.
>
> Of course they could not see him with their eyes, but they thought that by touching him they could learn just what kind of animal he was.
>
> The first one happened to put his hand on the elephant's side. "Well, well!" he said, "now I know all about this beast. He is exactly like a wall."
>
> The second happened to take hold of the elephant's trunk. "Both of you are wrong," he said. "Anybody who knows anything can see that this elephant is like a snake."
>
> The third reached out his arms, and grasped one of the elephant's legs. "Oh, how blind you are!" he said. "It is very plain to me that he is round and tall like a tree."
>
> The fourth was a very tall man, and he chanced to take hold of the elephant's ear. "The blindest man ought to know that this beast is not like any of the things that you name," he said. "He is exactly like a huge fan."
>
> The fifth was very blind indeed, and it was some time before he could find the elephant at all. At last he seized the animal's tail. "O foolish fellows!" he cried. "You surely have lost your senses. This elephant is not like a wall, or a spear, or a snake, or a tree; neither is he like a fan. But any man with a particle of sense can see that he is exactly like a rope."
>
> Then the elephant moved on, and the five blind men sat by the roadside all day, and quarreled about him. Each believed that he knew just how the animal looked; and each called the others hard names because they did not agree with him. People who have eyes sometimes act as foolishly.

In order to understand and treat aggressive behavior, we need to examine it from many angles. We have tried to show here how the research and theories developed by social psychologists can help in understanding the dynamics of psychodynamic therapy. A clear advantage of such a theory-based approach is that it suggests new directions for therapeutic techniques and offers the therapist greater insight into aggressive behavior.

Although this chapter is far from comprehensive, we hope that in taking a first step in uniting these two approaches we may have increased the awareness among both social psychologists and psychodynamic therapists that they do indeed deal with similar issues and face similar questions regarding aggression. Therefore, we view this chapter as an invitation to researchers and practitioners to combine their efforts toward creating a greater understanding of aggression and developing new techniques for dealing with it in therapy.

ACKNOWLEDGMENTS

We would like to thank Frances Worchel for her helpful comments on the chapter during its preparation.

REFERENCES

Alexander, F. Development of the fundamental concepts of psychoanalysis. In F. Alexander & H. Ross (Eds.), *Dynamic psychiatry*. Chicago: University of Chicago Press, 1952.

Allen, V., & Greenberger, D. Destruction and complexity: An application of aesthetic theory. *Personality and Social Psychology Bulletin*, 1980, *6*, 479–483. (a)

Allen, V., & Greenberger, D. Destruction and perceived control. In A. Baum & J. Singer (Eds.), *Advances in environmental psychology* (Vol. 2). Hillsdale, N.J.: Lawrence Erlbaum, 1980. (b)

Allport, G. *The nature of prejudice*. Reading, Mass.: Addison-Wesley, 1954.

Aronson, E. Some antecedents of interpersonal attraction. In W. J. Arnold & D. Levine (Eds.), *Nebraska Symposium on Motivation*. Lincoln: University of Nebraska Press, 1969.

Bach, G. R. *Aggression lab: The fair fight training manual* (1st ed.). Dubuque, Iowa: Kendall/Hunt, 1971.

Bandura, A. Behavior modification through modeling procedures. In L. Kranser & L. Ullman (Eds.), *Research in behavior modification*. New York: Holt, Rinehart & Winston, 1965.

Bandura, A. Psychotherapy based upon modeling principles. In A. E. Bergin & S. L. Garfield (Eds.), *Handbook of psychotherapy and behavior change*. New York: Wiley, 1971.

Bandura, A. *Aggression: A social learning analysis*. New York: Holt, Rinehart & Winston, 1973.

Bandura, A., & Walters, R. *Social learning and personality development*. New York: Holt, Rinehart & Winston, 1963.

Bandura, A., Ross, D. & Ross S. Transmission of aggression through imitation of aggressive models. *Journal of Abnormal and Social Psychology*, 1961, *63*, 575–582.

Bandura, A., Ross, D., & Ross, S. Imitation of film-mediated aggressive models. *Journal of Abnormal and Social Psychology*, 1963, *66*, 3–11. (a)

Bandura, A., Ross, D., & Ross, S. A comparative test of the status envy, social power, and secondary reinforcement theories of identification learning. *Journal of Abnormal and Social Psychology*, 1963, *66*, 527–534. (b)

Baron, R. *Human aggression*. New York: Plenum Press, 1977.

Berkowitz, L. The concept of aggressive drive: Some additional considerations. In L. Berkowitz (Ed.), *Advances in experimental social psychology* (Vol. 2). New York: Academic Press, 1965.

Bibring, E. Psychoanalysis and the dynamic psychotherapist. *Journal of American Psychoanalytic Association*, 1954, *2*, 745–770.

Birk, L. Psychoanalysis and behavioral analyis: Natural resonance and complementarity. *International Journal of Psychiatry*, 1973, *11*, 160–166.

Blackman, N., Weiss, J., & Lamberti, J. The sudden murderer. *Archives of General Psychiatry*, 1963, *8*, 289–294.

Bramel, D., Taub, B., & Blum, B. An observer's reaction to the suffering of his enemy. *Journal of Personality and Social Psychology*, 1968, *8*, 384–392.

Brenner, C. *An elementary textbook of psychoanalysis*. New York: International Universities Press, 1955.

Burnstein, E., & Worchel, P. Arbitrariness of frustration and its consequences for aggression in a social situation. *Journal of Personality*, 1962, *30*, 528–540.

Cloniger, C. R. The antisocial personality. *Hospital Practice*, August 1978, pp. 97–106.

Davanloo, H. *Basic principles and techniques in short-term dynamic psychotherapy*. New York: SP Medical and Scientific Books, 1978.

Dollard, J., & Miller, N. E. *Personality and psychotherapy*. New York: McGraw-Hill, 1950.

Dollard, J., Doob, L., Miller, N., Mowrer, O., & Sears, R. *Frustration and aggression*. New Haven: Yale University Press, 1939.

Eron, L. Prescription for reduction of aggression. *American Psychologist*, 1980, *35*, 244–252.

Fairweather, G. W. *Social psychology in treating mental illness: An experimental approach*. New York: Wiley, 1964.

Fancher, R. E. *Psychoanalytic psychology: The development of Freud's thought*. New York: W. W. Norton, 1973.

Feather, B. W., & Rhoads, J. M. Psychodynamic behavior therapy. I, II. *Archives of General Psychiatry*, 1972, *26*, 496–511.

Fenichel, O. *The psychoanalytic theory of neurosis*. New York: W. W. Norton, 1945.

Freud, A. *The ego and the mechanisms of defense*. New York: International Universities Press, 1946.

Freud, S. *Civilization and its discontents*. London: Hogarth Press, 1930.

Geen, R., & Quanty, M. The catharsis of aggression: An evaluation of a hypothesis. In L. Berkowitz (Ed.), *Advances in experimental social psychology* (Vol. 10). New York: Academic Press, 1977.

Halleck, S. L. *The treatment of emotional disorders*. New York: Jason Aronson, 1978.

Harbin, H. T. Children as initiators of family violence. *Frontiers of Psychiatry*, 1980, *1*, 14.

Harris, M. Mediators between frustration and aggression in a field experiment. *Journal of Experimental Social Psychology*, 1974, *10*, 561–571.

Hine, F. R. *Behavioral science: A selective view*. Boston: Little, Brown, 1972.

Hokanson, J., & Shelter, S. The effect of overt aggression on physiological arousal level. *Journal of Abnormal and Social Psychology*, 1961, *63*, 446–448.

Hokanson, J., Burgess, M., & Cohen, M. Effect of displaced aggression on systolic blood pressure. *Journal of Abnormal and Social Psychology*, 1963, *67*, 214–218.

Hovland, C., & Sears, R. Minor studies in aggression. VI. Correlation of lynchings with economic indices. *Journal of Psychology*, 1940, *9*, 301–310.

Kardiner, A., Karush, A., & Ovesey, L. A methodological study of Freudian theory. IV. The structural hypothesis, the problem of anxiety, and post-Freudian ego psychology. *Journal of Nervous and Mental Disease*, 1959, *129*, 341–356.

Kernberg, O. F., Burnstein, E. D., Coyne, L., Applebaum, A., Horowitz, L., & Voth, H. Psychotherapy and psychoanalysis: Final report of the Menninger Foundation Psychotherapy Research Project. *Bulletin of the Menninger Clinic*, 1972, *36*, 3–275.

Kulik, J., & Brown, R. Frustration, attribution of blame, and aggression. *Journal of Experimental Social Psychology*, 1979, *15*, 183–194.

Lazarus, A. A. Multimodal behavior therapy: Treating the "BASIC ID." *Journal of Nervous and Mental Disease*, 1973, *156*, 404–411.

Lovaas, O. I. A program for the establishment of speech in psychotic children. In J. K. Wing (Ed.), *Early childhood autism*. Oxford: Pergamon Press, 1966.

Lovaas, O. I., Koegel, R., Simmons, J. Q., & Long, J. S. Some generalization and follow-up measures on autistic children in behavior therapy. *Journal of Applied Behavior Analysis*, 1973, *6*, 131–166.

Maddi, S. R. *Personality theories: A comparative analysis* (3rd ed.). Homewood, Ill.: Dorsey Press, 1976.

Mason, W. A. Motivational factors in psychological development. In W. J. Arnold & M. M. Page (Eds.), *Nebraska Symposium on Motivation*. Lincoln: University of Nebraska Press, 1970.

McCarthy, J. Narcissism and the self in homicidal adolescents. *American Journal of Psychoanalysis*, 1978, *38*, 19–29.

McDougall, W. *An introduction to social psychology*. Boston: Luce, 1908.

Miller, N. Frustration-aggression hypothesis. *Psychological Review*, 1941, *48*, 337–342.

Moreno, J. L. *Psychodrama*. Boston: Beacon Press, 1946.

Nedelman, D., & Salzacher, S. I. Dicky, at 13 years of age: A long-term success following early application of operant conditioning procedures. In G. Semb (Ed.), *Behavior analysis and education*. Laurence: Follow-through Project, 1972.

Nichtern, S. Has psychiatry failed the acting out and violent adolescent? *Frontiers of Psychiatry*, 1981, 2(8), 4.

Paolino, T. J. *Psychoanalytic psychotherapy: Theory, technique, therapeutic relationship and treatability*. New York: Brunner/Mazel, 1981.

Pastore, N. The role of arbitrariness in the frustration-aggression hypothesis. *Journal of Abnormal and Social Psychology*, 1952, *47*, 728–731.

Perls, F. S. *Gestalt therapy verbatim*. Lafayette, Calif.: Real People Press, 1969.

Rada, R. T. The violent patient: Rapid assessment and management. *Psychosomatics*, 1981, 22 (2), 101–109.

Rado, S. *Psychoanalysis of behavior: Collected papers* (Vol. 1). New York: Grune & Stratton, 1956.

Rado, S. *Psychoanalysis of behavior: Collected papers* (Vol. 2). New York: Grune & Stratton, 1962.

Rado, S. *Adaptational psychodynamics: Motivation and control*. New York: Science House, 1969.

Salzman, L. *Developments in psychoanalysis*. New York: Grune & Stratton, 1962.

Seligman, M. *Helplessness*. San Francisco: W. H. Freeman, 1975.

Sheras, P., & Worchel, S. *Clinical psychology: A social-psychological approach*. New York: Van Nostrand, 1979.

Stone, L. Psychoanalysis and brief psychotherapy. *Psychoanalytic Quarterly*, 1951, *20*, 215–236.

Tardiff, K. J. A survey of psychiatrists in Boston and their work with violent patients. *American Journal of Psychiatry*, 1974, *131*, 1008–1011.

Thomas, M., Horton, R., Lippincott, E., & Drabman, R. Desensitization to portrayals of real-life aggression as a function of exposure to television violence. *Journal of Personality and Social Psychology*, 1977, *35*, 450–458.

Thompson, C. *Psychoanalysis: Evolution and development*. New York: Hermitage House, 1950.

Turner, C., & Berkowitz, L. Identification with film aggressor (covert role taking) and relations to film violence. *Journal of Personality and Social Psychology*, 1972, *21*, 256–264.

Turpin, J. P. Predicting violence before and after release. *Frontiers of Psychiatry*, 1982, *12* (2), 1–4.

Wheeler, L., Deci, E., Reis, H., & Zukerman, M. *Social influence* (2nd ed.). Boston: Allyn & Bacon, 1978.

Woody, R. H. *Psychobehavioral counseling and therapy: Integrating behavioral and insight therapies*. New York: Appleton-Century-Crofts, 1971.

Worchel, S., & Andreoli, V. Facilitation of social interaction through deindividation of the target. *Journal of Personality and Social Psychology*, 1978, *36*, 549–557.

Zeig, J. K. *A teaching seminar with Milton H. Erickson*. New York: Brunner/Mazel, 1980.

Zillman, D. Excitation transfer in communication-mediated aggressive behavior. *Journal of Experimental Social Psychology*, 1971, *7*, 419–434.

Zillman, D. *Hostility and aggression*. Hillsdale, N.J.: Lawrence Erlbaum, 1978.

Zillman, D., & Cantor, A. Effect of timing of information about mitigating circumstances on emotional responses to provocation and retaliatory behavior. *Journal of Experimental Social Psychology*, 1976, *12*, 38–55.

Zimbardo, P. The human choice: Individuation, reason and order versus deindividuation, impulse and chaos. In W. Arnold & D. Levine (Eds.), *Nebraska Symposium on Motivation*. Lincoln: University of Nebraska Press, 1970.

7

Attitude Change Theory, Research, and Clinical Practice

John H. Harvey, Gifford Weary, James E. Maddux, John S. Jordan, and Kathryn S. Galvin

In this chapter we shall focus on the relevance to clinical practice of social psychological theories of attitude formation and change. We shall argue that attitude change processes are important in all major approaches to psychotherapy and that three decades of social psychological research attention to attitude development and change provide an empirical foundation for the investigation of persuasion as a clinical treatment strategy. The general position that clinical practice should be informed by the substantial literature on attitudes and attitude change is not original (e.g., see S. S. Brehm, 1976; Goldstein & Simonson, 1971, Goldstein, Heller, & Sechrest, 1966; Strong, 1978). However, it is our objective to provide a new look at the literature to which researchers and practitioners should be especially attentive in considering the nature and importance of persuasion in psychotherapy. We shall begin by examining the idea that attitude change processes are fundamental to all

John H. Harvey and James E. Maddux • Department of Psychology, Texas Tech University, Lubbock, Texas 79409. Gifford Weary • Department of Psychology, Ohio State University, Columbus, Ohio 43210. John S. Jordan • Duke Medical School, Duke University, Durham, North Carolina 27706. Kathryn S. Galvin • Counseling Center, Texas A & M University, College Station, Texas 77840.

psychotherapies and shall then outline several attitude change conceptions that should have relevance to clinical practice. Finally, we shall present examples of recent research that examined the role of attitude change in clinical and counseling contexts.

ATTITUDE CHANGE AND PSYCHOTHERAPY: THEORETICAL RELATIONSHIPS

Psychotherapy has been defined as "an interpersonal process designed to bring about modification of feelings, cognitions, attitudes, and behavior which have proven troublesome to the person seeking help from a trained professional" (Strupp, 1978, p. 3). Psychotherapy is an interpersonal or social process, and most psychological problems, to a great extent, need to be viewed within the context of social and interpersonal relationships (Strong, 1978). Regardless of the therapist's theoretical orientation, much of the time and effort expended in therapy will be directed toward the goal of changing or eliminating certain essential dysfunctional attitudes and beliefs about the social environment held by the client. These beliefs may range from the unconsciously held taboos and sanctions of the superego, as posited by psychoanalysis, to the "basic mistakes" in the individual's "cognitive map" as viewed by Adlerian individual psychology (Mosak, 1979), to Beck's (1976) "cognitive triad" in the genesis and maintenance of depression, to the "irrational beliefs" of rational-emotive therapy (Ellis, 1973).

The procedures and techniques by which the therapist attempts to change or eliminate these dysfunctional beliefs and attitudes and replace them with beliefs and attitudes that are more adaptive will vary greatly with various theoretical schools. A traditional psychoanalyst may rely primarily on free association as an avenue for attitude exploration, insight, and change. The Rogerian or "person-centered" therapist (Meador & Rogers, 1979) may reflect to the client the attitudes, beliefs, and feelings expressed by the client, a procedure that results in clients' reexamination of their belief system and basic assumptions or attitudes about themselves and the world. The Adlerian therapist (Mosak, 1979) may rely more on "education" and direct encouragement in an effort to influence the client and eliminate the "basic mistakes" in the client's cognitive processes. The rational-emotive therapist (Ellis, 1973) is likely to engage in considerable direct challenging and confrontation of the client's irrational beliefs and dysfunctional attitudes, and thus elicit attitudinal change.

Despite differences in style and procedure, each of these therapists is involved to a great extent in a process of persuasion or attitude change.

The success of even the most explicitly behavioral interventions (e.g., aversion therapy, systematic desensitization, relaxation training) depends largely upon verbal activity and cognitive change (Beck, 1976). This point follows from the fact that such procedures are dependent upon persuading the client that a problem exists, that the therapist understands the problem, that the problem is treatable, that the intervention procedure proposed by the therapist will be successful, and that the client is capable of performing the recommended intervention. Modification of these beliefs may be seen as most crucial to the opening phase of behavioral therapies, but the success of any therapy depends on initially engaging the client's interest and optimism.

Murray and Jacobson (1978) have described four basic processes that are viewed as essential to the success of all therapeutic techniques: (1) the arousal of expectations in clients that they will be helped by psychotherapy, (2) the correction of nonadaptive or dysfunctional beliefs about the world, (3) the correction of nonadaptive or dysfunctional beliefs about the self, and (4) the development of competencies in dealing with social living. The first three of these basic processes fall squarely within the realm of attitudes and social cognition, the fourth factor is largely concerned with the relationship between attitudes and beliefs about the self and the world and actual social behavior. The contemporary researcher of the psychotherapy process is displaying an interest in central attitude change processes when questions such as the following are asked: What beliefs and attitudes about the world and the self are maladaptive and lead to psychological and behavioral difficulties? What beliefs and attitudes are adaptive and lead to adjustment and psychological well-being? How are both adaptive and maladaptive beliefs, attitudes, and social cognitions learned and developed? What are the most effective and efficient procedures for eliminating maladaptive beliefs and instilling adaptive, "healthy" beliefs? If we take the perspective that the "therapeutic process can be viewed in terms that apply to all social relationships" (Orlinksy & Howard, 1978), then the commonalities between the theoretical concerns of the attitude researcher and the psychotherapist become even more apparent.

Cognitive approaches to psychotherapy have been viewed as outgrowths of behavioral approaches to treatment (Mahoney & Arnkoff, 1978) and represent, perhaps, the clearest examples of "attitudinal therapy" (Allport, 1968). Cognitive psychotherapies have been defined as "all the approaches that alleviate psychological distress through the medium of correcting faulty conceptions and self-signals" (Beck, 1976, p. 214) and that focus largely on the process of "making the patient's attitudes explicit and helping him decide whether they are self-defeating" (Beck, 1976, p. 256). Notable cognitive approaches include Ellis's (1973) rational-emotive therapy and Beck's (1976) cognitive therapy for depression.

The idea that beliefs, attitudes, cognitions, and other internal events are important to understanding and changing psychological problems is not a recent innovation. Such an emphasis has existed since the advent of theories of psychotherapy. A principal tenet of traditional psychodynamic approaches to psychotherapy has been their emphasis upon the influence of unconscious determinants of behavior. This focus is contrasted with that of the newer cognitive theories that consider conscious or accessible experiences to be more salient and amenable to successful modification. With this contemporary approach, the relationship between theories of psychotherapy and theories of attitudes and attitude change has become easier to discern. In the next section, we shall examine some of the major attitude change conceptions and discuss their potential relevance to the clinical practice.

SOCIAL PSYCHOLOGICAL THEORIES OF ATTITUDE CHANGE

What is an attitude according to social psychological theory? A popular definition was provided by Thurstone (1946):

> I define attitude as the intensity of positive or negative affect for or against a psychological object. A psychological object is any symbol, person, phrase, slogan, or idea toward which people can differ as regards positive or negative affect. (p. 39)

We might also wish to include in the category of psychological object "groups of people" since clearly attitude research on prejudice is usually directed toward affect about more than one person.

Most of the theories of attitude change that will be briefly characterized were developed in the late 1950s, 1960s and early 1970s. This period represents the zenith of the development of cognitive theories in social psychology, and each theory to be depicted has as a central component an emphasis upon the individual's cognitions and cognitive processes. For a comprehensive treatment of this era of work, the reader is referred to Kiesler, Collins, and Miller's (1969) excellent textbook on attitude change.

LEARNING THEORY IN THE YALE COMMUNICATION AND ATTITUDE CHANGE PROGRAM

The work of Hovland and his colleagues at Yale University in the early 1950s was the precursor of two decades in research in social psychology concerned with attitude change from a straightforward learning perspective. These investigators (see Hovland, Janis, & Kelley, 1953) presented a program of work focusing on "who says what with whom with what effect." Hovland *et al.* assumed that attitudes are learned verbal answers that an individual gives in response to stimulus situations. In this conception, attitudes have drive value, and they persist unless the individual undergoes some new learning experience. As Kiesler *et al.* (1969) suggest, Hovland and associates probably assumed that the laws governing the learning of an overt response are quite similar to those that are involved in the "change" of an attitude. This type of reasoning no doubt was influenced by the tenor of the research zeitgeist at Yale in the first half of the twentieth century with such workers as Hull (1932) and Doob (1947) having a major impact with their quasi-behavioristic reinforcement conceptions.

For our purposes in this chapter, the principal contribution of Hovland *et al.* (1953) was in providing a theoretical perspective and accompanying research on aspects of how people learn messages. In this framework, a persuasive message, such as a therapist's advice, questions a recipient's initial attitude (e.g., the initial position of the client on a particular problem) and provides incentives (e.g., promise to reduce an unpleasant drive such as anxiety) for attending to, understanding, yielding to, and retaining the new rather than the initial attitude. Thus, the independent variables in Hovland *et al.*'s work were classes of source variables (such as the communicator's expertise or attractiveness), message variables (such as number of arguments presented and the comprehensibility of the arguments), and recipient variables (such as intelligence, self-esteem, and sex differences).

The rather simple, logical model of the Yale program has had considerable impact on research concerned with persuasion in clinical and counseling contexts. Indeed, in a recent review, Corrigan, Dell, Lewis, and Schmidt (1980) found many studies in the clinical and counseling literatures that examined the effects of source, message, and recipient factors on clients' attending, understanding, yielding, and retaining processes with classes of variables relevant to the source, message, and the recipient. The pervasiveness of the Hovland *et al.* line of logic will be apparent throughout the remainder of this chapter as we consider attitude conceptions relevant to practice.

COGNITIVE DISSONANCE THEORY

Dissonance theory may be a close second to the Yale learning approach in its potential usefulness for practitioners. As will be discussed later, the research of Cooper and colleagues has suggested direct ways in which dissonance principles may be used to effect therapeutic change. Cognitive dissonance theory was first proposed by Festinger (1957), and other important revisions were offered by J. W. Brehm and Cohen (1962) and Aronson (1968). Dissonance theory is probably the most important of the several cognitive consistency theories. It certainly has generated the most research and further theoretical analyses (see Wicklund & Brehm, 1976, for one of the most recent reexaminations of dissonance ideas). These theories posit that a negative drive state is aroused when cognitive elements (thoughts) are inconsistent (or dissonant). Dissonance supposedly increases as a function of the importance of the inconsistent elements. For example, the thoughts "I smoke and smoking causes cancer" would most likely be dissonant and of much importance for the individual. But, according to this view, the cognitions "I smoke, and it is snowing outside" would not stimulate dissonance since they do not represent a relevant linkage of thoughts for dissonance to occur.

Because cognitive dissonance is theoretically an unpleasant state, individuals are presumed to be motivated to reduce or eliminate it by modifying their attitudes or behavior so as to restore balance and consistency to the cognitive system. When it is impossible to change an attitudinal or behavioral element, alternative dissonance reduction strategy possibilities include denial and reducing the importance of the elements (e.g., the smoker's rebuttal, "I can only live so long anyway").

As has been suggested elsewhere (e.g., S. S. Brehm, 1976), the process of dissonance induction and reduction and the psychodynamic process of conflict development and resolution are similar in several ways. Both emphasize the importance of balance versus imbalance and consistency versus inconsistency within the individual's cognitive system. Both posit an "uncomfortable psychological state" that is an emotional or affective product of the imbalance and that is presumed to serve as a mediating force between the cognitive conflict and the individual's attempts to resolve the inconsistency. Finally, both emphasize the importance of cognitive and behavioral changes in the restoration of balance and consistency and thus in the alleviation of the unpleasant emotional state.

SELF-PERCEPTION THEORY

Bem (1967) argues that individuals' attitude statements may be viewed as inferences from observations of their own behavior and the context in which it occurs. This postulation was offered as an alternative to dissonance theory to account for attitude change. Bem suggests that his self-perception approach was superior to dissonance theory because it made no assumptions about drive state (which cannot be observed directly) and, hence, was more parsimonious. An illustration of the self-perception approach to attitude change is provided by the so-called foot-in-the-door technique for inducing compliance. In this technique, individuals are induced initially to take some action that may not be consistent with their attitudes (e.g., inducing college students to circulate a petition advocating higher tuition). Subsequently, the individuals often show more atittudinal endorsement of the position indicated by the action than if they had not been induced to take the action. According to Bemian logic, they may think, "If I sent that petition around to my friends, then I must believe there are good reasons to increase tuition."

The foot-in-the-door effect may also be explained by dissonance theory if one assumes that the act of committing oneself to a position—especially a counterattitudinal position—arouses dissonance. We do not need to be concerned too much here with the relative cogency of dissonance versus self-perception approaches. But the possible value of getting clients to act or make commitments (e.g., to particular remedial activities) is well known, and either of these theoretical processes may be seen as providing helpful interpretations of why action–commitment sequences are sometimes effective.

The Bemian self-perception approach also bears a strong similarity to so-called cognitive relabeling or self-statement modification (Craighead & Craighead, 1980). Self-statements, in effect, become attitudes about self that the client is encouraged to believe. The general premise is that if a client is induced to make the self-statements with little external pressure or reward, the person may come to perceive these statements as highly representative of personal internal attitudes and beliefs.

Theory of Reasoned Action

In a recent formulation, Fishbein and Ajzen (1981) have proposed a theory of reasoned action that specifies in a fairly precise way the relationships among beliefs, attitudes, and behaviors. According to this theory, the immediate determinant of a person's overt behavior is the person's intention to perform (or not to perform) that behavior. Thus, the knowledge of a person's intention with regard to some object or person would be an important piece of information to be used in predicting future behavior. Fishbein and Ajzen suggest that when we do not know a person's intentions, two factors can be used to deduce intention, namely: (1) the person's attitude toward the behavior, and (2) the person's subjective norm. The first component refers to the person's positive or negative sentiment toward engaging in that behavior. The second refers to the person's perceptions of the social pressures to perform or not to perform the behavior in question. It is argued that people will perform behaviors that they value highly and that are popular with others and will refrain from behaviors that they do not regard favorably and that are unpopular with others.

Although Fishbein and Ajzen's conception has not yet been linked directly to clinical practice questions, it has obvious relevance. The conception specifies with some degree of precision the elements of attitude change with which the therapist should be most concerned. Nonetheless, the task of pinpointing a client's attitude toward a behavior, subjective norm, and, consequently, the client's intention, may be as imposing as engineering significant behavioral change. Each type of determination may require considerable investigation of the individual and his or her idiosyncratic viewpoints.

Cognitive Response Theory

The so-called cognitive response theory is also of recent vintage and derives from the writings of Greenwald (1968), Petty, Ostrom, and Brock (1981), and Petty and Cacioppo (1981). In the cognitive response conception, it is assumed that when a person anticipates or receives a persuasive communication, an attempt is made to relate the information in the message to the preexisting knowledge that the person has about the topic. The person is assumed to consider a substantial amount of information that is not found in the communication itself. These additional self-generated cognitive responses (thoughts) may agree with the proposals being made in the message, disagree, or be entirely irrelevant to the communication. If, for example, a therapist is trying to convince a client to adopt a particular position, then it is imperative to induce the client to develop cognitive responses that are supportive (proarguments) rather than antagonistic (counterarguments). Thus, the cognitive response approach holds that the cognitions elicited at the time of message transmission will determine the amount and direction of attitude change that is produced.

INTERACTION THEORY

This approach has been developed by Strong and his colleagues (e.g., Dell, 1973; Strong, 1968, 1978; Strong & Claiborn, 1982) and involves clear-cut suggestions for linking attitude change concepts and therapeutic approaches. In part, these theorists have focused on source factors (therapist expertness, attractiveness, credibility) that may enhance the therapist's ability to influence the client. A direct prediction, for example, from this line of work is that physically attractive therapists are most persuasive and hence most effective in their counseling. The evidence to support such assertions, however, is mixed (Strong, 1978).

At the heart of interaction theory is the assumption that the therapeutic relationship, like most interpersonal relationships, involves interdependence, identity negotiation, and attempts at interpersonal control. In this conception, symptomatic behaviors are tactical self-presentational devices used to influence the nature of the relationship between other individuals and the client, who avoids responsibility for failing to change by creating an aberrant social identity. One technique deriving from this position discussed by Strong and his associates involves paradoxical directives. These directives are given by the therapist when symptomatic behavior is displayed and involve directing the client to continue displaying the behavior. Presumably such a technique may influence clients to change their actions so as to adapt to this unexpected interpersonal event; in turn, the client may begin to use more socially acceptable styles of interaction and to feel a sense of responsibility for more effective psychological functioning.

A premise behind the paradoxical directive technique is found in reactance theory (S. S. Brehm, 1981). This theory assumes that people experience psychological discomfort and react to restore their freedoms if they feel that those freedoms are threatened or have been eliminated. It can be assumed that clients may refuse to obey paradoxical directives because such directives threaten their sense of freedom to change courses of action. Therapists who show eagerness to persuade, to maintain control, to establish their expertise, or to limit the client's choices may create reactance in the client which then may lead to the client's refusal to follow any directives from the therapist.

Although the interaction approach is rich in its linkage of social and clinical concepts, to date little empirical work has been done to evaluate its efficacy or to compare it to other approaches.

ATTITUDE CHANGE AND PSYCHOTHERAPY: ILLUSTRATIVE RESEARCH

In the area of attitude change, an exemplary integration of social and clinical psychology is the work of Cooper and his colleagues on a derivative of dissonance theory, the effort justification hypothesis. This hypothesis posits that the more effort a person puts into a counterattitudinal activity in the absence of external pressure, the more dissonance will be aroused. To reduce this dissonance created by the incongruent cognitions that one has freely engaged in an unpleasant activity, one may come to view the activity more positively. Applying this reasoning to therapy, Cooper and Axsom (1982) have argued that the extreme effort required of those in psychotherapy will, under certain conditions, create pressures toward therapeutic success (e.g., the goal of therapy may become more attractive) through the arousal and subsequent reduction of cognitive dissonance. Moreover, these investigators have argued that it makes little difference what type of effort is required by the various therapies (e.g., dream recall, modeling). It is sufficient for therapeutic improvement that clients perceive themselves as freely choosing to engage in difficult or unpleasant activity.

Cooper (1980) reports two conceptually and methodologically similar experiments that attempted to demonstrate that effort justification may account for therapeutic success. Using 75 snake-phobic participants recruited from a newspaper advertisement, Cooper in the first experiment compared the effects of two 40-min therapies engaged in under conditions of high and low choice. One therapy employed by Cooper was a conventional treatment for phobias, implosion therapy (Stampfl & Lewis, 1967), which is based on principles of extinction. The other therapy, called "effort therapy," consisted solely of physical exercises that the participants were to perform. (Pretesting revealed that both of these therapies were perceived as equally aversive and effortful.) All participants learned that the treatments could be very effortful and anxiety provoking, but only high-choice participants were told explicitly that they could terminate the experiment. Therapeutic success was measured by changes in participants' pre- and posttherapy ability to approach a caged boa constrictor.

Predictions based on the theoretical integrity of the therapies would suggest that implosion therapy should work, whereas "effort therapy" should not. Dissonance-based predictions, however, call for participants in high-choice conditions to improve, whereas low-choice condition participants should not, regardless of the therapy type. Consistent with a dissonance-based, effort justification hypothesis, only a significant main effect for choice emerged. When choice to engage in the therapy was high, both of the therapies resulted in improvement, irrespective of differences in their content; when perceived choice was low, neither therapy was effective.

Although these results seem to provide strong support for the notion that therapeutic improvement results from dissonance arousal and reduction, they may be subject to the potential alternative interpretation of experimental demand. Under high-choice conditions, participants may have perceived the experimenter as especially fair or sympathetic and may have tried harder to approach the snake in order to please her. In an attempt to rule out this possible alternative explanation, Cooper conducted a second study in which the effectiveness of two "treatments" for assertiveness—conventional behavioral rehearsal therapy (Wolpe, 1963) and the "effort therapy" employed in the previous experiment—was compared. Participants in this study, college students who had scored low on a measure of assertiveness, received one of the two therapies under conditions of high and low choice and then reported at the conclusion of the therapy session to a receptionist to receive their $2 promised for participation in the study. Participants' reactions when they received only $1 were rated by two hidden observers (blind to experimental conditions) on a scale of assertiveness (a low score was assigned to participants who accepted the dollar without question and a high score was given to subjects who expressed an intent to correct the receptionist's error). It was hoped that this measure of assertiveness was sufficiently unobtrusive so that a desire to please the experimenter would not be prominent. As in the first study, no effects of type of therapy were found. What was important was whether participants believed they had a high degree of choice to engage in the therapy. When choice was high, both therapies produced an improvement; when choice was low, neither therapy was effective.

A related experiment concerned with the durability of effort justification effects (Axsom & Cooper, 1981) involved weight loss as the target problem. In this study a single therapy consisting of various cognitive tasks was used, with the degree of effort expenditure (high or low) and choice to engage in therapy (high or low) manipulated. Although the choice manipulation was not successful in this study (all subjects perceived a high degree of choice), the results are consistent with dissonance theory predictions. The high-effort condition subjects lost more weight at the end of five "therapy" sessions, at a 6-month and 1-year follow-up, than did low-effort condition subjects.

These three studies, each focusing on a different target problem and employing different subject populations, indicated that conventional psychotherapies worked only to the extent that participants chose freely to engage in them. Moreover, these conventional therapies were no more effective than the "effort therapies," despite the vacuous nature of these latter treatment approaches. Effort justification, then, may be an important factor contributing to the success of many forms of psychotherapy.

EXTRAPOLATING ATTITUDE CHANGE PRINCIPLES AND PERSPECTIVES TO CLINICAL PRACTICE

The practitioner is advised to be extremely cautious in applying attitude change principles to issues that develop in the course of practice. Very little attitude change research has been conducted in clinical settings and/or with representative clinical populations. Furthermore, as several analysts (e.g., Harris & Harvey, 1978; Holmes, 1974) have suggested, many logical and empirical difficulties are often encountered in attempts to apply concepts deriving from basic social psychological research to the interpretation of psychopathology phenomena. Nonetheless, for each persuasion–clinical practice problem, there are probably some relatively clear extrapolations of principle and perspective that may be offered in the interest of facilitating treatment. An example will be used to discuss how certain extrapolations might hold for a particular situation.

A complex decision dilemma that many therapists have encountered is that involving a married client's request for help in connection with the handling of a surreptitiously conducted extramarital affair that threatens the client's marriage and feelings of self-worth and dignity. The client's decision may pertain to how and when to terminate the affair, or to how and when to terminate the marriage in order to develop the outside relationship. Over the course of therapy, the therapist may become convinced that the client's and others' welfare requires a particular kind of decision. The therapist may wish to try to persuade the client to adopt a particular attitudinal position about the affair and related issues so that the attitude will influence certain behavioral consequences. What advice derives from the study of attitude change processes for the therapist helping the client deal with this situation of momentous decision?

1. An obvious extrapolation from the attitude literature relates to the success of the therapist in establishing his or her trustworthiness and credibility as an advisor on the problem. Without the development of these basic ingredients of interaction, all additional approaches are apt to founder.

2. It is likely that clients in such situations may exhibit self-perception tendencies in inferences about self relevant to the affair. For example, they may begin to look back and evaluate self as immoral since the course of action may have highly negative consequences for others and since it was carried out in a volitional and secretive fashion. Whatever decision the therapist wishes the client to reach, it may be helpful for the client to become more aware of the potent persuasive influence of self-perception processes. This type of enlightenment may be especially useful as the client makes the decision and then starts to look back and evaluate the merits of the decision and how the decision may have an effect on self-esteem.

3. Whether or not the therapist chooses to try the interaction/paradoxical directives approach, caution is warranted regarding the client's possible reactance against perceived threats to personal freedom to act. Reactance may be especially likely to occur if the therapist early in the course of therapy tries to persuade the client to abandon the affair.

4. The decisional situation has qualities that make it amenable to some principles from dissonance theory. Dissonance theory would emphasize the import of the client's feeling choice in first seeking therapy and then in following the therapist's lead in arriving at a particular decision. Perceived choice about the decision will probably stimulate a sense of personal responsibility and control (Harvey & Harris, 1975). Also, as suggested by dissonance theory, a sense of having expended effort and being committed (as via commitments to the therapist) should enhance feelings of responsibility and control, and ultimately positively influence general well-being. The therapist again may do well to help the client anticipate the likely postdecisional experiences and conflicts.

The client may encounter not only rather pronounced dissonance, but also regret and even a strong desire to reverse the decision. Awareness that these processes are normal for consequential decisions may be helpful, as also may the therapist's ability to help the client decide how to reduce dissonance and generally come to grips with the decision and its consequences. Dissonance literature provides at least suggestive evidence that certain techniques (e.g., attempted avoidance of painful cognitions) may be less effective in this postdecisional phase for the client than other techniques (e.g., use of distracting cognitions).

Finally, such a decision situation, with its gravity, defies a ready, precise answer to the question of how to persuade a client to take some action. Common sense, clinical experience, wisdom and intuition, and knowledge of a variety of theoretical perspectives and research findings will be required for a thoughtful approach to the problem.

CONCLUSIONS

One of the most significant recent developments in research on psychotherapy processes is the growing interest in strategies of behavior change that may be common to all psychotherapies. For example, Strupp and Bergin (1969) have contended that there is a specifiable number of social influence factors that are operative in all of the major psychotherapies. Among these Strupp and Bergin include imitation, identification, empathy, warmth, interpretation, counterconditioning, extinction, discrimination, learning, reward, and punishment. In a similar vein, Goldfried (1980) argues that

> although commonalities across approaches may be found in the realm of specific techniques (e.g., role-playing, relaxation training), it is unlikely that such comparisons would reveal much more than trivial points of similarity. I would suggest, however, that the possibility of finding meaningful consensus exists at a level of abstraction somewhere between theory and technique which, for want of a better term, we might call clinical strategies. . . . In essence, such strategies function as clinical heuristics that implicitly guide our efforts during the course of feedback. (p. 994)

Goldfried identifies two such clinical strategies that may be common to all therapies: (1) providing the client with new, corrective experiences, and (2) giving the client direct feedback.

We have argued that processes of attitude change are important components of most (if not all) of the major psychotherapies. Indeed, part of the attractiveness of focusing on strategies of attitude change is that such an approach can facilitate the integration of diverse concepts and findings and offers the opportunity for viewing the similarities rather than the dissimilarities across different forms of psychotherapy. As we have indicated, research on attitude change as a clinical strategy can benefit from the substantial social psychological literature on persuasion.

Our treatment of the link between attitude change literature and clinical practice has a few notable limitations. First, it is obvious that persuasion goes both ways in the therapeutic encounter. We have too little evidence on how the client influences the therapist. To assume that the therapist does all of the influencing and is a pillar of resistance to the client's persuasion attempts would be folly. The persuasion from client to therapist may run the gamut from influence attempts aimed at changing the therapist's attitudes toward the client to changing significant world views that the therapist holds. A typical psychotherapy encounter probably involves a reciprocity of influence attempts, although the therapist's attempts may be more explicit and presented in a more formal fashion.

Second, implicit in our analysis is the idea that the therapist is often engaged in an attitude change attempt that will eventuate in a positive outcome for the client. But such attempts are not always nor necessarily felicitous. As Brodsky (1982) has suggested, therapists may communicate socially negative biases, or cultural stereotypes, to clients who may be prone to adopt some of these views. In addition, the therapist may subvert the process of therapy by inadvertently conveying dissatisfaction with some attributes or features of the client. For example, most therapists would agree to a preference for working with clients who fit the YAVIS prototype (young, attractive, verbal, intelligent, successful). As a result of this type of bias, a client who does not meet these standards may receive subtle cues that serve to reinforce self-perceptions of deficiency or inadequacy.

Finally, this chapter is directed mainly toward the practicing clinician. We have suggested that the practitioner may benefit from general scrutiny of the extensive literature in social psychology on attitudes and attitude change. But we would be quite negligent if no mention were made of the other side of the coin. The theorist and researcher can probably learn much from the practitioner's intuitions about attitude change in clinical settings and possibly by observing therapeutic encounters. These intuitions and observations will most likely suggest useful hypotheses. But even more likely, they will point toward the subtleties of social influence (such as reciprocal influence) that pervade the therapeutic interaction.

REFERENCES

Allport, G. *The person in psychology*. Boston: Beacon Press, 1968.

Aronson, E. Dissonance theory: Progress and problems. In R. P. Abelson, E. Aronson, W. J. McGuire, T. M. Newcomb, M. J. Rosenberg, & P. H. Tannenbaum (Eds.), *Theories of cognitive consistency: A sourcebook*. Chicago: Rand McNally, 1968.

Axsom, D., & Cooper, J. Reducing weight by reducing dissonance: The role of effort justification in inducing weight loss. In E. Aronson, (Ed.), *Readings about the social animal*. (3rd ed.). San Francisco: W. H. Freeman, 1981.

Beck, A. T. *Cognitive therapy and the emotional disorders*. New York: Meridian Books, 1976.

Bem, D. J. Self-perception: An alternative interpretation of cognitive dissonance phenomena. *Psychological Review*, 1967, *74*, 183–200.

Brehm, J. W., & Cohen, A. R. *Explorations in cognitive dissonance*. New York: Wiley, 1962.

Brehm, S. S. *The application of social psychology to clinical practice*. New York: Halsted Press, 1976.

Brehm, S. S., & Brehm, J. W. *Psychological reactance: A theory of freedom and control*. New York: Academic Press, 1981.

Brodsky, A. M. Sex, race, and class issues in psychotherapy research. In J. H. Harvey & M. M. Parks (Eds.), *The Master Lecture Series. Vol. 1. Psychotherapy research and behavior change*. Washington, D.C.: American Psychological Association, 1982.

Cooper, J. Reducing fears and increasing assertiveness: The role of dissonance reduction. *Journal of Experimental Social Psychology*, 1980, *16*, 199–213.

Cooper, J., & Axsom, D. Effort justification in psychotherapy. In G. Weary & H. L. Mirels (Eds.), *Integrations of clinical and social psychology*. New York: Oxford University Press, 1982.

Corrigan, J. D., Dell, D. M., Lewis, K. N., & Schmidt, L. D. Counseling as a social influence process: A review. *Journal of Counseling Psychology Monographs*, 1980, *27*, 395–441.

Craighead, L. W., & Craighhead, W. E. Implications of persuasive communication research for the modification of self-statements. *Cognitive Therapy and Research*, 1980, *4*, 117–134.

Dell, D. M. Counselor power base, influence attempts, and behavior change in counseling. *Journal of Counseling Psychology*, 1973, *20*, 399–405.

Doob, L. W. The behavior of attitudes. *Psychological Review*, 1947, *54*, 135–156.

Ellis, A. *Humanistic psychotherapy: The rational-emotive approach*. New York: McGraw-Hill, 1973.

Festinger, L. *A theory of cognitive dissonance*. Stanford, Calif.: Stanford University Press, 1957.

Fishbein, M., & Ajzen, I. Acceptance, yielding, and impact: Cognitive processes in persuasion. In R. E. Petty, T. M. Ostrom, & T. C. Brock (Eds.), *Cognitive responses in persuasion*. Hillsdale, N.J.: Lawrence Erlbaum, 1981.

Goldfried, M. R. Toward the delineation of therapeutic change principles. *American Psychologist*, 1980, *35*, 991–999.

Goldstein, A. P., & Simonson, N. R. Social psychological approaches to psychotherapy research. In A. E. Bergin & S. L. Garfield (Eds.), *Handbook of psychotherapy and behavior change*. New York: Wiley, 1971.

Goldstein, A. P., Heller, K., & Sechrest, L. B. *Psychotherapy and the psychology of behavior change*. New York: Wiley, 1966.

Greenwald, A. G. Cognitive learning, cognitive response to persuasion, and attitude change. In A. G. Greenwald, T. C. Brock, & T. M. Ostrom (Eds.), *Psychological foundations of attitudes*. New York: Academic Press, 1968.

Harris, B., & Harvey, J. H. Social psychological concepts applied to clinical processes: On the need for precision. *Journal of Consulting and Clinical Psychology*, 1978, *46*, 326–328.

Harvey, J. H., & Harris, B. Determinants of perceived choice and the relationship between perceived choice and expectancy about feelings of internal control. *Journal of Personality and Social Psychology*, 1975, *31*, 101–106.

Holmes, D. S. Investigations of repression: Differential recall of material experimentally or naturally associated with ego threat. *Psychological Bulletin*, 1974, *81*, 632–653.

Hovland, C. T., Janis, I. J., & Kelley, H. H. *Communication and persuasion*. New Haven: Yale University Press, 1953.

Hull, C. L. The goal-gradient hypothesis and maze learning. *Psychological Review*, 1932, *39*, 25–43.

Kiesler, C. A., Collins, B. E., & Miller, N. *Attitude change*. New York: Wiley, 1969.

Mahoney, M. J., & Arnkoff, D. Cognitive and self-control therapies. In S. L. Garfield & A. E. Bergin (Eds.), *Handbook of psychotherapy and behavior change* (2nd ed.). New York: Wiley, 1978.

Meador, B. D., & Rogers, C. R. Person-centered therapy. In R. Corsini (Ed.), *Current psychotherapies*. Itasco, Ill.: F. E. Peacock, 1979.

Mosak, H. H. Adlerian psychotherapy. In R. Corsini (Ed.), *Current psychotherapies*. Itasco, Ill.: F. E. Peacock, 1979.

Murray, E. J., & Jacobson, L. I. Cognition and learning in traditional and behavioral psychotherapy. In S. L. Garfield & A. E. Bergin (Eds.), *Handbook of psychotherapy and behavior change* (2nd ed.). New York: Wiley, 1978.

Orlinsky, D. E., & Howard, K. I. The relation of process to outcome in psychotherapy. In S. L. Garfield & A. E. Bergin (Eds.), *Handbook of psychotherapy and behavior change* (2nd ed.). New York: Wiley, 1978.

Petty, R. E., & Cacioppo, J. T. *Attitudes and persuasion: Classic and contemporary approaches*. Dubuque, Iowa: William C. Brown, 1981.

Petty, R. E., Ostrum, T. M., & Brock, T. C. (Eds.). *Cognitive responses in persuasion*. Hillsdale, N.J.: Lawrence Erlbaum, 1981.

Stampfl, T., & Lewis, D. Essentials of implosive therapy: A learning theory based psychodynamic behavior therapy. *Journal of Abnormal Psychology*, 1967, *72*, 496.

Strong, S. R. Counseling: An interpersonal influence process. *Journal of Counseling Psychology*, 1968, *15*, 215–224.

Strong, S. R. Social psychological approach to psychotherapy research. In S. L. Garfield & A. E. Bergin (Eds.), *Handbook of psychotherapy and behavior change* (2nd ed.) New York: Wiley, 1978.

Strong, S. R., & Claiborn, C. D. *Change through interaction*. New York: Wiley, 1982.

Strupp, H. H. Psychotherapy research and practice: An overview. In S. L. Garfield & A. E. Bergin (Eds.) *Handbook of psychotherapy and behavior change* (2nd ed.). New York: Wiley, 1978.

Strupp, H. H., & Bergin, A. E. Some empirical and conceptual bases for coordinated research in psychotherapy. *International Journal of Psychiatry*, 1969, *7*, 18–90.

Thurstone, L. L. Comment. *American Journal of Sociology*, 1946, *52*, 39–40.

Wicklund, R. A., & Brehm, J. W. *Perspectives on cognitive dissonance*. Hillsdale, N.J.: Lawrence Erlbaum, 1976.

Wolpe, J. *The conditioning therapies*. New York: Holt, Rinehart, & Winston, 1963.

8

Social Stereotypes

ROBIN B. SHAFRAN

The occurrence of stereotyping in psychotherapy traditionally has been minimized. This is particularly true within the more psychoanalytically oriented paradigms in which the dynamics of the individual patient are emphasized, and the patient–therapist relationship is explored within the framework of transferential and countertransferential processes. However, as examination of the psychotherapy process has expanded, its essential characteristics have been seen to include those of any interpersonal, dyadic interaction. Within this context, the question of whether social stereotypes are operative in the psychotherapeutic situation has become an issue. Conceptualizations regarding the manner in which stereotypes shape our views of those with whom we interact have been extended to the psychotherapy relationship.

The literature on stereotyping in psychotherapy covers wide areas, including racial, ethnic, socioeconomic, and gender stereotyping. A discussion of so extensive a subject matter is beyond the scope of this chapter. Accordingly, in an effort to highlight the relevance of a particular area of psychological research to the psychotherapeutic situation, one aspect of stereotyping, sex-role or gender stereotyping, will be the focus of discussion. This chapter, then, will first present a brief discussion of stereotyping in general and sex-role stereotyping in particular. The empirical literature specifically directed to the occurrence of sex-role

ROBIN B. SHAFRAN • 262 Central Park West, Suite 1-B, New York, New York 10024.

stereotyping in psychotherapy will be sampled as well. A discussion of related clinical issues will follow the presentation of empirical data.

STEREOTYPES: AN OVERVIEW

The study of stereotyping, its occurrence and its impact, has been a focus of social psychological research for more than half a century. Writing in 1922, Lippmann distinguished between "the world outside and the pictures in our heads" (p. 1), and defined stereotypes as oversimplified pictures of the world that serve to make it more comprehensible. According to Lippmann, stereotypes are broad, simple generalizations about categories of people and are, by definition, incorrect.

Reviewing the literature on racial stereotypes, Brigham (1971) highlights the lack of consensus regarding the nature of stereotyping, and describes the research literature as "fragmented and of limited value" (p. 15). He points to insufficient data regarding such issues as the learning and modification of stereotyping, and the relation between stereotypes, and attitudes and behavior toward stereotyped groups.

Ashmore and DelBoca (1979) also call for increased conceptual clarity regarding stereotyping. Using the framework of implicit personality theory, they define a stereotype as "a structured set of inferential relations that link a social category with personal attributes" (p. 225). Ashmore and DelBoca stress that stereotyping is a cognitive process that is presumed to influence social behavior. Stereotyping can be seen as a means of organizing experiences, and as creating expectancies that allow the holder of the stereotype to anticipate behavior. All stereotypes have a specified referent, a particular social category, such as gender or race. Ashmore and DelBoca emphasize that perceivers (stereotypers) play an active role in dividing their social world into these categories. Although cultural factors obviously influence this, they believe that stereotypes are not simply absorbed. Rather, they suggest that there are significant differences among members of the same culture regarding social category definitions and actual stereotypic content.

Ashmore and DelBoca believe that an individual's social category system, on which stereotypes are based, is a major influence on the way in which another person is classified. The characteristics of target persons present another major source of input into their classification by another, and the tendency to ascribe traits on the basis of social category membership alone is believed to decline with the availability of additional information. Finally, the situational context may be considered in determining the salience of various social categories in the assumptions made about another's personality.

Ashmore and DelBoca also address the issue of the stability of stereotypic attitudes and beliefs. To the extent that stereotypes organize experience and can influence the behavior of the stereotyped individual, they seem them as self-fulfilling. Furthermore, information that confirms a stereotype may be selectively attended to and differentially retained. There are also motivational factors that contribute to the persistence of stereotypes, resisting change in the implicit structures that guide perception and behavior.

In a recent review article, McCauley, Stitt, and Segal (1980) suggest a shift in the conceptualization of stereotyping and a change in research methodology. Rather than assessing stereotypes by asking perceivers to characterize the stereotyped group according to lists of traits, a common research strategy, McCauley *et al.* advocate understanding stereotypes as probabilistic predictions, and investigating them as such. They define stereotypes as

> those generalizations about a class of people that distinguish that class from others. In other words, stereotyping is differential trait attribution or differential prediction based on group membership information. (p. 197)

Stereotypes are not seen as absolute generalizations, but as guidelines for the possible presence of a character trait.

McCauley *et al.* (1980) concur with Ashmore and DelBoca (1979), viewing stereotypes as a means of cognitively structuring new events and experiences, and caution that they can have negative impacts. A stereotype can be misused if its predictions are relied upon too heavily, perhaps to the exclusion of other information. Furthermore, they warn that stereotypic expectations can be self-fulfilling in their effects on the memory of the stereotyper and the behavior of the stereotyped person.

These considerations are also raised by Snyder, Tanke, and Berscheid (1977), who consider social stereotyping to be a special case of interpersonal perception. These researchers provide empirical documentation that social stereotypes can significantly impact upon contact and communication between individuals. They demonstrate that stereotypically based attributes of a perceiver may shape interaction such that the stereotyped individual behaves in such a way as to confirm the stereotype.

SEX ROLES AND SEX-ROLE STEREOTYPING

In a discussion of sex-role-related issues in psychotherapy, Kaplan (1979) comments that "our culture makes clear distinctions between the feeling, behaviors, and role expectations that are assumed to be associated with each sex" (p. 112). She believes that such stereotypes are present throughout society, within families, school systems, the media, and work roles. Kaplan believes such sex-role stereotypes are

> internalized by us as children, and remain with us as adults as conscious and unconscious components of our personalities, whether or not our political beliefs support the efficacy of differential traits for males and females. (p. 112)

As such, sex-role-related assumptions would, by definition, dictate differential expectations regarding what is appropriate for men and women in our society.

In a review of their own widely cited studies, Broverman, Vogel, Broverman, Clarkson, and Rosenkrantz (1972) reach the conclusion that sex-role stereotypes do exist and do influence the manner in which people perceive themselves and others. A Sex-Role Stereotype Questionnaire, consisting of 41 stereotypic items in bipolar form, and based upon the differential attribution of the items to men and women, was developed in the first of the studies done by the Broverman group (Rosenkrantz, Vogel, Bee, Broverman, & Broverman, 1968). This instrument has been used extensively in subsequent research (e.g., Garske, 1975; Kravetz, 1976; Tolor, Kelley, & Stebbins, 1976; Zanna & Pack, 1975).

Rosenkrantz *et al.* (1968) found that significant differences in self-concept exist between the sexes, and that traits ascribed to men are more frequently highly valued. Men and women incorporate both the positive and negative traits of their stereotypes into their self-concepts, but more female traits are held in low regard, suggesting that women, who are more likely to internalize these negative characteristics, are handicapped by sex roles. Individuals' self-concepts were found to be less extreme in terms of stereotypes (i.e., subjects rated themselves as less feminine or masculine than were their stereotypic responses for their own sex), and self-concept responses correlated highly with social desirability.

Spence, Helmreich, and Stapp (1975) also had male and female college students rate their own sex-role identity and make judgments about sex-role-related differences in typical male and female college students. Ratings of the typical male and female students were more stereotypic than would be predicted by subjects' sex-role-related descriptions of themselves. However, self-ratings of sex-role identity also frequently conformed to established stereotypes.

In a series of three well-designed and executed studies to assess the reactions people have to male dependence and passivity and to female aggression and self-assertion, Costrich, Feinstein, Kidder, Maracek, and Pascalle (1975) consistently found that individuals are penalized for these sex-role reversals. These results emphasize the consequences of stereotypic expectations, in channeling behavior in the direction appropriate to the given stereotype.

Zanna and Pack (1975) looked at the situational effects of sex-role stereotypes, proposing that an individual's conformity to stereotypic expectations will be influenced by the stereotypic attitudes held by an attractive other. These authors found that their female subjects modified both stated sex-role attitudes and a sex-role-related behavior about which they were given no prior information, as a result of feedback regarding the sex-role attitudes of an attractive male partner. These findings again support the cautions of both McCauley *et al.* (1980) and Snyder *et al.* (1977) that the stereotypic attitudes held by one person in an interaction modify the behavior of the other, thereby inducing behavior that is either consistent or inconsistent with the stereotypic assumption.

Most paradigms in sex-role-stereotyping research ask subjects to ascribe characteristics to either "typical men" or "typical women." Garske (1975) predicted that stimulus variation would affect the attribution process and asked subjects to rate, on the Rosenkrantz *et al.* (1968) scale, stimulus persons whom he described in terms of sex and role (undergraduate student, graduate student, or adult). He reports signficant differences between ratings of female and male stimulus persons in the graduate student and adult roles, and no difference for the undergraduate role. Garske sees these results as supporting the situation-specific quality of stereotypes. He points out that male and female undergraduates have similar roles, thereby mitigating stereotypic attribution.

Ashmore and DelBoca (1979) take issue with the idea that there is a general consensus about what is meant by sex-role stereotyping, and define sex roles as "the structured set of inferential relations that link personal attributes to the social categories female and male" (p. 225). They discuss two of their own studies that use two of the methods of implicit personality theory, "personality description" and "trait difference," to assess stereotypes. In the first of these studies, subjects, using a sorting procedure, consistently described female "targets" in terms of "soft" traits (e.g., naïve, sentimental) and male "targets" in terms of "hard" traits (e.g., scientific, critical). Sex of target was found to be a more central or basic construct for the male than for the female subjects. In their second study, using the trait inference method, females were described more frequently in terms reflecting communion (e.g., generosity), and males were rated higher in traits reflecting intellectual desirability. These results, then, using different methodologies than are typically seen in this research area, corroborate both basic stereotypic conceptions of men and women and provide additional confirmation for the findings that men tend to rely on stereotypes to a greater extent than do women.

Locksley, Borgida, Brekke, and Hepburn (1980) presented two experiments that confirm the findings reported by Garske, namely, that with the inclusion of additional information the occurrence of sex-role stereotyping decreases markedly. In the first experiment, these investigators hypothesized that knowledge of an individual's sex and a sampling of his or her behavior should affect sex-role-related trait attributions to that person. Using "assertiveness" as the sex-role stereotypic trait, Locksley *et al.* suggest that since the trait is stereotypically masculine, assertive behavior is more likely to be attributed to an enduring personality characteristic in men than would the same behavior in a women.

In this study, subjects were given an account of a college student's behavior through ordinary problem situations, varying the student's sex and behavior (assertive or passive). The subjects were then asked to make predictions about the student's behavior in four novel situations and to rate the student on the Bem Sex-Role Inventory (1974). Locksley *et al.* found that the behavior predictions and BSRI ratings bore a closer resemblance to the target's actual behavior than to his or her sex. In discussing their results, the authors comment that "a strong implication of this research is that a minimal amount of diagnostic target case information would be sufficient to eliminate the impact of stereotypic beliefs or judgments of the target individual" (p. 826).

In the second experiment, Locksley *et al.* examined the impact of case information (sex of subject, relevant and irrelevant behavioral information) on judgments regarding the assertiveness of the target individual. In the absence of any additional information, or with nonrelevant information, subjects considered male targets to be assertive more frequently than female targets. With the provision of relevant information, judgments about male and female target persons were similar, to such an extent that sex-role stereotypic beliefs did not appear to play any part in the judgments.

> Perhaps the most striking implication of this line of reasoning is that social stereotypes may not exert as pervasive or as powerful an effect on social judgments as has been traditionally assumed. Social stereotypes may affect judgments of individuals about whom little else is known besides their social category. But as soon as individuating, subjectively diagnostic characteristics of a person are known, stereotypes may have minimal, if any, impact on judgments of that person. (p. 830)

The literature suggests, then, that stereotypes exist regarding characteristics associated with "typical" men and "typical" women. With the provision of additional data, however, the tendency to stereotype diminishes markedly. These findings, true of college students serving as research subjects, should be no less true of practicing psychotherapists, who, one hopes, would be at least as sensitive to the bases for their impressions and judgments about their patients. I shall now turn to the literature on sex-role stereotyping in psychotherapy, in order to ascertain whether sex-role stereotyping plays a significant role in the psychotherapeutic interaction and, if so, how.

SEX-ROLE STEREOTYPING AND PSYCHOTHERAPY

The issue of sex-role stereotyping and psychotherapy, that is, the concern that stereotypic attitudes held by psychotherapists affect the treatment of the patients they see, has received considerable attention in recent years. Since most psychotherapy patients are female, and since stereotypically feminine traits are more frequently considered to be less desirable (Broverman *et al.*, 1972), the focus of concern has generally been whether female patients are adversely affected by the stereotypic attitudes of their therapists. This is an area charged with considerable emotion, and accusations about the sexist nature of psychotherapy are frequently heard. Feminist therapy has developed to provide an alternative, and women are routinely warned to be wary of the dangers inherent in a psychotherapeutic situation that is likely to perpetuate societally held stereotypes about the nature of women and femininity.

The major focus in the controversy over sex roles and psychotherapy has been the stereotypic attitudes held by therapists. It is the therapists who potentially influences their patients, possibly with considerable impact. However, in a full exploration of the influence of stereotypic attitudes on the psychotherapy process, a discussion of patient beliefs must be included as well. There is literature indicating that attitudes and expectations patients have regarding their therapists significantly influence the treatment process (Goldstein, 1962; Gurman, 1977; Wilkins, 1973; Ziennelis, 1974). Furthermore, such patient-held beliefs may well be involved in whether a psychotherapeutic relationship is initially established. The psychotherapeutic relationship is, essentially, a dyadic relationship in which attitudes held by each member can shape the course of treatment. Accordingly, literature dealing with both therapist and patient stereotyping will be surveyed.

Accusations about the sexist nature of psychotherapy in general, and on treatment paradigms based on Freudian theory in particular, began to be made with increasing frequency with the growth of the women's movement. Broverman, Broverman, Clarkson, Rosenkrantz, and Vogel (1970) published their now classic study "demonstrating" a bias against women by both male and female psychotherapists. Broverman and her associates asked clinicians to rate either a "healthy man," "healthy female," or "healthy adult" using the Sex-Role Stereotype

Questionnaire (Rosenkrantz *et al.*, 1968). They report strong agreement among male and female clinicians on the characteristic attributes of each (male, female, adult). Furthermore, whereas healthy men and healthy adults were viewed as possessing similar characteristics, healthy women were characterized as possessing fewer socially desirable characteristics. It was this result that provided empirical documentation for the cry that psychotherapy was sexist.

> These results, then, confirm the hypothesis that a double standard of health exists for men and women, that is, the general standard of health is actually applied only to men, while healthy women are perceived as considerably less healthy by adult standards. (p. 6)

Although the Broverman study was published over a decade ago, it continues to be regarded as the most striking and influential illustration of sex-role stereotyping in psychotherapy. As such, it is frequently cited by those who argue that psychotherapy hampers, rather than facilitates, psychological growth for women.

The extensive literature on sex-role stereotyping among psychotherapists has been the subject of several major review articles in the past 5 years. Stricker (1977) devotes considerable attention to the Broverman *et al.* (1970) study. He points out certain methodological weaknesses in the Sex-Role Stereotype Questionnaire, including the means by which the bipolar end points were labeled. Stricker notes, for example, that if men were rated more "logical" than women, the male pole for that characteristic was labeled "logical" and the female pole "illogical," language he sees as "slanting the use of the scale for all further investigations" (p. 17). He sees the absence of raw data, the lack of individual item analyses, and the assumption of significant differences without statistical evidence as major criticisms substantially lessening the impact of the results. Stricker considers the conclusions reached by the investigators to be "unsubstantiated and overdramatic" and the accusation of a "double standard of mental health" premature and insufficiently documented.

Stricker also enumerates some of the major difficulties in adequately investigating stereotypic practices in psychotherapy, raising concerns about the validity of analog designs in psychotherapy research and emphasizing that psychotherapists work with unique individuals, not "typical" men and women. Generalizing from what therapists say about "typical" people to what they do with their own unique patients may be a questionable practice.

After reviewing the research in this area, Zeldow (1978) concludes that, generally, the values and norms of mental health professionals reflect those of society. However, he suggests that "differential judgment does not automatically prove differential treatment, nor do double standards of optimal health necessarily imply differential assessments of mental illness" (p. 89). Similarly, citing data indicating that female therapists may be more empathic and male therapists more authoritarian, Zeldow goes on to suggest that such differences in therapeutic style have not been related to gender-associated differences in treatment outcomes.

Whitley (1979), in a review of the research literature, criticizes the Broverman *et al.* work as well, also noting that the extent to which attitudinal bias leads to discriminatory behavior was not addressed. Whitley comments that subsequent research has "generally found no such discrimination and casts doubt on the generality of the bias" (p. 1309). As is often the case in controversial areas, those who believe that psychotherapy may offer something of value to women attend to more of these criticisms, whereas those who fear that the treatment may be worse than the initial problem emphasize the initial results to a greater extent.

In his review, Whitley evaluates the data indicating that there are differential, sex-role-related standards of mental health for men and women, inquires whether violations of sex-role characteristics are associated with negative judgments regarding mental health, and explores the frequency with which therapy goals are related to sex-role attributes. He found empirical support for the existence of differential standards of mental health, but in a research area plagued by serious methodological flaws. Significant effects of violation of sex roles were found in the judgments of nonprofessional, but not professional, therapists. He did not find therapy goals to be related to sex-role stereotypic characterizations of men and women. Again, Whitley underscores methodological difficulties—in subject samples, choice of assessment instruments, and particularly in the overreliance on the Sex-Role Stereotype Questionnaire used in 10 of the 12 studies he reviewed.

Smith (1980) also reviewed the research on counselor attitudes, stereotype, and behavior. She did not find evidence for the existence of counselor sex bias, and concludes that empirical support for sexism in psychotherapy is weak, with the evidence leaning toward bias balanced by contradictory data.

> Stripped of the selectivity, motivation, and rhetoric, the body of evidence looks both different and more clear: Counselor sex bias has not been demonstrated despite a dozen years of attempts to do so. (p. 407)

Sherman (1980) reviewed therapist sex-role stereotypic attitudes toward women and concludes that the data support the contention that therapist sex-role attitudes are operative during counseling, that there are sex-role-related mental health standards, and that behavior that violates sex-role norms is judged more negatively. Sherman does believe that there is evidence that stereotyping is decreasing.

Sherman's assessment of the status of stereotyping in psychotherapy is discussed by Stricker and Shafran (1983), who suggest alternative interpretations of the data she presents. For example, Sherman cites results reported by Rabkin (1977) that significant numbers of psychiatrist respondents believed that psychiatrists have stereotypes of "normal healthy females" and "normal healthy males" and that psychiatrists indeed have sex-role-related goals for their patients. Sherman interprets these results as indicating that psychiatrists hold stereotypes about their male and female patients. Stricker and Shafran offer an alternative explanation, namely, that the results may more accurately indicate stereotypes psychiatrists have of their colleagues. Such alternative explanations of the same data are seen by Stricker and Shafran as illustrative of the persistence of stereotyping, despite the lack of clearly unambiguous data.

The question of a double standard in assessing the mental health of men and women was addressed more recently by Marwit (1981). Discussing the limitations of existing research, Marwit points out that with one exception, sex-stereotyping of patients by therapists has been demonstrated via sex-role questionnaires that conceptualize masculinity and femininity as bipolar, mutually exclusive opposites. Marwit draws upon both Bem's (1974) challenge to such a model, her conceptualization of androgyny as a successful integration of both masculine and feminine traits, and Spence *et al.*'s (1975) findings of a positive relationship between androgyny and mental health in suggesting that the occurrence of sex-role stereotyping among psychotherapists might be more effectively assessed using the BSRI.

Marwit asked practicing male and female psychologists to use the BSRI to describe either a "mature, socially competent" adult male, adult female, or adult. Differential characterizations of male and female adults were obtained when masculinity, femininity, and androgyny were defined both by scale scores (simple difference scores) and by median splits, supporting previous contentions of stereotyping among psychotherapists. Interestingly, when the traits "masculine" and "feminine" were removed from the scales and the data were reanalyzed, relevant significant effects were not obtained—findings that throw the use of the BSRI into question.

In their assessment of the empirical literature on sex-role stereotyping in psychotherapy, Stricker and Shafran (1983) emphasize that the available data are not sufficient to draw definite conclusions. They believe that assertions of demonstrated effects of stereotyping are more likely "anecdotal or political" than based in data. Stricker and Shafran emphasize the importance of realizing that the "crucial findings will not concern attitudinal characteristics of therapists, but the effect of those attitudes on treatment interventions and outcomes" (p. 198). In their estimation, sex-role stereotypes are likely to affect the psychotherapy situation in more complex interaction with other moderator variables.

Although there are few studies assessing patient stereotypes regarding male and female therapists, there are data regarding patient preferences for a therapist of a particular sex. Turkel (1976) sees the two as related and suggests that societally based, stereotypic conceptions of men and women may influence preferences. Similarly, Davidson (1976), in a study of new patient attitudes and preferences regarding sex of therapist, interprets her findings as an indication that patients, especially women, may prefer male or female therapists for different reasons. Davidson suggests that female therapists may be perceived as possessing different personality characteristics than male therapists.

Interestingly, changes in attitudes toward women are clearly discernible in the literature on patient preferences. It had generally been found that, among patients with preferences, male therapists were preferred more frequently than female therapists (Davidson, 1976). However, there has been somewhat of a shift, with a growth of interest in female therapists particularly among women patients (Shafran, 1979). Additionally, once therapy has begun, women seeing female therapists report greater satisfaction than women working with male therapists (Orlinsky & Howard, 1976). Although the data reported by Olinsky and Howard do not directly assess stereotyping, the investigators relate the satisfaction reported to stereotypic issues rather than to the competence of male and female therapists involved. They see their results as directing "attention to the patients' more than to the therapists' attributes. Whether the therapist is able and sensitive may be less important than what a male or female therapist means to the patient" (p. 87).

Delk, Ryan, and Livingston (1978) looked at the relationships between patients' sex-role stereotypes, A-B status, and preferences for male and female therapists. They had subjects rate themselves and male and female therapists on an adaptation of the Sex-Role Stereotypes Questionnaire. The authors judge significant and "sizable" differences in ratings of male and female therapists as indicating that patients "endorse culturally loaded male and female stereotypes which tend to denigrate the female therapist" (p. 5). Stereotyping occurs more frequently among male patients than female patients. A-status patients of either sex more frequently described male and female therapists along stereotypic lines and more frequently expressed preferences for male therapists. B patients tended to prefer female therapists. The investigators emphasize their findings that, with the exception of those female patients who preferred male therapists, patients generally perceived themselves as more similar (in sex-role identity) to the sex of the therapist they preferred. Delk *et al.* conclude that people tend to seek therapists perceived as similar to themselves. However, Delk *et al.*'s design did not include a no-preference category in choosing gender of therapist. Patients were asked to rank order their preferences, and, for those subjects who chose the no-preference option, their second choices were substituted in the data analysis. Therefore, no comparison of patients with preferences and those without preferences is possible.

Shafran (1979) looked at preferences for male and female therapists and differential sex-role-related expectations that patients have of male and female therapists. In order to assess the degree to which therapists of the preferred sex are believed to possess characteristics considered important, the relationship between preference and characteristics associated with an "ideal" (sex unspecified) therapist was also investigated.

Shafran found that more men than women endorsed the stereotypic expectations of differences between male and female therapists. Where both male and female subjects expected differences, those differences fell in the same, stereotypic direction. Male and female patients expressed strong agreement in their assessment of characteristics important in an "ideal" therapist, and the majority of the characteristics considered important were female-valued traits.

Of interest in the findings reported by Shafran are the relationships between expectancy and preference. The male patients sampled were more likely to expect differences between male and female therapists, and were less likely than female patients to express preferences. Those male patients who had preferences were divided between preferring male and female therapists. In contrast, women were less likely to make differential attributions to male and female therapists but were more likely to express preferences. More often the preference was to see a woman.

The literature suggests that an increasing number of women prefer and report better experiences with female therapists. The data reported by Shafran suggest that the preferred therapist, regardless of sex, has at least some feminine characteristics. The relationship of preferences and sex-role-based expectancies to actual treatment is, as yet, undemonstrated. Stricker and Shafran (1983) point out that, as in the case of therapist attitudes, information regarding patient stereotypes must also be considered in conjunction with other variables.

Evaluating the literature regarding sex-role stereotyping and psychotherapy, it becomes clear that the data, standing alone, do not convincingly demonstrate that sex-role stereotypes necessarily interfere with the psychotherapy process. It also becomes clear that the same data, viewed through the eyes of either believers or nonbelievers in the impact of stereotyping, can be interpreted to support the preferred stance.

As pointed out by Stricker and Shafran (1983), five out of the six major reviews they discuss conclude that clear-cut stereotyping in psychotherapy has not been demonstrated. Methodological weaknesses are a frequent basis upon which results demonstrating stereotyping are minimized or discredited (Stricker, 1977; Stricker & Shafran, 1983; Whitley, 1979). Overreliance on questionnaires assessing masculinity and femininity as bipolar opposites and in particular on the Sex-Role Stereotype Questionnaire raises questions about reported results. Whitley points out the lack of convergent validity in this area, dominated as it is by one major research instrument. Two other issues raised in examining data regarding stereotyping concern the identifying data provided (healthy male, female, and adult), and the question of the relationship between attitudes and behavior.

The more general literature regarding stereotyping may be helpful in addressing both of these considerations. As pointed out earlier in this chapter, stereotypes are most often relied upon in the absence of additional information, and their prominence declines as additional information is made available. Social categories are used to shape expectations when there is little else to go on. As Stricker (1977) points out, when asked to rate healthy adult men and women, subjects have little choice but to fall back upon what is "known" about those generic categories. Once confronted with an actual individual, reliance on stereotypic conceptions decreases. Thus, the occurrence of stereotyping in the cited research may be, at least partially, an artifact of experimental design.

Ashmore and DelBoca's (1979) suggestion that the context in which one meets a new person influences the salience of characteristics attributed on the basis of social category is also relevant here. There is a clear possibility that in the psychotherapy situation, what is expected stereotypically of men and women may be less prominent than stereotypes elicited by the category "patient" or "therapist." Role expectations not necessarily related to gender may here influence the new relationship. Maxfield (1976) cites previous research suggesting that with the introduction of a role, "sex stereotyping decreases and role stereotyping increases" (p. 100).

The relationship between attitude and behavior is more controversial. The belief that stereotypic conceptions influence the treatment process is implicit in the idea of a "double standard of mental health," in that therapist judgments regarding the characteristics of healthy men and women are assumed to be reflected in treatment goals. This view is questioned by Zeldow (1978). Although he agrees that the empirical literature supports the contention that therapists share the stereotypes of society, he suggests that attitudes do not necessarily influence behavior.

The social psychological literature may again be useful in delineating the factors involved. Insofar as stereotypes shape expectations and lead to selective attention (and also inattention) and perhaps selective responses to information provided by a patient, the stereotypic conceptions of men and women may well influence therapist behavior. To the extent that therapists' stereotypes may actually foster stereotypically consistent patient behavior, any therapist stereotyping is potentially detrimental to the psychotherapeutic process. These possibilities are extensions of results reported in the literature but have not been demonstrated empirically in regard to psychotherapy.

The more general literature, then, may be useful in suggesting how to assess the occurrence of sex-role stereotyping in psychotherapy. It is clear that studying therapist judgments regarding typical patients (male, female, or adult) is not sufficient to demonstrate that stereotyping influences the treatment process. To be more convincing, research needs to demonstrate that stereotypic attitudes persist in the face of more information about the patient to be rated, and that such attitudes affect the manner in which therapists listen to and respond to their patients, and consequently the patient behavior they elicit.

Discussion

Social psychologists view stereotypes as a framework for organizing and structuring information about other people. To the extent that stereotyping is operative in psychotherapy, it can occur on the part of either patient or therapist, both of whom are initially confronted with and expected to relate to an unfamiliar person about whom little may be known. Although the literature on sex-role stereotyping in psychotherapy has not conclusively demonstrated a general effect of such stereotyping, it has raised some provocative questions and outlined areas for further study. This section of the chapter will explore some of those issues and attempt to illustrate some of the previously discussed ideas. The discussion will be extended beyond stereotyping to the closely related issue of preferences, and then move to the areas of transference–countertransference and resistance, in an effort to differentiate between what is a stereotypic assumption by patient or therapist and what is more truly a reflection of the treatment process itself.

Reliance on stereotypes is most likely to occur in the initial phases of treatment, during which both patient and therapist have the least information about each other. Stereotypes may influence patient choice of a therapist of a particular sex, and expectations about the other person by both patient and therapist. The extent to which stereotypes influence the setting of goals for the treatment is an area of particular concern. The manner in which patient and therapist relate to each other can be influenced by stereotypic conceptions held by either participant in the psychotherapy dyad. It is in regard to this final point that transference–countertransference and resistance become relevant in clarifying the nature of the therapeutic interaction.

Sex-role stereotypic assumptions have been postulated to shape the expectations of both patients and therapists regarding the characteristics they are likely to find in the other. To the extent that such assumptions interfere with learning about the patient (or therapist), they are detrimental to the treatment process. Harry Stack Sullivan called stereotypes "inescapable handicaps in becoming acquainted with strangers" (1953, p. 304).

Stereotypic expectations are most likely to influence the treatment process at its inception. When held by the patient, they may enter into the choice of a therapist of a particular gender. Clara Thompson singled out the choice of the sex of analyst as "probably the most discussed part of the whole question of choice of analyst" (1938/1964, p. 128). Thompson focused her discussion on the defensive, or neurotic, aspects of patient preferences, but did point out that culturally reinforced beliefs—for example, about the superiority of the male analyst, who is characterized as strong and more capable of assuming responsibility—are often the basis upon which a therapist of a particular sex is chosen. The increasing awareness of the role such stereotypes play in influencing beliefs about men and women has led to a heightened interest in the types of beliefs to which Thompson referred.

Thompson's comments remain relevant. New psychotherapy patients often have clear preferences for male or female therapists, and, as she pointed out, it is difficult to determine the antecedents of these preferences. For example, a man in his late thirties requested a female therapist. He based his choice on the belief that a woman would be more sensitive and supportive, a better listener. Further discussion revealed his fear that if he were to work with a male therapist, the competition between them would hamper the progress of the treatment. Similarly, a young female professional also specifically sought a female therapist, saying that a woman was more likely to understand her. When pressed, she added that it would be easier for her to disagree with a woman, and ignore her advice or recommendations. Were she to see a male therapist, he would be such an authority that she would give up her autonomy and find it difficult to determine her own solutions.

Although both these individuals offered stereotypically consistent explanations for their choices, it became clear in the exploration of their preferences that other factors were involved. For the man, a combination of a dominant, forceful, and overshadowing father and a highly competitive work situation led him to shy away from involvement with any man with clear professional stature. The woman's choice masked both her contempt for her own sex and therefore for herself, and her experience of men as powerful and rigid in the imposition of their own opinions. In both cases, stereotypic characterizations were minor elements in a more complicated constellation of facts. Individual dynamics proved to be more salient issues.

For both participants in the psychotherapy relationship, stereotypic assumptions about the other may serve as part of the framework within which the initial meeting occurs. Sullivan saw psychiatrists' stereotypes as a potential source of difficulty.

> If you have a number of implicit assumptions that have not been questioned by you for twenty or twenty-five years about the alleged resemblance of the person before you to some stereotype that you have in your mind, you may find yourself greatly handicapped, for these stereotypes are often viciously incomplete or meaninglessly erroneous. (1953, p. 238)

The importance of seeing one's patient in a clear, unhampered manner cannot be overemphasized.

Patients' sex-role-related assumptions about their therapist are frequently discussed long after the treatment begins, especially when what is expected is seen as something negative. Although doubts are often heard about the "competence" of a woman or the "sensitivity" of a man during initial sessions, it is likely that if the concerns were sufficiently great the woman or man would have been rejected as a potential therapist from the outset. Once the treatment relationship is established, however, expectations harbored from its inception are often discussed more freely.

A highly intelligent and competitive male patient assigned to a female therapist at a clinic offering psychoanalytically oriented psychotherapy accepted the assignment without hesitation. Months into the treatment, he was able to discuss his surprise in finding a woman to be an intellectual equal, strong enough to meet him head-on. He admitted to having had some reservations about seeing a woman, fearing that he would be the stronger of the two and therefore be cast in the role of caretaker. As in the examples discussed regarding preferences for male and female therapists, it is difficult to delineate the degree to which this young man's reactions reflect the incorporation of stereotypic attitudes about women, and how much they are manifestations of individual psychodynamics.

The extent to which sex-role stereotypes influence the goals therapists set for their patients is the area that has generated the greatest controversy. The cry of a "double standard of mental health" under which men are encouraged to be competent, independent achievers and women are expected to incorporate conforming, passive, and dependent qualities has served as the slogan for those who challenge the notion that psychotherapy is a viable option for women today. Although it is unlikely that many therapists in a society sensitized to women's issues would subscribe to such a view of women's mental health, it is important to explore the more subtle manifestations of such stereotyping. It is equally important to keep in mind the impact such stereotypic assumptions can have on male patients as well, a perspective less frequently addressed in the literature.

A particularly clear example of the operation of stereotypic assumptions in the setting of goals falls in the area of employment and professional success. Unemployment and dependence upon parents and/or spouses for financial support is frequently considered "less unhealthy" for intelligent, physically able young women than for similarly equipped young men. The stereotype negatively affects both women and men, if women are not urged as strongly to develop autonomy and men are not permitted to maintain their dependence. Men who are reminded that they have wives and families to support are penalized as strongly as women who are reassured that theirs is the family's second income.

A less frequently addressed consequence of increased sensitivity to sex-role stereotyping occurs when a therapist's need not to stereotype collides with a patient's stereotypically consistent goals. For example, an intelligent, educated wife and mother presented at a community mental health center with the explicit wish to be happier as a homemaker. The therapist to whom she was assigned could not accept that goal for her, unable to comprehend why a capable adult would choose a lifestyle revolving about children, husband, and the smooth running of a household. The therapist legitimately questioned the degree to which the patient's goals reflected a fear of moving beyond the protective boundaries of her home environment, but had difficulty understanding the equally legitimate possibility that for this particular woman, homemaking could be a source of satisfaction.

The impact of sex-role stereotypes on the therapy process itself is particularly difficult to assess, since once treatment is in progress, relational issues are likely to assume greater importance. Individual and idiosyncratic reactions must be distinguished from stereotypes, and, in an ongoing psychotherapy relationship, transferential and countertransferential processes become more prominent as the interplay between a particular patient and therapist becomes the focus of attention. Nonetheless, it is likely that societally reinforced styles of interaction may, to some extent, influence men and women in the psychotherapy relationship just as in any other. It is in this context that modes of relatedness may be stereotypic.

For example, a female therapist who is stylistically tough and confronting commented that her male and female patients related to her differently, the male patients in a consistently stereotypic fashion. The women in her practice were likely to acknowledge her competence and obvious ability to fend for herself. Her male patients, in contrast, frequently expressed interest in protecting and caring for her. In her view, their responses were consistent with traditional stereotypes. The male role requires that women, stereotypically less competent and capable, be cared for, despite the actual data suggesting the contrary.

Is this an example of male stereotyping of a female therapist? The therapist believed so. However, it is not known whether this particular therapist interacted with her male patients in such a way as to elicit the fantasies and wishes they then consistently reported. If so, her stereotypic behavior may have served to "create" the reciprocal responses by her patients. What is occurring may be more reflective of some perhaps unconscious needs on the part of the therapist than of stereotypic rigidity on the part of her male patients. In this case, the therapist's sex-role stereotypic patterns may be affecting the treatment, limiting the responses available to her male patients. Such an example highlights the importance of therapists' own personal psychotherapy, so that they can be more sensitive to their role in shaping the therapeutic interaction.

Thus the concept of stereotyping may prove to be of value in understanding therapist responses that occur consistently with patients of a particular sex. However, when looking at the mode of interaction between a therapist and a particular patient, or at the attitudes, feelings, reactions, and so on arising in either as psychotherapy proceeds, stereotyping becomes a less useful explanation. Transference and countertransference are frequently more useful concepts in studying the ongoing therapeutic relationship, in their emphasis on the unique characteristics each patient brings to the psychotherapy dyad.

For example, a male patient in his late twenties began treatment with a female therapist. As the therapy progressed, it became clear that his expectation was to be cared for and protected by a warm and nurturant woman. Although his expectations can be seen as consistent with stereotypic conceptions of female therapists, such an explanation is not especially helpful. Looking past the stereotype to the transference implications of his expectations, the patient was able to explore his experience of his mother as an ungiving, self-absorbed woman who had expected the patient, as the eldest of several children, to stand on his own. His wish in seeking a female therapist, similar to his goals in his relationship with the women he dated, was to replace his mother with a more caring model. Working within the framework provided by this transferential formulation put the patient in closer touch with his own personal history, and enabled him to explore more realistically what he might hope to obtain in his current relationships.

Stereotypes can be used by both patients and therapists as a form of resistance, in that such an understanding of an attitude or response may serve to minimize the need to look further. Attributing a male therapist's authoritarian behavior to his gender may spare both patient and therapist the task of examining their own roles in eliciting that attitude at a given point in the treatment process. Similarly, explaining a female patient's reluctance to directly express angry feelings as a stereotypically feminine hesitation eliminates the necessity of dealing with those factors in the therapeutic interaction that may perpetuate such inhibitions. In both situations, therapist and patient are drawing upon stereotypic conceptions of men and women in a collaborative effort to avoid looking at something going on between them.

It seems, then, that there are situations in the course of psychotherapy that may be interpreted as indications of sex-role stereotyping. The danger lies in not looking further to explore the additional, more personal meaning of the same situations for that patient or therapist, or their interaction. Explanations that go no further than identifying something as "stereotypic" fail to grasp and respect the richness and complexity of human relatedness.

CONCLUSIONS

Continuing sensitivity to the consequences of constricting sex-role stereotypes has encouraged mental health professionals to examine their own attitudes and behaviors, in order to exonerate themselves of charges that they perpetuate less than optimal growth and development for both women and men. This evaluation of the psychotherapy process, and of the notion of psychotherapy itself, has occurred on both an empirical level, in the form of investigation of stereotyping among psychotherapists, and on a more clinical level in the ongoing discussion and exploration of how sex-role stereotyping may be manifested in the psychotherapeutic situation.

Social psychological formulations regarding the phenomena of stereotyping in general, and sex-role stereotyping in particular, have been useful indicators of where stereotypic thinking may affect the psychotherapeutic process. Studies of nonpsychotherapy dyadic interactions are clearly relevant to clinical practice. Research that, in evaluating the strength of stereotypic attribution, highlights the importance of situational context or the presence or absence of additional information about the "target" of the stereotype is similarly useful in any thoughtful consideration of the impact of stereotypic attitudes on the therapy process. Such research findings can direct clinicians in their examination of where stereotyping may interfere in the uncluttered communication important for successful psychotherapy.

Integrating the results of empirical research with the clinical material underscores the importance of recognizing that human interactions are rarely, if ever, unidimensional. Although the identification of a given behavior or attitude as stereotypic may in itself be accurate, it may not be the most useful way of understanding what is going on. Furthermore, satisfaction with such an explanation on the part of either participant may serve to obscure additional meanings or to make further discussion appear unnecessary. In either case, the danger exists that by accepting the designation "stereotypic" as a satisfactory level of explanation, a more helpful understanding of the behavior or attitude may be unwittingly passed by.

Although the research reviewed does not convincingly demonstrate that sex-role stereotypes operate to hinder treatment, it has highlighted the complexity of the phenomenon and the need to look at it within the context of other variables. Social psychological conceptualizations, such as those of implicit personality theory, emphasize stereotypes as cognitive structures that are part of the way we all anticipate and organize experience. Stereotypes, then, are likely to be operative in the anticipation of a new person, providing the most rudimentary structure of what one expects upon meeting an unfamiliar "female" or "male." As such, stereotypes may not be detrimental. However, should the existence of a stereotypic conceptualization interfere with learning about and responding appropriately to the unique individual who also happens to be "female" or "male," it will clearly impair the course of treatment, as would any other interference in clearly perceiving another person. Social psychologists have provided a valuable framework for understanding and researching this phenomenon. It may be profitably studied and extended by clinicians interested in increasing their understanding of the psychotherapy process.

REFERENCES

Ashmore, R. D., & DelBoca, F. K. Sex stereotypes and implicit personality theory. *Sex Roles*, 1979, *5*, 219–248.

Bem, S. L. The measurement of psychological androgyny. *Journal of Consulting and Clinical Psychology*, 1974, *42*, 155–162.

Brigham, J. C. Ethnic stereotypes. *Psychological Bulletin*, 1971, *76*, 15–38.

Broverman, I. K., Broverman, D. M., Clarkson, F. E., Rosenkrantz, P. S., & Vogel, S. R. Sex-role stereotypes and clinical judgments of mental health. *Journal of Consulting and Clinical Psychology*, 1970, *34*, 1–7.

Broverman, I. K., Vogel, S. R., Broverman, D. M., Clarkson, F. E., & Rosenkrantz, P. S. Sex-role stereotypes: A current appraisal. *Journal of Social Issues*, 1972, *28*, 59–78.

Costrich, N., Feinstein, J., Kidder, L., Maracek, J., & Pascalle, L. When stereotypes hurt: Three studies of penalties for sex-role reversals. *Journal of Experimental Social Psychology*, 1975, *11*, 520–530.

Davidson, V. Patient attitudes toward sex of therapist: Implications for psychotherapy. In J. L. Claghorn (Ed.), *Successful psychotherapy*. New York: Brunner/Mazel, 1976.

Delk, J. L., Ryan, T. T., & Livingston, M. M. *Patients' sex-role stereotypes of psychotherapists and their preferences for male or female therapist*. Paper presented at the Southeastern Psychological Association Convention, Atlanta, March 1978.

Garske, J. P. Role variation as a determinant of attributed masculinity and femininity. *Journal of Psychology*, 1975, *91*, 31–37.

Goldstein, A. P. *Therapist-patient expectancies in psychotherapy*. New York: Pergamon Press, 1962.

Gurman, A. S. Therapist and patient factors influencing the patient's perception of facilitative therapeutic conditions. *Psychiatry*, 1977, *40*, 218–231.

Kaplan, A. Toward an analysis of sex-role related issues in the psychotherapeutic relationship. *Psychiatry*, 1979, *42*, 112–120.

Kravetz, D. F. Sex role concepts of women. *Journal of Consulting and Clinical Psychology*, 1976, *44*, 437–443.

Lippmann, W. *Public opinion*. New York: Harcourt, Brace, 1922.

Locksley, A., Borgida, E., Brekke, N., & Hepburn, C. Sex stereotypes and social judgment. *Journal of Personality and Social Psychology*, 1980, *39*, 821–831.

Marwit, S. J. Assessment of sex-role stereotyping among male and female psychologist practitioners. *Journal of Personality Assessment*, 1981, *45*, 593–599.

Maxfield, R. B. *Sex-role stereotypes of psychotherapists*. Unpublished doctoral dissertation, Adelphi University, 1976.

McCauley, C., Stitt, C. L., & Segal, M. Stereotyping: From prejudice to prediction. *Psychological Bulletin*, 1980, *87*, 195–208.

Orlinsky, D. E., & Howard, K. I. The effects of sex of therapist on the therapeutic experience of women. *Psychotherapy: Theory, Research, and Practice*, 1976, *13*, 82–88.

Rabkin, J. G. Therapists' attitudes toward mental illness and health. In A. S. Gurman & A. M. Razin (Eds.), *Effective psychotherapy: A handbook of research*. New York: Pergamon Press, 1977.

Rosenkrantz, P. S., Vogel, S. R., Bee, H., Broverman, I. K., & Broverman, D. M. Sex-role stereotypes and self-concepts in college students. *Journal of Consulting and Clinical Psychology*, 1968, *32*, 287–295.

Shafran, R. B. *Differential patient expectations and preferences regarding male and female therapists*. Paper presented at the American Psychological Association, New York, September 1979.

Sherman, J. Therapist attitudes and sex-role stereotyping. In A. Brodsky & R. Hare-Mustin (Eds.), *Women and psychotherapy: An assessment of research and practice*. New York: Guilford Press, 1980.

Smith, M. L. Sex bias in counseling and psychotherapy. *Psychological Bulletin*, 1980, *87*, 392–407.

Snyder, M., Tanke, E. D., & Berscheid, E. Social perception and interpersonal behavior: On the self-fulfilling nature of social stereotypes. *Journal of Personality and Social Psychology*, 1977, *35*, 656–666.

Spence, J. T., Helmreich, R., & Stapp, J. Ratings of self and peers on sex-role attributes and their relation to self-esteem and conceptions of masculinity and femininity. *Journal of Personality and Social Psychology*, 1975, *32*, 29–39.

Stricker, G. Implications of research for the psychotherapeutic treatment of women. *American Psychologist*, 1977, *32*, 14–22.

Stricker, G., & Shafran, R. B. Gender and psychotherapy: A review of the empirical literature. In J. Murray & P. Abramson (Eds.), *Bias in psychotherapy*. New York: Praeger, 1983.

Sullivan, H. S. *The interpersonal theory of psychiatry*. New York: W. W. Norton, 1953.

Thompson, C. Notes on the psychoanalytic significance of the choice of an analyst. In M. R. Green (Ed.), *Interpersonal psychoanalysis: The selected papers of Clara M. Thompson*. New York: Basic Books, 1964. (Originally published, 1938)

Tolor, A., Kelly, B. R., & Stebbins, C. A. Assertiveness, sex-role stereotyping, and self-concept. *Journal of Psychology*, 1976, *93*, 157–164.

Turkel, A. R. The impact of feminism on the practice of a woman analyst. *American Journal of Psychoanalysis*, 1976, *36*, 119–126.

Whitley, B. E. Sex roles and psychotherapy: A current appraisal. *Psychological Bulletin*, 1979, *86*, 1309–1321.

Wilkins, W. Expectancy of therapeutic gain: An empirical and conceptual critique. *Journal of Consulting and Clinical Psychology*, 1973, *40*, 69–77.

Zanna, M. P., & Pack, S. J. On the self-fulfilling nature of apparent sex differences in behavior. *Journal of Experimental Social Psychology*, 1975, *11*, 583–591.

Zeldow, P. Sex differences in psychiatric evaluation and treatment: An empirical review. *Archives of General Psychiatry*, 1978, *35*, 89–93.

Ziennelis, A. Effects of client preference and expectancy upon the initial interview. *Journal of Counseling Psychology*, 1974, *21*, 23–30.

9

Interpersonal Attraction

SUSAN SPRECHER AND ELAINE HATFIELD

At every stage of an intimate relationship, couples must deal with difficult and unsettling interpersonal problems. As a consequence, therapists often find themselves dealing with clients when the problems have become too difficult and too unsettling.

Different problems characterize different stages of a relationship. Early on, one must deal with the problems associated with initiating a relationship; then the problems associated with keeping as established relationship vibrant and growing; or the painful experiences associated with relationship breakdown and dissolution.

In this chapter, we shall review the social psychological research on interpersonal attraction, love, and intimacy, and attempt to point out its relevance to three stages of a relationship:

1. How relationships begin.
2. How relationships grow.
3. How and why they end (see Duck, 1982; Duck & Gilmour, 1981 a, b, c).

We conclude by discussing some differences between men and women in what they expect from their intimate relationships.

SUSAN SPRECHER • Department of Sociology, University of Wisconsin, Madison, Wisconsin 53706. ELAINE HATFIELD • University of Hawaii at Manoa and King Kalakaua Center for Psychotherapy, Honolulu, Hawaii 96822.

How Relationships Begin

"Who am I likely to fall in love with?" "Where can I meet someone?" These questions are often posed by people who are having trouble initiating relationships. Therapists may be more adequately equipped to advise them if they are aware of how couples generally meet and what factors are most important in initial attraction.

Determinants of Initial Attraction

Social psychologists have conducted a great deal of research to determine what factors are most important in determining how attracted people will be to one another (see Berscheid & Walster [Hatfield], 1978; Hatfield & Walster, 1978). Among these are proximity, physical attractiveness, and similarity.

Proximity

In a song from *Finian's Rainbow* there is a line, "When I'm not near the one I love, I love the one I'm near." Numerous studies have demonstrated that sheer proximity between two people is critical in determining how attracted they will be to each other.

In one classic study, Festinger, Schachter, and Back (1959) examined the development of friendships in a new housing project for married students. They found that proximity in apartment residences was the major factor in determining who became friends with whom. Although the linear distance between two apartment residences was important, the functional distance was even more important. They found, for example, that residents whose apartments were located near entrances or exits, or near utility rooms, were more likely to interact and become friends with other residents than were those who did not live near common areas. Other studies provide additional documentation that proximity matters. Co-workers become closer to those who work right next to them than to those who work farther away (Gullahorn, 1952; Kipnis, 1957; Zander & Havelin, 1960), and students who sit near each other in class are more likely to become friends than those who sit farther away (Byrne, 1961; Segal, 1974).

More interesting, however, is the apparent importance of proximity in mate selection. Studies have demonstrated that people are more likely to marry those who live closer than those who live at greater distances. For example, in an examination of 5,000 Philadelphia marriage licenses,

Bossard (1932) found that many of the couples were already living at the same address or within a few blocks at the time they applied for their license (see also, Clarke, 1952). After reviewing all the studies on the importance of proximity in mate selection, Kephart (1961) states: "Cherished notions about romantic love notwithstanding, it appears that when all is said and done, the 'one and only' may have a better than 50-50 chance of living within walking distance" (p. 269).

Operationalizing proximity as the actual or functional distance between two people at one point in time may be an outdated approach in a society that is becoming increasingly mobile. As suggested by Monge and Kirste (1980), a more appropriate conceptualization of proximity may be psychological distance, which can be thought of as "the opportunity to communicate." Viewed in such a way, proximity can be considered a phenomenon that fluctuates over time. Future studies that examine the importance of proximity for both the initiation and maintenance of relationships may need to take a more flexible approach to proximity.

A distinction has also been made between two different kinds of settings that encourage close physical proximity, a "closed field" and an "open field" (Murstein, 1970). In a closed field, people have little choice about with whom they interact. For example, the typist in the typing pool has little choice about who her co-workers are. However, when she leaves work and goes to a party, she is entering an open field. In an open field there are many opportunities to interact and begin relationships, and thus choices have to be made. It is possible that people differ on whether they are more successful socially in a closed versus open field. For example, as we shall discuss, people are initially attracted to others who are physically attractive. Thus, the attractive may well have the greatest advantage in open fields. The unattractive, on the other hand, may have better luck developing social relations in closed fields, where others are forced to get to know them over time.

The simple notion behind the proximity effect is the following: When people are thrown together, they are likely to interact; when people interact, they are likely to become attracted to each other. This has implications for those desiring to expand their social networks or to find someone to love. It suggests that people must arrange things so that they will be physically proximate to others—for example, by joining exercise clubs, by taking night courses, or by attending dances. Furthermore, less attractive people might do well to place themselves in settings that are closed.

Physical proximity can be considered a necessary but insufficient condition for attraction to occur between two people. It is an important precursor for other personal and situational factors to have an influence on the attraction process.

Physical Attractiveness

People seem to work harder on their physical appearance when they are seeking a new lover or spouse. They suddenly lose several pounds, adopt a new hairstyle, and double the size of their wardrobe. They know that looks *do* matter; they, too, are searching for an attractive catch.

Considerable research has been devoted to the importance of attractiveness in heterosexual attraction, date selection, and marriage. The two questions that have received the most attention are: (1) Do people seek to date the most attractive persons? or (2) do they try to date others of the same level of physical attractiveness? The data indicate that although people would like to date the most attractive persons, they are generally realistic—they generally settle for someone near their own level of attractiveness.

In a classic study that demonstrated the importance of physical attractiveness, Walster (Hatfield), Aronson, Abrahams, and Rottman (1966) held a "college computer dance" and paired men and women randomly (with the only requirement that the man be taller than the woman). Each individual's personality, intelligence, social skill, and physical attractiveness were measured. It was assumed that men and women would be attracted to others of approximately the same level of social desirability—and that all of the preceding factors would be important in determining people's worth. The researchers found, however, that only physical attractiveness mattered! How much people liked their dates, wanted to see them again, and actually tried to see them again was influenced only by how attractive the date was. Similar results were found in other computer dating studies (Brislin & Lewis, 1968; Tesser & Brodie, 1971). There was no evidence that daters were willing to "settle" for "appropriate" partners.

These early computer dating studies, however, are different in a fundamental way from most dating situations. With the exception of arranged marriages in traditional societies, people are rarely "assigned" to date or marry particular others. Instead, people have to ask (and possibly be rejected) or wait to be asked (and possibly never be asked). When people are required to *choose* a dating partner and the possibility for rejection is made salient, people often do tend to choose someone of about the same level of physical attractiveness. This has been demonstrated in computer dating studies by Berscheid, Dion, Walster (Hatfield), and Walster (1971), Huston (1973), and Stroebe, Insko, Thompson, and Layton (1971). A high degree of similarity in physical attractiveness of couples has also been found in natural settings (Murstein, 1972; Silverman, 1971). In summary, people aspire to date those who are most physically attractive, but assortive mating is pervasive—that is, people end up with someone of about the same level of physical attractiveness.

Why are the physically attractive preferred? There are several possible reasons. It may simply be that people have an innate preference for what is aesthetically pleasing (Valentine, 1962). What is aesthetically pleasing, however, seems to be greatly influenced by cultural standards. Attractive persons may also be preferred because it is assumed that beauty is *more* than skin-deep. Much experimental evidence exists to support the physical attractiveness stereotype, "What is beautiful is good." People assume that those who are physically attractive possess many other desirable qualities as well (K. K. Dion, Berscheid, & Walster, 1972; Miller, 1970). Attractive people may be preferred for yet another reason, namely, that people hope that their physical attractiveness will rub off. Many years ago, Waller (1937) stated that there is a prestige value in being seen with attractive people. This has more recently been verified by an experiment demonstrating that men are most likely to be rated positively by outsiders when they are accompanied by a good-looking woman, and most likely to be rated negatively when they are accompanied by an unattractive woman (Sigall & Landy, 1973).

Therapists counsel attractive persons, unattractive persons, and persons of average appearance. Although unattractive persons probably have their share of interpersonal problems stemming from physical appearance (e.g., not being able to get dates, having low self-confidence), they do not have a monopoly on appearance-related problems. For example, very physically attractive persons (particularly women) may be more likely to attribute success in heterosexual relations to their appearance, which might result in a lowered overall self-evaluation (K. K. Dion, 1982). They may also have problems in being accepted by others on personal dimensions unrelated to physical attractiveness, which can lead to disappointing, superficial relationships.

Similarity

People who are similar on physical attractiveness often end up together. Similarity on other personal characteristics, such as social background, attitudes, and values, is also important in determining initial attraction. In one classic study, Newcomb (1961) examined the development of friendships in a group of college men who were given free housing in return for completing questionnaires on their friendship choices and attitudes. After the men had time to become acquainted, it was found that mutual liking was greatest among those who had similar attitudes and beliefs. In a succession of laboratory experiments conducted by Byrne (1971) and his colleagues, it was found that the greater the proportion of attitudes expressed by a stranger that were similar to the subject's, the more liking the subject expressed for the stranger. In addition, evidence in naturalistic settings suggests that friends and married couples tend to come from similar backgrounds, similar family relationships, and have similar attitudes and values (Burgess & Wallin, 1943; Kandel, 1978).

Why are people attracted to similar others? There is one very simple explanation: People are more likely to live, work, and play near others who share their background, values, attitudes, hobbies, and so on—in other words, proximity. Another explanation is that similarity is reinforcing because it provides validation for people's beliefs and attitudes. Still another explanation is that similarity affects people's beliefs about the probability of being liked—people assume that similar others will like them.

It is also possible for similarity to be a consequence rather than a cause of attraction—that is, attraction can lead to actual and/or perceived similarity. Evidence does exist to support the attraction–perceived similarity link. In fact, it has been found that husbands and wives tend to overestimate how similar they actually are to each other (Byrne & Blaylock, 1963; Levinger & Breedlove, 1966). Whether husbands and wives actually do become more similar to each other, however, is less clear. Early studies found either no evidence that this occurred (Hoffeditz, 1934; Hunt, 1935) or mixed evidence (Newcomb & Svehla, 1937; Schooley, 1936). However, more recent studies suggest that actual similarity between husband and wife may increase for at least some couples. In a reanalysis of an earlier longitudinal study by Kelly (1955) of husbands and wives, it was found by Uhr (cited in Barry, 1970) that happy couples became more similar whereas unhappy couples became more dissimilar over 18 years of marriage. In a cross-sectional study by Ferreira and Winter (1974), couples who had been married for 3 years or more were more likely to express spontaneous agreement on various issues than couples who had been married less than 3 years. This was found, however, only for couples who did not have emotional or psychiatric problems. There is also evidence that couples may come to be more similar to each other in neuroticism (Kreitman, Collins, Nelson, & Troop, 1970; Nelson, Collins, Kreitman, & Troop, 1970).

Although the evidence suggests that people generally seem to prefer to associate with similar others, people may sometimes prefer dissimilar others. In marriage, people may be attracted to those who complement their needs and personalities (Kerckoff & Davis, 1962; Winch, 1952). For example, dominant people may be attracted to those with submissive personalities. People may also differ in their need to be around similar versus dissimilar others. Snyder and Fromkin (1980) have proposed that people differ in their need to be distinguished from other people, and have developed a scale to measure this desire for uniqueness. They define uniqueness as "a positive striving for differences relative to other people" (Snyder & Fromkin, 1977, p. 518). In one experiment, they found that those who were high in need for uniqueness were more likely to react negatively to finding out that they were similar to other college students (Snyder & Fromkin, 1980).

Other Factors Important in Determining Attraction

There are many other factors in addition to proximity, physical attractiveness, and similarity that are important in determining how attracted people are to others. It helps, for example, if the others are competent (Solomon & Saxe, 1977) and have pleasant personalities (M. F. Kaplan & Anderson, 1973). But perhaps one of the most important determinants of attraction is reciprocal liking—people like others who like them (Mettee & Aronson, 1974; Newcomb, 1961). This is knowledge that anyone who feels deprived of love can use to his or her benefit. As the philosopher Hecato put it in the second century B.C.:

> I will show you a love potion without drug or herb or any witch's spell; if you wish to be loved, love.

INITIATING THE RELATIONSHIP

We have discussed several important determinants of initial attraction. But how do relationships actually get started? How and where do people meet? These questions have been explored only recently.

Evidence suggests that many of our relationships begin with an introduction provided by a same-sex friend. In a study of East Carolina University students, Knox and Wilson (1981) found that one-third of the college students met their dating partners through a friend. This suggests that the best way of meeting someone of the opposite sex is to establish relationships with same-sex friends. Those couples who were not introduced often met in such commonplaces as parties, work, and classes.

Similar results were found in a study by Marwell, McKinney, Sprecher, Smith, & DeLamater (1982) at the University of Wisconsin.

The researchers asked a random sample of college sophomores to think of their most recent "couple" experience with someone of the opposite sex. The students were then asked a set of questions about when they first saw the other, how they met him or her, and what activities they were involved in at the time. Results indicate that couples often met at such common places as classes, parties, or work—or were introduced by a friend. Marwell and his colleagues argue that what is critical for the initiation of heterosexual relationships are *legitimating factors*—factors such as sharing a class or being introduced by someone. These legitimating factors help to overcome socially specified barriers to interactions between strangers.

Sometimes, however, people do meet opposite-sex individuals in ways considered somewhat less legitimate—for example, they may be picked up at single bars. Perper and Fox (1981) spent 2 years observing how people meet for the first time in singles bars, college pubs, or other public settings. For a total of 350 hours they observed over 500 couples make initial contacts. They also interviewed some men and women on what happened when they either tried to pick someone up or were picked up. The researchers found that the pickup can be characterized by steps of escalating flirtatious behaviors, often initiated by the female. *Approach* is the first step. One person (usually the female) acknowledges the presence of the other. If the other is also interested, the acknowledgment is returned. Then the two people begin to *synchronize* their body movements—for example, they make take on identical postures or lift drinks at the same time. Finally, *touching* occurs. This is often initiated by the woman, and is the most obvious sign of interest. The touching is either acknowledged by the other, or the interaction begins to wane. From the personal interviews, the researchers discovered that women were more aware than men of what goes on during the pickup.

How Relationships Grow

Once people get involved in relationships, their problems are far from over. Many people complain of relationships that lack honest and open communication, relationships full of conflict, strife, and frustration, relationships that do not seem to go anywhere. Relationships can simply exist, or they can be exciting intersections between two unique individuals. According to some theorists (Kelley, Berscheid, Christensen, Harvey, Huston, Levinger, McClintock, Peplau, & Peterson, 1983), interdependence is critical for close relationships. We shall explore those actions that can lead the lives of two people to become more and more intertwined.

SELF-DISCLOSURE

A central theme of many love stories is that lovers attempt to "know" each other. Pope (1980) describes how important knowledge of each other is for lovers:

> They sought a knowledge of not only what made the beloved human but also of what made him or her different from all other humans. They acquired a knowledge not only of the head but also of the heart and body. (p. 8)

To gain knowledge about the other is important not only for lovers, but also for close friends and family members. Mutual self-disclosure is the means by which this knowledge is gained. Self-disclosure has been defined as "the act of making yourself manifest, showing yourself so others can perceive you" (Jourard, 1971).

Altman and Taylor (1973) have written about the importance of self-disclosure for the growth of intimate relationships. They describe the personality as an onion, with different layers of skins that need to be peeled in the process of self-disclosure. The outer layers are superficial items, such as biographical information. The inner layers contain that which is most critical to the person, involving "basic core feelings about life, trust in others, and the nature of one's self-image" (p. 17).

According to these theorists, as relationships grow, self-disclosure is likely to increase in both breadth and depth. Breadth refers to the number of surface areas that something is revealed, and depth refers to how far the penetration proceeds to the basic core feelings in each area. People move to greater self-disclosure and intimacy in relationships only if it is rewarding to do so.

Altman and Taylor (1973) describe the stable exchange of self-disclosure achieved in very few intimate relationships:

> Stable exchange: Achieved in only a few relationships, stable exchange continues to reflect openness, richness, spontaneity, and so on in public areas. . . . Dyad members know one another well and can readily interpret and predict the feelings and probable behavior of the other. . . . For the first time, perhaps, there is a considerable richness of communication in the central core areas and a high degree of mutual spontaneity, permeability, and dyadic uniqueness. In addition to verbal levels, there is a great deal of exchange of nonverbal and environmental behaviors, movements, touching, and so one. They are more willing to allow each other to use, have access to, or know about very private apparel and belongings. (pp. 136–141)

Much research has documented that self-disclosure tends to be reciprocal (for a review, see Chelune, 1979). One relationship for which this may be an exception, however, is that between the therapist and the client. The client is expected to self-disclose, the therapist to listen and give advice. (For exceptions to this model, see Doster & Nesbitt, 1979.)

The norm of reciprocity in self-disclosure may change at different stages of the relationship. Altman (1973) has suggested that the obligation to immediately reciprocate self-disclosure is more important in early stages of a relationship when mutual trust and respect are being developed. At later stages of the relationship, when mutual trust is well established, self-disclosure reciprocity can become less important. In support of this contention, it has been found that reciprocity of self-disclosure is higher among strangers than spouses (Morton, 1978) or friends (Derlega, Wilson, & Chaikin, 1976).

Altman and Taylor's self-disclosure theory assumes that self-disclosure precedes love. However, it is also possible that love generates greater self-disclosure. A recent study explored whether love is more predictive of the development of self-disclosure than is self-disclosure of the development of love. In the two-wave panel study of over 300 college students, it was found that romantic love led to self-disclosure for women, but self-disclosure led to romantic love for men (Adams & Shea, 1981). In general, and disregarding the causal direction, evidence indicates that self-disclosure in the relationship is related to love feelings (Rubin, Hill, & Peplau, 1980) and to satisfaction with the relationship (Blood & Wolfe, 1960; Jorgenson & Gaudy, 1980; Hendrick, 1981; Komarvsky, 1962; Levinger & Senn, 1967).

There is some partial evidence to suggest that too much self-disclosure may not always be good for relationships. When people are first getting acquainted, the person who confides information about skeletons in the family closet or about personal problems may be seen to be very peculiar indeed (for a review, see Kleinke, 1979). Later in the relationship, too much self-disclosure may result in difficulties in a relationship because partners become bored with each other or begin taking each other for granted. Evidence for a curvilinear relationship between self-disclosure and satisfaction in marriage has been found by Cutler and Dyer (1965), Gilbert (1976), and Shapiro and Swensen (1969).

One interesting question therapists often face is: Should intimates disclose things that might upset their relationship? For example, should a husband admit to his wife that he had a casual affair while on a business convention? Many theorists have argued that some things are best kept undiscussed. Goffman (1959) suggests that there are things that should not be revealed because it might be too costly to the self-discloser and/

or might hurt the recipient of the information. However, compelling data suggest that unhappiness in intimate relationships is related to suppressed negative feelings (Gilbert, 1976). There is obviously no straightforward answer, but in practice we have found that in families there are no "secrets"—but there is sometimes confusion. When people disclose their deepest feelings, the disclosures usually turn out not to be so lethal as they had supposed. Self-disclosure is risky. In casual encounters one must often be false to oneself. But in intimate relationships, it is a risk worth taking.

EVERYDAY EXCHANGES

As two people become closer, they also become increasingly interdependent. Characteristic of close relationships is a high degree of exchange. Intimates exchange not only confidences but a wide variety of things including love, sex, money, helping favors, and emotional support. Recently, social psychologists have tried to examine the exchange process in intimate relationships. In studies by Hatfield, Traupmann, Sprecher, Greenberger, & Wexler (1981), couples were asked about the level of give-and-take in the following areas.

Personal concerns: How attractive were they and their partners? How sociable? How intelligent? *Emotional concerns:* How much liking did they express for one another? How much loving, understanding, and acceptance? How much sexual pleasure did they give and get? Were they faithful? Were they committed to one another? Did they respect their partner's needs for freedom? *Day-to-day concerns:* How much of the day-to-day maintenance of the house did they and their partners do? How were finances handled? What was the situation with companionability; conversation; decision making; remembering special occasions? Did they fit in with one another's friends and relatives? *Opportunities gained and lost:* How much did they gain simply from going together or being married? (For example, how much did they appreciate the chance to be married; to be a parent or a grandparent; or to have someone to grow old with?) What opportunities had to be forgone?

In these studies, men and women were asked how fair they thought their relationships were. Were they getting more than they felt they deserved, just about what they deserved, or less than they thought they had coming from the relationship?

It has been found that, in general, equitable (fair) relationships are happy, satisfying, and growing relationships. However, when the exchange is unbalanced and participants report that things are grossly unfair, both participants are likely to be unhappy with the state of affairs.

Those who are getting more than they feel they deserve experience, not delight, but guilt. Those who are getting far less than they deserve feel angry. In general, both are motivated to set things right.

CONFLICT

A distraught newlywed woman cries to her therapist, "My husband doesn't love me anymore. He yelled at me this morning. What am I to do?" The therapist may well reassure her that conflict is an integral part of intimacy. In close relationships, people have intense feelings about each other—both positive and negative. Having negative feelings or experiencing conflict is not necessarily a sign that the relationship is about to end, but may actually suggest the contrary. Most theorists state that the opposite of love is not hate, but indifference.

That negative feelings are an integral part of intimate relationships is evidenced by several recent studies. In one study, college students were asked to report their most recent outbreak of anger. The first most frequently mentioned source of outbreak was a family member or relative, and the second most frequently mentioned was a lover (Fitz & Gerstenzang, 1978). In another study, lovers were most likely to be named as the source of feelings of depression (Berscheid & Fei, 1977).

Braiker and Kelley (1979) found conflict to be typical among married couples. Husbands and wives were asked to independently complete a questionnaire about four different stages of their relationship: casual dating, serious dating, engagement, and marriage. The questions were about attitudes, feelings, and behaviors they may have experienced in each stage of the relationship. The researchers found that as the relationship developed over time, the partners became increasingly interdependent at several different levels—emotionally, behaviorally, normatively, and even in personal characteristics and attitudes. The researchers found not only a large increment in interdependence as the couple became more serious, but also an increase in conflict. The couples were more likely to feel angry, communicate negative feelings, argue with each other, and try to change bothersome aspects of the other in the serious dating stage and later than in the early, casual dating stage. Interestingly, the researchers found no correlation between the level of love and interdependence and the level of conflict. That is, couples who reported conflict were just as likely to feel love and commitment toward each other as those who did not report conflict.

Kelley (1976) and others have pointed out the importance of the attribution process during interpersonal conflict. According to attribution theory (Heider, 1958; E. E. Jones & Davis, 1965; Kelley, 1967), people attribute their own behavior and the behavior of others to various sources—to the person, to circumstances, to the situation. Evidence suggests that people are generally biased in their attributions, and this tendency may be particularly aggravated in interpersonal conflict. More specifically, people are likely to attribute the actions of the other to personality characteristics and dispositions, but attribute their own actions to situational causes (Heider, 1958; Ross, 1977).

This actor–other discrepancy has been found in investigations of intimate relationships. Orvis, Kelley, and Butler (1976) found that members of close relationships tended to attribute their partner's behavior to such stable dispositions as traits, whereas they attributed their own behavior to such unstable causes as judgments or environmental states. Perhaps if couples become more aware of this bias in how they make attributions, they can prevent conflict from escalating to unmanageable levels.

In an insightful article on conflict in close relationships, Braiker and Kelley (1979) discuss ways in which conflict can, in fact, facilitate growth in close relationships. Some of these ways include: (1) Expressing conflict may lead the couple to devote thought to the relationship. This increased thought about the relationship may actually result in the couple coming to realize that they are more important to each other than originally realized. (2) Expressing conflict may also allow the couple to come to a new definition of the relationship, with new norms and goals. (3) In the process of resolving the conflict, the couple may come to see their relationship as unique and special.

The authors admit, however, that conflict may not always be a constructive experience; it may also have many destructive effects on the relationship.

> The escalation that is characteristic of open conflict often lays bare basic disagreements—differences at the higher levels of interdependence—that might have remained implicit had the interaction remained at the concrete level of specific behaviors. The two persons may realize that they have not only specific conflicts of interest but also basic and perhaps irreconcilable differences in moral principles and human values. (p. 163)

HOW AND WHY RELATIONSHIPS END

Sometimes relationships fail. Conflict may escalate until it appears that there can be no reconciliation in the relationship. Or, the imbalance of exchange may become so inequitable that the participants feel indifferent and lack motivation to try to set things right again. Every day people fall out of love, break engagements, and get divorces. Therapists are frequently faced with couples who are deciding whether or not to finally break up their relationship. However, other relationships exist that endure and continue to grow over a lifetime. Why do some relationships work, whereas others do not?

To examine breakups among dating couples, Rubin and his colleagues (Hill, Rubin, & Peplau, 1976; Rubin, Peplau, & Hill, 1981; Stewart & Rubin, 1976) interviewed 231 dating couples in the Boston area, and then followed them up after a 2-year period. By that time, 103 couples had split up, 65 were still dating, 9 were engaged, 43 were married, and 11 could not be reached. In examining the dating partners' original questionnaires, it was found that those couples who reported higher love feelings for their partner at Time 1 (1972) were more likely to be together at Time 2 (1974). However, whether the couple had engaged in sexual intercourse or had lived together did not affect whether the couple was still living together 2 years later.

The study did find that similarity of the two partners on some dimensions was related to whether the couple remained together. In particular, similarity on age, education, intelligence, and physical attractiveness was important in predicting the success of the relationship. However, similarity (or dissimilarity) on religion, sex-role attitudes, and desired family size was not seen to matter.

One intriguing factor that was found to be related to breakups was the man's need for power. Couples in which the man had a high need for power (as measured on TAT cards) were much more likely to break up than couples in which the man had a lower need for power. On the other hand, need for power in women was not found to be associated with breaking up (Stewart & Rubin, 1976).

Most of the breakups were not mutual. Furthermore, people were likely to say that they were more interested than their partner in breaking off the relationship. This can be interpreted as a strategy to protect one's self-esteem. By combining the men's and women's independent reports, the researchers determined that women were more interested in breaking up in 51% of the couples, the men in 42%, and that there was mutual interest in 7% of the couples. Other evidence also suggests that females are more likely to initiate breakups (Hill, 1974).

The researchers had speculated that in our society men tend to fall in love more readily than women, and that women fall out of love more readily than men. Thus, it was expected that breaking up would be a more traumatic experience for men than for women. Indeed, it was found that men were more likely to feel depressed and lonely after the breakup than were women. As suggested by Goethals (1973; cited in Rubin *et al.,* 1981):

> The notion that the young adult male is by definition a heartless sexual predator does not bear examination. In point of fact some of the most acute cases of depression I have ever had to deal with occurred in attempting to help young men with their betrayal by a young woman in which they had invested a great deal and who had, as the relationship developed, exploited them rather ruthlessly. (p. 94)

Whether the man or the woman initiated the breakup also influenced whether or not the couple remained friends. If the man broke up the relationship, the couple was likely to remain casual friends. On the other hand, if the woman broke up the relationship, this was much less likely to be so.

Other than the study by Rubin and his colleagues, very little research has examined the process of relationship breakdown. However, some theorists have speculated on what happens during relationship dissolution. Taking a social psychological perspective on marital dissolution, Levinger (1979) suggests that relationships break up when there is a shift in the perceived rewards and costs of the relationship. That is, the rewards in the relationship become fewer, the costs greater. Furthermore, attractive alternative states (e.g., being alone, beginning a new relationship) can be a threat to the relationship.

More recently, Duck (1982) discusses several reasons why relationships can break down.

1. *Personal factors.* There may be certain characteristics of individuals that make them unsuccessful at maintaining relationships. For example, the personality trait of neuroticism has been found to be related to high levels of marital disruption (Bloom, Asher, & White, 1978). Age and sex have also been found to be important. Females seem more likely than males to fall out of love (Hill *et al.,* 1976; Rubin *et al.,* 1981), and young people are more likely to have unstable relationships than older people (Duck, Miell, & Gaebler, 1980; Foot, Chapman, & Smith, 1980).

2. *Exterior influences.* These include circumstantial events (such as the relocation of one partner) or the effect of other people (a rival) on the relationship. These exterior influences can actually be the cause of the relationship breakdown, or may just be the last straw or an "excuse" to dissolve a relationship that is actually falling apart for other reasons.

3. *Processual/behavioral/management features.* In order for a relationship to exist, the participants must have a certain amount of social skill. For example they must be able to express their feelings, maintain an equitable exchange, and self-disclose an appropriate amount. To the

extent that these features break down in the relationship, the relationship may also be likely to break down.

4. *Emergent properties of the relationship.* Sometimes the relationship can break down because of relationship properties. For example, a relationship could be too intense for one or both partners—or not intense enough. The partners may find that there is not enough similarity between the two of them—or too much similarity. Or, they may disagree on the future of the relationship.

LONELINESS

We have described how relationships begin, grow, and dissolve. At every stage of a relationship, loneliness can be experienced. Loneliness can be the impetus for people to begin relationships, but people enmeshed in relationships are certainly not immune to this unpleasant experience. It may be during particularly stressful conflicts in a relationship that each partner, focusing on his or her own pain and bitterness, feels the most loneliness. Indeed, the evidence indicates that dissatisfaction with one's intimate relationships is positively related to feeling lonely. In one study of senior citizens, it was found that marital dissatisfaction was related to feeling loneliness (Perlman, Gerson, & Spinner, 1978). In another study conducted with freshman college students (Cutrona, 1982), dissatisfaction with friendships, dating relationships, and family relationships were all associated with loneliness experienced during the first year on a large college campus. Dissatisfaction with current friendships was found to be the most important predictor of student loneliness.

Loneliness has only recently been subject to theorizing and empirical research.

WHAT IS LONELINESS?

Loneliness has been defined by Perlman and Peplau (1981) as an unpleasant experience that results when there is a discrepancy between one's expectations about social relationships and what actually occurs. In particular, when there is a deficit in social relations, either quantitatively or qualitatively, the person may experience loneliness. Perlman and Peplau (1981) further define loneliness as a subjective phenomenon (being alone is not synonymous with being lonely) that is unpleasant and distressing. The most frequently used scale to measure loneliness is the UCLA Loneliness Scale (Russell, Peplau, & Ferguson, 1978). It consists of 20 items such as "I feel starved for company," "I lack companionship," and "I am unhappy being so withdrawn." Research indicates that the scale is reliable and valid.

WHO ARE THE LONELY?

Everyone can experience loneliness at some point. Research on loneliness, in fact, is starting to gain momentum because of the awareness of just how widespread it is.

According to recent research, however, there may be certain types of people who are prone to experience loneliness. Although the stereotype is of the lonely elderly person, large-scale surveys suggest that loneliness may actually be more common in youth. Adolescents tend to report the highest levels of loneliness and older age groups the lowest. This was found in a study by Rubenstein and Shaver (1982) of persons throughout the country ages 18 to 87, and in other studies as well (Fidler, 1976; Parlee, 1979; Woodward & Visser, 1977).

It is not yet fully understood why loneliness is more common among younger people, since they generally have more social opportunities. One hypothesis is that they may have high and unrealistic expectations about social relations that would lead them to experience loneliness. Another possible explanation is that younger people may be more willing to acknowledge such negative feelings as loneliness (Peplau *et al.*, 1982).

Other research has also disproved common notions about who the lonely are. In a survey by Rubenstein and Shaver (1982), geographic mobility was not found to be related to loneliness. In a study of loneliness in middle-age women (Paloutzian & Ellison, 1982), it was expected that women living in large urban cities would experience greater loneliness than women living in rural areas, and that women who were homemakers would experience greater loneliness than working women. Surprisingly, loneliness was found to be unrelated to both the size of the city the women lived in and whether or not she worked.

THE EXPERIENCE OF LONELINESS

What is it like to be lonely? Rubenstein and Shaver (1980) published a questionnaire on loneliness in several newspapers around the country. The respondents were asked to indicate how loneliness feels by indicating to what degree they experienced 27 different possible negative feelings such as desperate, sad, empty, uneasy, angry, and insecure. After factor analyzing the responses, the researchers found the following major factors: (1) desperation; (2) depression; (3) inpatient boredom; and (4) self-deprecation.

Other investigators have also found that the experience of loneliness is related to negative feelings about the self or others. In a study of adolescents, Brennan and Auslander (1979) found that loneliness was related to self-pity, feeling rejected by parents, and feeling unpopular among peers. In a study of college students, Russell, Peplau, and Cutrona (1980) found that lonely students had low self-esteem, were unassertive, and were sensitive to rejection. Several other studies have also found loneliness to be related to low self-evaluation and to such negative feelings as pessimism, social alienation, hostility, mistrust of others, and lack of control (Barrett & Becker, 1978; Jones, Freemon, & Goswick, 1981; Jones, Hansson, & Smith, 1980).

How do people react to their loneliness? Rubenstein and Shaver (1980), in the study described earlier, also examined this question. They asked in their newspaper questionnaire, "When you feel lonely, what do you usually do about it?" Twenty-four possible behaviors were listed. It was found that the most common responses were reading, listening to music, and calling a friend. A factor analysis of the responses yielded four factors. The first, called "sad passivity" by the researchers, contained behaviors related to self-pity: cry, sleep, sit and think, do nothing, overeat, take tranquilizers, watch television, drink, or get "stoned". The second, called "active solitude", contained such behaviors as study or work, listen to music, and go to movies. The third factor involved behavior related to spending money, and the final was seeking out social contact.

Precipitating Factors

In addition to the personal factors (i.e., age) mentioned earlier as being related to loneliness, several antecedent factors of loneliness have been defined. In fact, a distinction has been made between *predisposing factors,* such as characteristics of the individual, and *precipitating factors,* which are events that actually lead to the onset of loneliness (Perlman & Peplau, 1981). We shall describe some of these precipitating factors as discussed by Perlman and Peplau (1981).

Precipitating factors to the experience of loneliness usually involve *changes* for the individual—either in the individual's actual, or desired, social relations.

Changes in one's actual social relations can occur because close relations terminate, either through death or divorce. Studies indicate that widowhood, divorce, and even the breakup of dating relationships are associated with feelings of loneliness (Gordon, 1976; Hill *et al.*, 1976; Lopata, 1969; Weiss, 1973). Changes can also occur when there are physical separations. Some research indicates that people who move are subject to feelings of loneliness (Weiss, 1976), although evidence also indicates that people adjust quite rapidly and make new friends (Rubenstein, Shaver, & Peplau, 1979). Changes in one's actual social relations can also occur because of a status change. When people go through significant role changes—losing their job or retiring, getting a promotion, having children leave home, entering parenthood—they can experience loneliness (see Bart, 1979; Dickens & Perlman, 1981; Rubenstein, 1979).

Changes occur not only in one's actual social relations, but also in the desire for social relations. An increase in desire for social relations that is not matched by actual changes is likely to precipitate the feeling of loneliness. There are several situational factors that can influence an individual's desire for social relations. It has been found that people seem very desirous of being around others when they are anxious. In a classic experiment by Schachter (1959), it was found that when subjects believed they were going to receive very painful electric shocks (high-anxiety condition), they were more likely to want to wait with others (vs. alone) than when they thought they were going to receive only painless shocks (low-anxiety condition). People may also want to be around people when they need to compare themselves to others in order to evaluate themselves on attitudes, values, and skills (Festinger, 1954).

People who experience loneliness also try to determine the precipitating factors—that is, they tend to make attributions for their loneliness. The types of attributions that are formed are important for how the individuals react to their loneliness.

Researchers applying attribution theory to loneliness have used the theoretical work of Weiner and his colleagues (Weiner, Russell, & Lerman, 1978). According to Weiner's model, causal attributions can be classified on two major dimensions: locus of causality (internal vs. external) and stability (stable vs. unstable). More recently, Weiner has added a third dimension, controllability—that is does the individual have control over his or her behavior? Michela, Peplau, and Weeks (1981) examined the reasons students gave for their loneliness. When the students attributed loneliness to stable and internal (personal) characteristics (e.g., being physically unattractive) or to stable characteristics of the situation, they were more likely to feel pessimistic and hopeless about the future.

It has also been found that when attributions are made to stable individual characteristics, the individuals are likely to feel depression (Bragg, 1979; Peplau, Russell, & Heim, 1979).

Coping with Loneliness

How can therapists help distressed individuals deal with their loneliness? First, there can be intervention into the attributions made for loneliness. It has been found that people are more likely to attribute their loneliness to personal factors than to the situation (Peplau *et al.*, 1979). This may result in lowered self-esteem and a helplessness about the future. Thus, lonely people should be taught to focus on situational or cultural factors that might have caused their loneliness.

If the individual is suffering from a social deficit, the therapist can help the client pinpoint what kind of deficit it is. Are close romantic relationships needed, or a wider social network of casual friends? Although lonely people are most likely to say that they need "one special person" (Rubenstein & Shaver, 1982), a loneliness treatment proposed by Young (1982) suggests that individuals should go through stages of commitment to others. In particular, he suggests six stages:

1. To overcome anxiety and sadness about spending time alone.
2. To engage in activities with a few casual friends.
3. To engage in mutual self-disclosure with a trustworthy friend.
4. To meet a potentially intimate, appropriate partner (usually of the opposite sex).
5. To begin to develop intimacy with an appropriate partner, usually through disclosure and sexual contact.
6. To make an emotional commitment to an appropriate partner for a relatively long period of time. (p. 391)

If the actual social contact cannot be altered, the therapist can help the client change the desired level of social contact.

Gender Differences in Love and Intimacy

Often men's and women's descriptions of what goes wrong in their relationships can be reduced to a single problem—men and women seem to want very different things out of love and intimate relationships. Here we shall review what social psychologists know about such gender differences, focusing on gender differences in four areas: (1) concern with love; (2) concern with sex; (3) desire for intimacy; and (4) desire for control.

CONCERN WITH LOVE

According to folklore, women are more concerned than men with love. Theorists seem to agree that women are intensely concerned with love, whereas men experience more muted feelings (see, e.g., Langhorn & Secord, 1955; Parsons, 1959; Parsons & Bales, 1955).

What do the facts say? The data suggest that the facts are more complicated than one might expect: Men and women seem to differ in how much they love, but the direction of the difference seems to depend on how love is conceptualized and measured.

Hatkoff and Lasswell (1979) argue that men and women differ in the way they conceptualize love. They interviewed over 500 males and females from several different ethnic and age groups. They concluded that men were more likely to experience romantic love *(eros)* and self-centered love *(ludis)*. Women, on the other hand, were more likely to experience obsessive, dependent love *(mania)*, companionate love *(storge)*, and logical, sensible love *(pragma)*.

Other studies have focused on gender differences in the experience of one specific type of love. Romantic love has frequently been examined. The finding of Hatkoff and Lasswell (1979) that men are more romantic than women has also been verified in several other studies. For example, Hobart (1958) asked 923 men and women to complete a romanticism scale. The scale included such items as: "Lovers ought to expect a certain amount of disillusionment after marriage" (scored, of course, in a negative direction) and "To be truly in love is to be in love forever." Hobart found that men had a considerably more romantic view of heterosexual relationships than did women. More recently, social psychologists have replicated Hobart's work and found much the same thing—men are still more romantic than women (K. L. Dion & Dion, 1973, 1979; Knox & Sporakowski, 1968; Rubin, 1973). Most of these studies, however, dealt with young dating couples. In one study of married couples (Reedy, Birren, & Schaie, 1976), it was found that married women were somewhat more romantic than married men.

A romanticism scale is a way of measuring a set of ideological beliefs. Other researchers have asked people directly about their romantic love experiences—how soon in relationships they fall in love, how frequently they fall in love, and so on. For example, Kanin, Davidson, & Scheck (1970) interviewed 700 young lovers. "How early," they asked, "did you become aware that you loved the other?" Whereas 20% of the men fell in love before the fourth date, only 15% of the women fell in love that early. At the other extreme, only 30% of the men, but a full 43% of the women, were not sure if they were in love by the 20th date.

Men seemed willing to fall headlong into love; women seemed far more cautious about getting involved. In another study, Hill *et al.*, (1976) asked dating partners to rate how important various factors had been as reasons for entering the dating relationship. Men were more likely than women to rate "desire to fall in love" as important. Kephart (1967) asked 1,000 college students the following question: "If a man (woman) had all other qualities you desired, would you marry this person if you were not in love with him (her)?" A full 65% of the men said no, but 24% of the women said no. However, in a more recent study (Adler & Carey, 1980), 86% of the men and 80% of the women said they would not marry without love. Final evidence that men may be more romantic than women comes from a computer dance study by Coombs and Kendell (1966). It was found that women were less impressed than men with the date to whom they had been randomly matched. Fifty-two percent of the women and 38% of the men reported absolutely no romantic attraction for their computer-matched date. Conversely, 19% of the men, but only 7% of the women, reported strong romantic attraction.

Other researchers have focused on interviewing men and women about the intensity of love feelings they actually experience in their relationships. This research has focused more on mania, or passionate love. Contrary to the evidence presented here that suggests that men may be more romantic than women (at least before marriage), researchers have found that women experience the euphoria and agony of love and romance more intensely than do men. For example, Kanin *et al.* (1970) asked men and women to describe how they felt when they were in love—for example, to what extent they experienced such love reactions as "felt like I was floating on a cloud," "had trouble concentrating," and "felt giddy and carefree." In this study, it was found that women appeared to be more passionate. Women experienced the symptoms of passionate love with greater intensity than did men. Similar results were also found by K. L. Dion and Dion (1973). But although women seem to experience more of the "symptoms" of passionate love, both men and women seem equally likely to say that they "passionately love" their partner (Sprecher, 1980; Traupmann & Hatfield, 1981).

Many researchers have measured romantic love via the Rubin Love Scale. In the original study by Rubin (1973), no differences were found between men and women in how much romantic love was expressed. Other studies have also found no differences between men and women on romantic love measured on Rubin's scale (Black & Angelis, 1974; Cunningham & Antill, 1980; Sprecher, 1980).

Finally, how much companionate love is experienced in the relationship has been examined. In studies by Sprecher (1980) and

Traupmann, Hatfield & Wexler, (1983), it has been found that women companionately love more than men.

In summary, there seems to be no simple answer to the question whether men or women are more loving. Men tend to have a more romantic view of love; women seem to experience the euphoria and agony of love more intensely. Both men and women seem to report equal amounts of passionate love, but women seem to love more companionately.

CONCERN WITH SEX

According to several theoretical perspectives, men and women differ in their enthusiasm for sex—with men being the more enthusiastic gender. According to the sociobiological perspective (which argues that sexual activity is determined primarily by genes, hormones, and anatomy), men are genetically programmed to seek out sexual activity and women to set limits on it. According to a cultural-contingency perspective (which argues that sexual behavior is learned), because it is a man's world, men have been encouraged to express themselves sexually whereas women have been punished for doing so. But regardless of theorists' debates as to *why* men and women may differ in their enthusiasm for sex, there is general agreement that they do. But what do the data say?

In the earliest sex research, scientists found fairly sizable gender differences. In more recent research, investigators find that although some gender differences still exist, they are no longer very strong. Gender differences have begun to narrow or disappear.

Gender Differences in Responsiveness to Erotica

Early research supported the traditional assumption that it is men who are most interested in erotica. Kinsey (Kinsey, Pomeroy, & Martin, 1948; Kinsey, Pomeroy, Martin, & Gebhard, 1953) found that the women in his sample were unlikely even to have been exposed to erotica. Even when both sexes were familiar with such literature, men reported being more aroused by it than women. For example, 47% of the men had been aroused by erotic stories; only 14% of the women reported such arousal.

Recently, however, researchers have begun both to ask men and women about their feelings and to get objective measures of their physiological arousal in response to pornography. In such studies, researchers generally find that there are few, if any, gender differences in responsiveness. For example, Veitch and Griffitt (1980) found no gender differences in response to literary erotica. In fact, some data suggest that explicit portrayals of sexual activity may evoke equal or greater erotic responsiveness in women than in men. Heiman (1977) observed that both men and women found audiotapes of exclusively "romantic" encounters less arousing than audiotapes describing either romantic-erotic or exclusively erotic sexual encounters. Women actually rated the explicitly erotic audiotapes as more arousing than the men did. Heiman found no sex differences on physiological measures of sexual arousal.

Eagerness to Initiate Sexual Activity

In Kinsey's day, a double standard existed. Men were allowed, if not encouraged, to get sex whenever and wherever they could. Women were supposed to save themselves for marriage. In light of such double standards, it was not surprising that men were more likely to initiate sex, and women to resist sexual advances (see Baker, 1974; Ehrmann, 1959; Kaats & Davis, 1970; Reiss, 1967; Schofield, 1965; Sorenson, 1973).

Recent evidence suggests that traditional standards, although changing, are not yet dead. Contemporary college students reject a sexual double standard (Hopkins, 1977; Komarovsky, 1976; Peplau, Rubin, & Hill, 1976). Yet the existence of a new single standard does not seem to have changed things much—it is still men who are the sexual initiators, women who are the limit setters (McCormick, 1979). In a recent study, Peplau, Rubin & Hill (1977) found that, among unmarried students, the man has more to say than the woman about the type and frequency of sexual activity—except when the dating couple has decided to abstain from sexual intercourse, in which case the woman's veto is the major restraining influence.

Gender Differences in Sexual Experience

There is, however, compelling evidence that men and women are becoming very similar with regard to sexual experience.

In the classic studies of sexuality, researchers found that society's double standard influences sexual experience. For example, Kinsey and his colleagues (Kinsey *et al.*, 1948; Kinsey *et al.*, 1953) tried to assess comparative sexual activity of men and women throughout their lives. They found that (1) indeed, men did seem to engage in more sexual activity than did women, and (2) men and women had strikingly different sexual histories. At 18, it was usually the man who pushed to have sex. Kinsey and his associates found that most men were as sexually expressive at 15 as they would ever be. In fact, according to Masters and Johnson (1966, 1970), 25% of men are impotent by age 65, 50% by age 75.

Women's experience was markedly different. Most women were slow to begin sexual activity. At 15, most women were quite inactive. Sometime between the ages of 16 and 20, they slowly shed their inhibitions and began to feel more enthusiastic about sexual exploration. They continued their high rates of sexual activity for fully two decades. It was not until their late forties that sexual behavior began to ebb.

> In commenting on women's sexual histories, Kinsey *et al.*, (1953) observed: One of the tragedies which appears in a number of the marriages originates in the fact that the male may be most desirous of sexual contact in his early years, while the responses of the female are still underdeveloped and while she is still struggling to free herself from the acquired inhibitions which prevent her from participating freely in the marital activity. But over the years most females become less inhibited and develop an interest in sexual relations, which they may then maintain until they are in their fifties or even sixties. But by then the responses of the average male may have dropped so considerably that his interest in coitus, and especially in coitus with a wife who has previously objected to the frequencies of his requests, may have sharply declined. (pp. 353–354)

Does this still hold true? Since Kinsey's day, researchers (DeLamater & MacCorquodale, 1979; Ehrmann, 1959; Reiss, 1967; Schofield, 1965; Sorenson, 1973) have continued to interview samples of young people about their sexual behavior: Had they ever necked? At what age did they begin? Had they fondled their lover's breasts or genitals or had their own genitals fondled? Had intercourse? Oral-genital sex? When we compare the data from these studies, we find that a sexual revolution is indeed occurring. In the early studies, in general, men were far more experienced than women. By the 1980s, these differences have virtually disappeared. As DeLamater and MacCorquodale (1979) observe,

> There are virtually no differences in the incidence of each of the behaviors. Unlike most earlier studies which generally reported lower frequencies of more intimate activities among females, we find that women are as likely as men to have ever engaged in these behaviors. The only exception occurs with coitus, which women . . . are less likely to have experienced. *(Among students,* 75% of men and 60% of women had had intercourse. *Among non-students,* 79% of men and 72% of women had had intercourse.) . . . Thus, the gender differences in lifetime behavior which were consistently found in studies conducted in the 1950s and 1960s have narrowed considerably. This is also an important finding; it suggests that those models which have emphasized gender as an explanatory variable are no longer valid. (p. 58)

When men and women are together in a close loving relationship, they seem equally likely to desire to engage in sexual activity. There is only one type of situation in which scientists find women are still more reserved than men—if men and women are offered a chance to participate in uncertain, unconventional, or bizarre sexual activities, men will be more willing to take the risk than will women (Clark & Hatfield, in press).

Other scientists have documented that even today men are more eager to have sex with a variety of partners, in a variety of ways, and so on. For example, Hatfield *et al.* (1981) interviewed casually dating and newlywed couples about their sexual preferences. They assessed desire for variety via such questions as: (1) "I wish my partner were much more unpredictable about *when* he/she wants to have sex." (2) "I wish my partner were much more variable about *where* we have sex."

The authors predicted that men would be more interested in exciting, diverse experiences than would women; this was exactly what they found. It was the men who wished their sex lives were a little more exciting, whereas women tended to be slightly more satisfied with the status quo.

In summary, then, recent evidence suggests that although some gender differences remain in men's and women's concern with sex, a sexual revolution is occurring. Gender differences in responsiveness to pornography, willingness to initiate sex, and sexual experience are rapidly disappearing. Recent studies indicate that women and men are becoming increasingly similar in their sexual preferences and experiences.

DESIRE FOR INTIMACY

We conceptualize intimacy not as a static state but as a process. Thus, we would define intimacy as a process by which a couple—in the expression of thought, emotion, and behavior—attempts to move toward more complex union.

Theorists from a variety of disciplines have agreed with Klimek (cited in Hatfield *et al.*, 1981) that "nearly any observer of family life and of the human condition in general has probably suspected—or has at least amassed subjective evidence that—women generally have a higher capacity for intimacy than men" (p. 248).

According to many clinicians, one of the major tasks facing people is to achieve a separate identity while at the same time achieving a deeply intimate relationship with others (Erikson, 1968; Kantor & Lehr, 1975; L. J. Kaplan, 1978). Both separateness and intimacy are generally considered to be basic human needs (see Freud, 1922; Maslow, 1954). Kaplan suggests that adults spend much of their lives resolving the dilemma between achieving a sense of self while at the same time establishing close, nurturant relations with others.

According to family therapists, men have the easiest time achieving an independent identity; women have the easiest time achieving closeness with others. Napier (1977) describes two types of people (Type 1 and Type 2) who seem, with uncanny accuracy, to attract one another. Type 1 (usually a woman) is only minimally concerned with maintaining her independence. What she cares about is achieving emotional closeness. She seeks "fusion with the partner," "oneness," "we-ness" in the marriage; she invests much energy into planning "togetherness" activities. What Type 1 fears is rejection and abandonment.

Type 1's partner, Type 2 (usually a man) is most concerned with maintaining his sense of self and personal freedom and autonomy. He feels a strong need to establish his territory within the common household—to have "his study," "his workshop," "his car." What he fears is being "suffocated," "stifled," or "engulfed," or in some manner intruded on by the spouse.

Napier observes that men's and women's efforts to reduce their anxieties make matters worse. Women—seeking more closeness—clasp tightly at their mates, thereby contributing to the men's anxiety. The men—seeking more distance—retreat further, which increases their wives' panic, inducing further clasping.

Theorists agree that women are far "better" at intimacy than men. Family therapists take it for granted that women are very comfortable with intimate relationships and men are not, and that this is a common cause of marital friction. There are literally dozens of books exhorting men to share their feelings. Considering these facts, it is startling that there has been so little research devoted to gender differences in intimacy.

Worse yet, it is difficult to draw any conclusions from the research that does exist. If we were forced to make a guess as to what future research will reveal, it would be as follows: Women's complaint that men refuse to share their deepest feelings is a legitimate one. In general, women are indeed more comfortable with intimacy than men. But, paradoxically, even though women complain about men's lack of intimacy in their love relationships, it is in a love affair that male–female differences are smallest. Women find it fairly easy to be intimate with their lovers, with men friends, with other women, and with children. For many men, it is only with their lovers that they can be intimate. It is here that they reveal most of themselves—not as much as their lovers might like, but far more than they share with anyone else. It is most difficult for men to be close to other men.

These overgeneralizations are based on sparse data. A few social psychologists have explored male–female differences in willingness to get close to others. Generally, such researchers have defined intimacy as a willingness to disclose one's ideas, feelings, and day-to-day activities to lovers, friends, or strangers . . . and to listen to their disclosures in return. Jourard (1964) developed one of the most commonly used measures of intimacy, the Jourard Self-Disclosure Questionnaire. Consisting of 60 questions in all, the JSDQ asks people to think about how much they typically disclose to others in six different areas of life: attitudes and opinions about sex, religion, and politics; tastes and interests; concerns about work (or studies); concerns about money; revelations about their personalities; and their feelings about their bodies.

In self-disclosure research, three findings on differences between men and women have emerged.

1. In their deeply intimate relationships, men and women often differ little, if at all, in how much they are willing to reveal to one another. For example, Rubin *et al.* (1980) asked dating couples via the JSDQ how much they had revealed themselves to their partners. The authors found that it was a small minority of traditional men and women who differed on emotional sharing. More egalitarian couples were more likely to disclose themselves fully to one another. Overall, men and women did not differ in how much they were willing to confide in their partners.

There was a difference, however, in the sorts of things men and women were willing to share with each other. Men were more willing to share their views on politics and their pride in their strengths; women to disclose their feelings about other people and their fears. Interestingly enough, Rubin *et al.* found that the stereotyped form of communications is not common with traditional couples.

Some authors have observed that neither men nor women may be getting exactly the amount of intimacy they would like. Women may want more intimacy than they are getting; men may want far less (there is evidence that couples do tend to negotiate a level of self-disclosure that is bearable to both—ensuring, in the words of *My Fair Lady*, that "neither really gets what either really wants at all"; (see Derlega & Chaiken, 1975).

2. In less intimate relationships, women disclose far more to others than do men (see Cozby, 1973; Jourard, 1971). As Rubin *et al.* (1980) point out:

> The basis for such differences appears to be in socialization practices. Whereas women in our culture have traditionally been encouraged to show their feelings, men have been taught to hide their feelings and to avoid displays of weakness.

Kate Millett (1975) puts it simply: "Women express, men repress" (p. 306).

3. Finally, women receive more disclosures than men. This is not surprising, in view of the fact that the amount of information people

reveal to others has a great impact on the amount of information they receive in return (see Altman, 1973; Davis & Skinner, 1974; Jourard, 1964; Jourard & Friedman, 1970; Marlatt, 1971; Rubin, 1975; Worthy, Gary, & Kahn, 1969).

There does seem to be evidence, then, that women feel slightly more comfortable with intense intimacy in their love relationships than men, and are far more comfortable revealing themselves in more casual relationships than men are.

Tradition dictates that women should be the "intimacy experts." Today, women are indeed more comfortable sharing their ideas, feelings, and behavior than men. But what would happen if this were to change? (Rubin *et al.*, 1980, suggest that such changes have already begun.)

The prognosis is mixed. Young women usually say they would be delighted if the men they love could be intimate. As a family therapist, one is titled to skepticism that it will be so easy. Change is always difficult. We have seen more than one man complain that when he finally dared to reveal his weaker aspects to a woman, he soon discovered that she was shocked by his lack of "manliness." Family therapists such as Napier have warned that the struggle to find both individuality and closeness is a problem for everyone. As long as men were fleeing from intimacy, women could safely pursue them. Now that men are turning around to catch them, women may well find themselves taking flight.

The change, on the whole, should be a healthy one. As Rubin *et al.* (1980) observe:

> Men and women should have the freedom to decide for themselves when they will reveal themselves—and when they will listen to another's revelations. "Full disclosure" need not be so full that it eliminates all areas of privacy, even within the most intimate relationships. . . . Especially when contemplating marriage, it is valuable for women and men to be able to share rather fully—and equally—their thoughts and feelings about themselves, each other, and their relationship. . . . It is encouraging to discover that a large majority of the college students we studied seem to have moved, even if incompletely and sometimes uneasily, toward the ethic of openness. (p. 316)

DESIRE FOR CONTROL

Traditionally, men are supposed to be in control—of themselves, other people, and the environment. The "ideal" man carefully controls his thoughts. He is "objective," "logical," unemotional. He hides his feelings, or if he does express any, he carefully telescopes the complex array of human feelings into a single emotion—anger. Men are supposed to be "dominant"; women "submissive."

There is considerable evidence that most men and women hold these stereotypes. Broverman, Vogel, Broverman, Clarkson, and Rosenkrantz (1972), in a study of sex-role stereotypes, found that women are perceived as expressive and nurturant, whereas men are perceived as in control and instrumental.

The purpose of all this control seems to be achievement. It may sound strange that some people may view intimate relations—the one area where people can be themselves, fully relaxed, confident that they will be accepted no matter what—as yet another arena for achievement. Yet apparently, some people do.

In reviewing the male sexual myths, Zilbergeld (1978) observes that even in their most intimate relationships men are more goal oriented than women:

> As boys and men we socialized in . . . the three A's of manhood: Achieve, Achieve, Achieve. . . . Give us a job with a goal, some job specificaitons, and perhaps a time limit, and we are in business. . . . It is understandable that we should bring this performance orientation to sex. . . . How else could we, given our training, handle such an anxiety-laden experience. . . . We make work of sex. It becomes businesslike and mechanical, another job to be done, another goal to be achieved. Rather than seeing sex as a way for two people to relate and have fun, and asking how much pleasure and closeness there was, we view it as a performance and ask how hard the erection was, how long we lasted, and how many orgasms she'd had. The goals—usually intercourse and orgasm—are the only important factors. (pp. 35–37)

According to theorists, then, there are marked gender differences in concern about being in control, dominating others, and achieving at love. As yet, however, there is no evidence for these speculations.

REFERENCES

Adams, G. R., & Shea, J. A. Talking and loving: A cross-lagged panel investigation. *Basic and Applied Social Psychology*, 1981, 2, 81–88.

Adler, J., & Carey, J. The science of love. *Newsweek*, 1980, *95*, 49–50.

Altman, I. Reciprocity of interpersonal exchange. *Journal of the Theory of Social Behavior*, 1973, 3, 249–261.

Altman, I., & Taylor, D. A. *Social penetration: The development of interpersonal relationships*. New York: Holt, Rinehart & Winston, 1973.

Aristotle. *Nicomachean ethics* (Martin Ostwald, Trans.). New York: Bobbs-Merrill, 1962.

Baker, M. J. *The effects of inequity on heterosexual behavior: A test for compensation in inequitable dating relationships*. Unpublished master's thesis, University of Wisconsin, 1974.

Barrett, C. J., & Becker, R. M. *The prediction of adjustment to widowhood from social and demographic data*. Paper presented at the meeting of the Western Social Science Association, Denver, April 1978.

Barry, W. A. Marriage research and conflict: An integrative review. *Psychological Bulletin*, 1970, *73*, 41–54.

Bart, P. The loneliness of the long-distance mother. In J. Freeman (Ed.), *Women: A feminist perspective* (2nd ed.). Palo Alto, Calif.: Mayfield, 1979.

Berscheid, E., & Fei, J. Romantic love and sexual jealousy. In G. Clanton & L. G. Smith (Eds.), *Jealousy*. Englewood Cliffs, N.J.: Prentice-Hall, 1977.

Berscheid, E., & Walster (Hatfield), E. *Interpersonal attraction* (2nd ed.). Reading, Mass.: Addison-Wesley, 1978.

Berscheid, E., Dion, K., Walster (Hatfield), E., & Walster, G. W. Physical attractiveness and dating choice: A test of the matching hypothesis. *Journal of Experimental Social Psychology*, 1971, *7*, 173–189.

Black, H., & Angelis, V. B. Interpersonal attraction: An empirical investigation of platonic and romantic love. *Psychological Reports*, 1974, *34*, 1243–1246.

Blood, R. O., & Wolfe, D. M. *Husbands and wives: The dynamics of married living*. Glencoe, Ill.: Free Press, 1960.

Bloom, B. L., Asher, S. J., & White, S. W. Marital disruption as a stressor: A review and analysis. *Psychological Bulletin*, 1978, *85*, 867–894.

Bossard, J. H. S. Residential propinquity as a factor in mate selection. *American Journal of Sociology*, 1932, *38*, 219–224.

Bragg, M. E. *A comparative study of loneliness and depression*. Unpublished doctoral dissertation, University of California, Los Angeles, 1979.

Braiker, H. B., & Kelley, H. H. Conflict in the development of close relationships. In R. L. Burgess & R. L. Huston (Eds.), *Social exchange in developing relationships*. New York: Academic Press, 1979.

Brennan, T., & Auslander, N. *Adolescent loneliness: An exploratory study of social and psychological predisposition and theory*. Boulder, Col.: Behavioral Research Institute, 1979.

Brislin, R. W., & Lewis, S. A. Dating and physical attractiveness: A replication. *Psychological Reports*, 1968, *22*, 976.

Broverman, I. K., Vogel, S. R., Broverman, D. M., Clarkson, F. E., & Rosenkrantz, P. S. Sex role stereotypes: A current appraisal. *Journal of Social Issues*, 1972, *28*, 59–78.

Burgess, E. W., & Wallin, P. Homogamy in social characteristics. *American Journal of Sociology*, 1943, *49*, 109–124.

Byrne, D. The influence of propinquity and opportunities for interaction on classroom relationships. *Human Relations*, 1961, *14*, 63–70.

Byrne, D. *The attraction paradigm*. New York: Academic Press, 1971.

Byrne, D., & Blaylock, B. Similarity and assumed similarity of attitudes between husbands and wives. *Journal of Abnormal and Social Psychology*, 1963, *67*, 636–640.

Byrne, D., & Byrne, A. *Exploring human sexuality*. New York: Thomas Y. Crowell, 1977.

Chelune, G. J. (Ed.). *Self-disclosure*. San Francisco: Jossey-Bass, 1979.

Clark, R. D., III, & Hatfield, E. *Gender differences in receptivity to sexual offers*. (Unpublished manuscript, 1981.) Cited in E. R. Allgeier & N. B. McCormick (Eds.), *Changing boundaries: Gender roles and sexual behavior*. Palo Alto, Calif.: Mayfield, 1982, pp. 106–132.

Clarke, A. C. An examination of the operation of residential propinquity as a factor in mate selection. *American Sociological Review*, 1952, *17*, 17–22.

Coombs, R. H., & Kendell, W. F. Sex differences in dating aspirations and satisfaction with computer-selected partners. *Journal of Marriage and the Family*, 1966, *28*, 62–66.

Cozby, P. C. Self-disclosure: A literature review. *Psychological Bulletin*, 1973, *79*, 73–91.

Cunningham, J. D., & Antill, J. K. Love in developing romantic relationships. In S. Duck & R. Gilmour (Eds.), *Personal relationships* (Vol. 2). New York: Academic Press, 1981.

Cutler, B. R., & Dyer, W. G. Initial adjustment processes in young married couples. *Social Forces*, 1965, *44*, 195–201.

Cutrona, C. E. Transition to college: Loneliness and the process of social adjustment. In L. A. Peplau & D. Perlman (Eds.), *Loneliness: A sourcebook of current theory, research, and therapy*. New York: Wiley, 1982.

Davis, J. D., & Skinner, A. E. Reciprocity of self-disclosure in interviews: Modeling of social exchange. *Journal of Personality and Social Psychology*, 1974, *29*, 779–784.

DeLamater, J., & MacCorquodale, P. *Premarital sexuality: Attitudes, relationships, behavior*. Madison: University of Wisconsin Press, 1979.

Derlega, V. J., & Chaikin, A. L. *Sharing intimacy: What we reveal to others and why*. Englewood Cliffs, N.J.: Prentice-Hall, 1975.

Derlega, V. J., Wilson, M. D., & Chaikin, A. L. Friendship and disclosure reciprocity. *Journal of Personality and Social Psychology*, 1976, *34*, 578–582.

Dickens, W. J., & Perlman, D. Friendships over the life cycle. In S. W. Duck & R. Gilmour (Eds.), *Personal relationships: Vol. 2. Developing personal relationships*. New York: Academic Press, 1981.

Dion, K. K. Physical attractiveness, sex roles, and heterosexual attraction. In S. Duck & R. Gilmour (Eds.), *Personal relationships* (Vol. 4). New York: Academic Press, 1982.

Dion, K. K., Berscheid, E., & Walster, E. What is beautiful is good. *Journal of Personality and Social Psychology*, 1972, *24*, 285–290.

Dion, K. L., & Dion, K. K. Correlates of romantic love. *Journal of Consulting and Clinical Psychology*, 1973, *41*, 51–56.

Dion, K. L., & Dion, K. K. Personality and behavioral correlates of romantic love. In M. Cook & G. Wilson (Eds.), *Love and attraction*. New York: Pergamon Press, 1979.

Doster, J. A., & Nesbitt, J. G. Psychotherapy and self-disclosure. In G. J. Chelune (Ed.), *Self-disclosure*. San Francisco: Jossey-Bass, 1979.

Duck, S. W. *Personal relationships: Vol. 4. Dissolving personal relationships*. New York: Academic Press, 1982.

Duck, S., & Gilmour, R. *Personal relationships: Vol. 1. Studying personal relationships*. New York: Academic Press, 1981. (a)

Duck, S. W., & Gilmour, R. *Personal relationships: Vol. 2. Developing personal relationships*. New York: Academic Press, 1981. (b)

Duck, S. W., & Gilmour, R. *Personal relationships: Vol. 3. Personal relationships in disorder*. New York: Academic Press, 1981. (c)

Duck, S. W., Miell, D. K., & Gaebler, H. C. Attraction and communication in children's interactions. In H. C. Foot, A. J. Chapman, & J. R. Smith, *Friendships and social relations in children*. New York: Wiley, 1980.

Ehrmann, W. *Premarital dating behavior*. New York: Holt, Rinehart & Winston, 1959.

Erikson, E. H. *Childhood and society* (Rev. ed.). New York: W. W. Norton, 1968.

Ferreira, A. J., & Winter, W. D. On the nature of marital relationships: Measurable differences in spontaneous agreement. *Family Process*, 1974, *13*, 355–369.

Festinger, L. A theory of social comparison processes. *Human Relations*, 1954, *7*, 117–140.

Festinger, L., Schachter, S., & Back, K. *Social pressures in informal groups: A study of human factors in housing*. New York: Harper, 1950.

Fidler, J. Loneliness: The problems of the elderly and retired. *Royal Society of Health Journal*, 1976, *96*, 39–41, 44.

Fitz, D., & Gerstenzang, S. *Anger in everyday life: When, where and with whom?* St. Louis: University of Missouri, 1978. (ERIC Document Reproductive Service No. ED 160-966)

Foot, H. C., Chapman, A. J., & Smith, J. R. *Friendship and social relations in children.* New York: Wiley, 1980.

Foucault, M. *The order of things.* New York: Vintage Books, 1973.

Freud, S. Certain neurotic mechanisms in jealousy, paranoia, and homosexuality. In *Collected papers* (Vol. 2). London: Hogarth Press, 1922.

Gibson, I. *The English vice: Beating sex and shame in Victorian England and after.* London: Duckworth, 1978.

Gilbert, S. J. Empirical and theoretical extensions of self-disclosure. In G. R. Miller (Ed.), *Explorations in interpersonal communication* (Vol. 5, Sage Annual Review of Communication Research). Beverly Hills, Calif.: Sage Publications, 1976.

Goffman, E. *The presentation of self in everyday life.* Garden City, N.Y.: Anchor Books, 1959.

Gordon, S. *Lonely in America.* New York: Simon & Schuster, 1976.

Gullahorn, J. Distance and friendship as factors in the gross interaction matrix. *Sociometry,* 1952, *15,* 123–134.

Hatfield, E., & Walster, G. W. *A new look at love.* Reading, Mass.: Addison-Wesley, 1978.

Hatfield, E., Utne, M. K., & Traupmann, J. Equity theory and intimate relationships. In R. Burgess & T. L. Huston (Eds.), *Social exchange in developing relationships.* New York: Academic Press, 1979.

Hatfield, E., Walster, G. W., & Traupmann J. Equity and premarital sex. In M. Cook & G. Wilson (Eds.) *Love and attraction: An international conference.* Oxford: Pergamon Press, 1979. (Reprinted from *Journal of Personality and Social Psychology,* 1978, *37,* 81–92)

Hatfield, E., Traupmann, J., Sprecher, S., Greenberger, D., & Wexler, P. *Male/female differences in concern with intimacy, variety, and powers in the sexual relationship.* Unpublished manuscript, 1981.

Hatkoff, T. S., & Lasswell, T. E. Male-female similarities and differences in conceptualizing love. In M. Cook & G. Wilson (Eds.), *Love and attraction: An international conference.* Oxford: Pergamon Press, 1979.

Heider, F. *The psychology of interpersonal relations.* New York: Wiley, 1958.

Heiman, J. A psychophysiological exploration of sexual arousal patterns in females and males. *Psychophysiology,* 1977, *14,* 266–274.

Hendrick, S. S. Self-disclosure and marital satisfaction. *Journal of Personality and Social Psychology,* 1981, *40,* 1150–1159.

Hill, C. T. *The ending of successive opposite-sex relationships.* Unpublished doctoral dissertation, Harvard University, 1974.

Hill, C. T., Rubin, Z., & Peplau, A. Breakups before marriage: The end of 103 affairs. *Journal of Social Issues,* 1976, *32,* 147–167.

Hobart, C. W. The incidence of romanticism during courtship. *Social Forces,* 1958, *36,* 362–367.

Hoffeditz, E. L. Family resemblances in personality traits. *Journal of Social Psychology,* 1934, *5,* 214–227.

Hopkins, J. R. Sexual behavior in adolescence. *Journal of Social Issues,* 1977, *33,* 67–85.

Hunt, A. M. A study of the relative value of certain ideals. *Journal of Abnormal and Social Psychology,* 1935, *30,* 222–228.

Huston, T. L. Ambiguity of acceptance, social desirability and dating choice. *Journal of Experimental Social Psychology,* 1973, *9,* 32–42.

Jones, E. E., & Davis, K. E. From acts to dispositions: The attribution process in person perception. In L. Berkowitz (Ed.), *Advances in experimental social psychology* (Vol. 2). New York: Academic Press, 1965.

Jones, W. H., Hansson, R. O., & Smith, T. G. *Loneliness and love: Implications for psychological and interpersonal functioning*. Unpublished manuscript, University of Tulsa, 1980.

Jones, W. H., Freemon, J. A., & Goswick, R. A. The persistence of loneliness: Self and other determinants. *Journal of Personality*, 1981, *49*, 27–48.

Jorgensen, S. R., & Gaudy, J. C. Self-disclosure and satisfaction in marriage: The relation examined. *Family Relations*, 1980, *29*, 281–287.

Jourard, S. M. *The transparent self*. Princeton, N.J.: Van Nostrand, 1964.

Jourard, S. M. *Self-disclosure: An experimental analysis of the transparent self*. New York: Wiley, 1971.

Jourard, S. M., & Friedman, R. Experimenter-subject distance in self-disclosure. *Journal of Personality and Social Psychology*, 1970, *15*, 278–282.

Kaats, G. R., & Davis, K. E. The dynamics of sexual behavior of college students. *Journal of Marriage and the Family*, 1970, *32* (3), 390–399.

Kandel, D. B. Similarity in real-life adolescent friendship pairs. *Journal of Personality and Social Psychology*, 1978, *36*, 306–312.

Kanin, E. J., Davidson, K. D., & Scheck, S. R. A research note on male/female differentials in the experience of heterosexual love. *Journal of Sex Research*, 1970, *6*, 64–72.

Kantor, D., & Lehr, W. *Inside the family*. San Francisco: Jossey-Bass, 1975.

Kaplan, L. J. *Oneness and separateness: From infant to individual*. New York: Simon & Schuster, 1978.

Kaplan, M. F., & Anderson, N. H. Information integration theory and reinforcement theory as approaches to interpersonal attraction. *Journal of Personality and Social Psychology*, 1973, *28*, 301–312.

Kelly, H. H. Attribution theory in social psychology. In D. Levine (Ed.), *Nebraska Symposium on Motivation* (Vol. 15). Lincoln: University of Nebraska Press, 1967.

Kelley, H. H. An application of attribution theory to research methodology for close relationships. In G. Levinger & H. L. Raush (Eds.), *Close relationships: Perspectives on the meaning of intimacy*. Amherst, Mass.: University of Massachusetts Press, 1977.

Kelley, H. H. *Personal relationships: Their structures and processes*. Hillsdale, N.J.: Lawrence Erlbaum, 1979.

Kelley, H. H., Berscheid, E., Christensen, A., Harvey, J., Huston, T., Levinger, G., McClintock, E., Peplau, A., & Peterson, D. *The psychology of close relationships*. San Francisco: W. H. Freeman, 1983.

Kelly, E. L. Consistency of the adult personality. *American Psychologist*, 1955, *10*, 659–681.

Kephart, W. M. *The family, society, and the individual*. Boston: Houghton Mifflin, 1961.

Kephart, W. M. Some correlates of romantic love. *Journal of Marriage and the Family*, 1967, *29*, 470.

Kerckhoff, A. C., & Davis, K. E. Value consensus and need complementarity in mate selection. *American Sociological Review*, 1962, *27*, 295–303.

Kinsey, A. C., Pomeroy, W. B., & Martin, C. E. *Sexual behavior in the human male*. Philadelphia: W. B. Saunders, 1948.

Kinsey, A. C., Pomeroy, W. B., Martin, C. E., & Gebhard, P. H. *Sexual behavior in the human female*. New York: Pocket Books, 1953.

Kipnis, D. M. Interaction between members of bomber crews as a determinant of sociometric choice. *Human Relations*, 1957, *10*, 263–270.

Kleinke, C. L. Effects of personal evaluations. In C. J. Chelune (Ed.), *Self-disclosure*. San Francisco: Jossey-Bass, 1979.

Knox, D. H., & Sporakowski, M. J. Attitudes of college students toward love. *Journal of Marriage and the Family*, 1968, *30*, 638–642.
Knox, D. H., & Wilson, K. Dating behaviors of university students. *Family Relations*, 1981, *30*, 255–258.
Komarovsky, M. *Blue-collar marriage*. New York: Random House, 1962.
Komarovsky, M. *Dilemmas of masculinity*. New York: W. W. Norton, 1976.
Krafft-Ebing, R. Von. *Psychopathia sexualis*. New York: Pioneer, 1953.
Kreitman, N., Collins, J., Nelson, B., & Troop, J. Neurosis and marital interaction. I Personality and symptoms. *British Journal of Psychiatry*, 1970, *117*, 33–46.
Langhorn, M. C., Secord, P. Variations in marital needs with age, sex, marital status, and regional locations. *Journal of Social Psychology*, 1955, *41*, 19–37.
Lee, J. A. *The colors of love*. New York: Bantam Books, 1977.
Levinger, G. A social exchange view on the dissolution of pair relationships. In R. L. Burgess & T. L. Huston (Eds.), *Social exchange in developing relationships*. New York: Academic Press, 1979.
Levinger, G., & Breedlove, J. Interpersonal attraction and agreement: A study of marriage partners. *Journal of Personality and Social Psychology*, 1966, *3*, 367–372.
Levinger, G., & Senn, D. J. Disclosure of feelings in marriage. *Merrill-Palmer Quarterly*, 1967, *13*, 237–249.
Lopata, H. Z. Loneliness: Forms and components. *Social Problems*, 1969, *17*, 248–261.
Marlatt, G. A. Exposure to a model and task ambiguity as determinants of verbal behavior in an interview. *Journal of Consulting and Clinical Psychology*, 1971, *36*, 268–276.
Marwell, G., McKinney, K., Sprecher, S., Smith, S., & DeLamater, J. *Legitimizing factors in the initiation of heterosexual relationships*. Paper presented at the International Conference on Personal Relationships, Madison, Wisconsin, July 1982.
Maslow, A. H. *Motivation and personality*. New York: Harper, 1954.
Master, W. H., & Johnson, V. E. *Human sexual response*. Boston: Little, Brown, 1966.
Masters, W. H., & Johnson, V. E. *Human sexual inadequacy*. Boston: Little, Brown, 1970.
McCormick, N. B. Come-ons and put-offs: Unmarried students' strategies for having and avoiding sexual intercourse. *Psychology of Women Quarterly*, 1979, *4*, 194–211.
Mead, M. *Sex and temperament in three primitive societies*. New York: Dell Books, 1969.
Mettee, D. R., & Aronson, E. Affective reactions to appraisal from others. In T. L. Huston (Ed.), *Foundations of interpersonal attraction*. New York: Academic Press, 1974.
Michela, J. L., Peplau, L. A., & Weeks, D. G. *Perceived dimensions and consequences of attributions for loneliness*. Unpublished manuscript, Columbia University, 1981.
Miller, A. Role of physical attractiveness in impression formation. *Psychonomic Science*, 1970, *19*, 241–243.
Millett, K. The shame is over. *Ms. Magazine*, January 1975, pp. 26–29.
Monge, P. R., & Kirste, K. K. Measuring proximity in human organization. *Social Psychology Quarterly*, 1980, *43*, 110–115.
Morton, T. L. Intimacy and reciprocity of exchange: A comparison of spouses and strangers. *Journal of Personality and Social Psychology*, 1978, *36*, 72–78.
Murstein, B. I. Stimulus-value-role: A theory of marital choice. *Journal of Marriage and the Family*, 1970, *32*, 465–481.
Murstein, B. I. Physical attractiveness and marital choice. *Journal of Personality and Social Psychology*, 1972, *22*(1), 8–12.
Napier, A. Y. *The rejection-intrusion pattern: A central family dynamic*. Unpublished manuscript, School of Family Resources, University of Wisconsin, Madison, 1977.
Nelson, B., Collins, J., Kreitman, N., & Troop, J. Neurosis and marital interaction. II Time sharing and social activity. *British Journal of Psychiatry*, 1970, *117*, 47–58.

Newcomb, T. M. *The acquaintance process*. New York: Holt, Rinehart & Winston, 1961.

Newcomb, T., & Svehla, G. Intra-family relationship in attitude. *Sociometry*, 1937, *1*, 180–205.

Orvis, B. R., Kelley, H. H., & Butler, D. Attributional conflict in young couples. In J. Harvey, W. Ickes, & R. Kidd (Eds.), *New directions in attribution research* (Vol. 1). New York: Wiley, 1976.

Paloutzian, R. F., & Ellison, C. W. Loneliness, spiritual well-being and the quality of life. In L. A. Peplau & D. Perlman (Eds.), *Loneliness: A sourcebook of current theory, research, and therapy*. New York: Wiley, 1982.

Parlee, M. B. The friendship bond: PT's survey report on friendship in America. *Psychology Today*, October 1979, pp. 42–54, 113.

Parsons, T. The social structure of the family. In R. N. Anshen (Ed.), *The family: Its function and density*. New York: Harper, 1959.

Parsons, T., & Bales, R. F. *Family, socialization, and interaction process*. New York: Free Press, 1955.

Peplau, L. A., Rubin, Z., & Hill, C. The sexual balance of power. *Psychology Today*, November 1976, pp. 142–147, 151.

Peplau, L. A., Rubin, Z., & Hill, C. Sexual intimacy in dating relationships. *Journal of Social Issues*, 1977, *33*, 86–109.

Peplau, L. A., Russell, D., & Heim, M. The experience of loneliness. In I. H. Frieze, D. Bar-Tal, & J. S. Carroll (Eds.), *New approaches to social problems: Applications of attribution theory*. San Francisco: Jossey-Bass, 1979.

Perlman, D., & Peplau, L. A. Toward a social psychology of loneliness. In S. Duck & R. Gilmour (Eds.), *Personal relationships* (Vol. 3). New York: Academic Press, 1981.

Perlman, D., Gerson, A. C., & Spinner, B. Loneliness among senior citizens: An empirical report. *Essence*, 1978, *2*, 239–248.

Perper, T., & Fox, V. S. *Flirtation behavior in public settings*. Final report to the Harry Frank Guggenheim Foundation, 1981.

Pleck, J. H., & Sawyer, J. (Eds.). *Men and masculinity*. Englewood Cliffs, N.J.: Prentice-Hall, 1974.

Pope, K. S. Defining and studying romantic love. In K. S. Pope & Associates (Eds.), *On love and liking*. San Francisco: Jossey-Bass, 1980.

Reedy, M. N., Birren, J. E., & Schaie, K. W. *Love in adulthood: Beliefs versus experience*. Paper presented at the meeting of the American Psychological Assocation, 1976.

Reiss, I. L. *The social context of premarital sexual permissiveness*. New York: Holt, Rinehart & Winston, 1967.

Ross, L. The intuitive psychologist and his shortcomings: Distortions in the attribution process. In L. Berkowitz (Ed.), *Advances in experimental social psychology* (Vol. 10). New York: Academic Press, 1977.

Rubenstein, C. *A questionnaire study of adult loneliness in three U.S. cities*. Unpublished doctoral dissertation, New York University, 1979.

Rubenstein, C., & Shaver, P. Loneliness in two northeastern cities. In J. Hartog, J. R. Andy, & Y. A. Cohen (Eds.), *The anatomy of loneliness*. New York: International Universities Press, 1980.

Rubenstein, C., & Shaver, P. The experience of loneliness. In L. A. Peplau & D. Perlman (Eds.), *Loneliness: A sourcebook of current theory, research, and therapy*. New York: Wiley, 1982.

Rubenstein, C., Shaver, P., & Peplau, L. A. Loneliness. *Human Nature*, 1979, *2*, 59–65.

Rubin, Z. *Liking and loving: An invitation to social psychology*. New York: Holt, Rinehart & Winston, 1973.

Rubin, Z. Disclosing oneself to a stranger: Reciprocity and its limits. *Journal of Experimental Social Psychology*, 1975, *11*, 233–260.

Rubin, Z., Hill, C. T., Peplau, L. A., & Dunkel-Schetter, C. Self-disclosure in dating couples: Sex roles and the ethic of openness. *Journal of Marriage and the Family*, 1980, *42*, 305–317.

Rubin, Z., Peplau, L. A., & Hill, C. T. Loving and leaving: Sex differences in romantic attachments. *Sex Roles*, 1981, *7*, 821–835.

Russell, D., Peplau, L. A., & Ferguson, M. L. Developing a measure of loneliness. *Journal of Personality Assessment*, 1978, *42*, 290–294.

Russell, D., Peplau, L. A., & Cutrona, C. E. The revised UCLA loneliness scale: Concurrent and discriminant validity evidence. *Journal of Personality and Social Psychology*, 1980, *39*, 153–166.

Schachter, S. *The psychology of affiliation*. Stanford, Calif.: Stanford University Press, 1959.

Schofield, M. *The sexual behavior of young people*. Boston: Little, Brown, 1965.

Schooley, M. Personality resemblances among married couples. *Journal of Abnormal Social Psychology*, 1936, *31*, 340–347.

Segal, M. W. Alphabet and attraction: An unobtrusive measure of the effect of propinquity in a field setting. *Journal of Personality and Social Psychology*, 1974, *30*, 654–657.

Shapiro, A., & Swensen, C. Patterns of self-disclosure among married couples. *Journal of Counseling Psychology*, 1969, *16*, 178–180.

Sigall, H., & Landy, D. Radiating beauty: The effects of having a physically attractive partner on person perception. *Journal of Personality and Social Psychology*, 1973, *28*, 218–224.

Silverman, I. Physical attractiveness and courtship. *Sexual Behavior*, September 1971, pp. 22–25.

Snyder, C. R., & Frankin, H. L. Abnormality as a positive characteristic: The development and validation of a scale measuring need for uniqueness. *Journal of Abnormal Psychology*, 1977, *86*, 518–527.

Snyder, C. R., & Frankin, H. L. *Uniqueness: The human pursuit of difference*. New York: Plenum Press, 1980.

Solomon, S., & Saxe, L. What is intelligent, as well as attractive, is good. *Personality and Social Psychology Bulletin*, 1977, *3*, 670–673.

Sorenson, R. C. *Adolescent sexuality in contemporary America*. New York: World, 1973.

Sprecher, S. *Men, women, and intimate relationships: A study of dating couples*. Unpublished master's thesis, University of Wisconsin, Madison, 1980.

Stewart, A. J., & Rubin, Z. The power motive in dating couples. *Journal of Personality and Social Psychology*, 1976, *34*, 305–309.

Stroebe, W., Insko, C. A., Thompson, V. D., & Layton, B. D. Effects of physical attractiveness, attitude similarity, and sex on various aspects of interpersonal attraction. *Journal of Personality and Social Psychology*, 1971, *18*, 79–91.

Traupmann, J., & Hatfield, E. Love: Its effects on mental and physical health. In J. March, S. Kiesler, R. Fogel, E. Hatfield, & E. Shanas (Eds.), *Aging: Stability and change in the family*. New York: Academic Press, 1981.

Traupmann, J., & Hatfield, E. How important is fairness over the lifespan? *International Journal of Aging and Human Development*, 1983, *17*, 89–101.

Traupmann, J., & Hatfield, E. The importance of "fairness" for the marital satisfaction of older women. Unpublished manuscript, 1981.

Traupmann, J., Hatfield, E., & Wexler, P. Equity and sexual satisfaction in dating couples. *British Journal of Social Psychology*, 1983, *22*, 33–40.

Valentine, C. W. *The experimental psychology of beauty*. London: Methuen, 1962.

Veitch, R., & Griffitt, W. The perception of erotic arousal in men and women by same- and opposite-sex peers. *Sex Roles*, 1980, *6*, 723–733.

Waller, W. The rating and dating complex. *American Sociological Review*, 1937, *2*, 727–734.

Walster (Hatfield), E., Aronson, V., Abrahams, D., & Rottman, L. The importance of physical attractiveness in dating behavior. *Journal of Personality and Social Psychology*, 1966, *4*, 508–516.

Weiner, B., Russell, D., & Lerman, D. Affective consequences of causal ascriptions. In J. Harvey, W. J. Ickes, & R. F. Kidd (Eds.), *New directions in attribution research* (Vol. 2). Hillsdale, N.J.: Lawrence Erlbaum, 1978.

Weiss, R. S. *Loneliness: The experience of emotional and social isolation*. Cambridge, Mass.: MIT Press, 1973.

Weiss, R. S. Transition states and other stressful situations: Their nature and programs for their management. In G. Caplan & M. Killilea (Eds.), *Support systems and mutual help: Multidisciplinary explorations*. New York: Grune & Stratton, 1976.

Winch, R. F. *The modern family*. New York: Holt, 1952.

Woodward, J. C., & Visser, M. J. Loneliness. *Farm, Ranch and Home Quarterly*, Fall 1977, pp. 5–6.

Worthy, M. A., Gary, L., & Kahn, G. M. Self-disclosure as an exchange process. *Journal of Personality and Social Psychology*, 1969, *13*, 63–69.

Young, J. E. Loneliness, depression and cognitive therapy. Theory and application. In L. A. Peplau & D. Perlman (Eds.), *Loneliness: A sourcebook of current theory, research, and therapy*. New York: Wiley, 1982.

Zander, A., & Havelin, A. Social comparison and interpersonal attraction. *Human Relations*, 1960, *13*, 21–32.

Zilbergeld, B. *Male sexuality*. Boston: Little, Brown, 1978.

10

The Experience of Injustice
Social Psychological and Clinical Perspectives

JANICE M. STEIL AND JOYCE SLOCHOWER

Justice is the first virtue of social institutions.
(Rawls, 1971, p. 3)

The first requisite of civilization . . .
(Freud, 1933, p. 42)

The uniting function in the individual man and in the social group . . .
(Plato, in Tillich, 1954, p. 55)

INTRODUCTION TO THE CHAPTER[1]

Despite our universal social need for justice, the presence of injustice is ubiquitous between people, groups, and societies. How the injustice is perceived and interpreted will determine how the individual responds

[1]This chapter is in two parts. The first reviews the social psychological literature on justice and was written by Janice M. Steil. The second part presents the clinical implications of the research and was written by Joyce Slochower.

JANICE M. STEIL • Institute of Advanced Psychological Studies, Adelphi University, Garden City, New York 11530. JOYCE SLOCHOWER • Department of Psychology, Hunter College, City University of New York, New York, New York 10021.

to it. On a social level, the individual's perception of the source of injustice may result in attempts to rescue or make restitution to the victim or, alternatively, in attempts to blame the victim. On an internal, emotional level, the experience of oneself or of others as the victims or perpetrators of injustice has emotional sequelae—including, at times, an overall sense of the self as an innocent or responsible victim or harmdoer. These self-perceptions will, in turn, alter the individual's social behavior toward others who are themselves victims or harmdoers.

In this chapter, we shall first examine the experience of injustice from a social psychological perspective, reviewing the empirical literature on injustice. That literature largely focuses on the experience of the individual rather than on larger social groups' responses to injustice.

In the second half of the chapter, the experience of injustice will be discussed in terms of its emotional impact from a psychoanalytic perspective. The clinical and treatment implications of the research on the victim and victimizer are then considered and, finally, are illustrated with a clinical case.

THE RESPONSE TO INJUSTICE: THE SOCIAL PSYCHOLOGICAL PERSPECTIVE[2]

DEFINITIONS AND ISSUES

Injustice is a psychological experience. What is required is "a dissatisfied state of mind; a felt discrepancy between what is perceived to be and what is perceived should be" (Adams, 1965, p. 272). It is relative rather than absolute because it comes from comparing our position or outcomes to the position or outcomes to which we aspire or to those that similar others enjoy (Crosby, 1976, 1982; Deutsch, 1976). What is perceived as unjust by one may be perceived as just by another depending on their respective aspiration levels or comparison choices.

The perception of injustice is distressing for all parties. The source and extent of the distress, however, varies with position (Walster, Walster, & Berscheid, 1978). Harmdoers are believed to be distressed

[2]I acknowledge and appreciate the efforts of Patricia Noel in the early stages of the chapter and thank Diane Sholomskas for her perceptive comments.

because of fear of retaliation and/or threatened self-esteem (Walster *et al.*, 1978). Observers are distressed because the victimization of others makes salient their own vulnerability (Lerner, 1977; Shaver, 1970; Walster, 1966). Victims are more sensitive to and more aggrieved by an injustice than are the other parties (Austin & Walster, 1974; Deutsch, 1976; Homans, 1961; Leventhal, Weiss, & Long, 1969; Walster *et al.*, 1978). Victims are deprived of desired outcomes, suffer a loss of self-esteem, and are often derogated and stigmatized in their own eyes as well as in the eyes of significant others.

Injustice is not only distressing but motivating (Adams, 1963, 1965; Austin & Walster, 1974; Steil, Tuchman, & Deutsch, 1978). As with all moral values, justice has an obligatory or "ought" component (Heider, 1958). We must neither inflict nor suffer unjust treatment. Furthermore, we must respond to an injustice when it is inflicted on any member of our moral community. Failure to do so jeopardizes the inviolability of the social norms and leaves us similarly vulnerable. Justice demands that we remedy or prevent injustice (Cahn, 1949).

Why then is injustice so pervasive? The answer lies in the very nature of justice itself. It is one of the most demanding, most complex, and most paradoxical of human values. It is a duty and an aspiration. It is relative and absolute. It is particularistic as well as universalistic. It is heteronomous and autonomous. It requires restrictions on individual liberty, and yet it is freeing. Although the moral demands require that we respond to injustice, restitution may be contrary to our immediate interests, costly, or futile. Not to respond, however, evokes psychic costs to our own self-regard. We are faced with the essential moral dilemma.

There is, however, an escape. By using the relative and particularistic aspects of justice, we can redefine our way out. There are, then, two ways to respond to injustice. We can restore justice objectively by seeking compensation, or we can restore justice psychologically by distortion (Walster *et al.*, 1978). The essential question, then, becomes not, Why is injustice so pervasive? but, What are the conditions that facilitate acknowledgment and restitution as compared to denial and distortion? An understanding of these conditions is essential to those who work with the victims of social injustice, of violent crime, and those who suffer severe accidents or trauma. Identification of the conditions and motivations that affect the response to injustice can further our understanding of the psychic costs of victimization, the potential conflict in the needs of the victim and the needs of those who may be the primary sources of support (e.g., family, friends, or professionals in the helping community), and, finally, the reasons why people often respond to their own victimization in seemingly self-defeating ways.

THE HARMDOER

A number of studies have shown not only that people feel guilty for their transgressions, but that it is surprisingly easy to elicit feelings of guilt and responsibility. Ross and DiTecco (1975, p. 75) cite a number of studies that show that laboratory subjects will assume responsibility for the consequences of their actions even when the consequences are not only somewhat "bizarre but unintended and unforeseeable." Furthermore, those in the responsibility-inducing conditions are far more likely to comply with subsequent requests for diverse kinds of help from the (confederate) victim than are controls. (Brock & Becker, 1966; Cialdini, Darby, & Vincent, 1973; Freedman, Wallington, & Blass, 1967). People, it seems, will fairly readily assume responsibility for the negative outcomes of others, feel distressed by the perception of themselves as harmdoers, and seek to alleviate the distress by helping to set things right.

It has also been shown, however, that rather than helping, harmdoers will respond to their transgressions by derogating the victim (Davis & Jones, 1960; Sykes & Matza, 1957) and denying responsibility for their acts (Brock & Buss, 1962, 1964; Sykes & Matza, 1957). Thus, it might equally be said that people are distressed by the perception of themselves as harmdoers and distort reality to eliminate that perception.

When do harmdoers compensate and when do they distort? Distortions, it has been asserted, must be credible. For this reason, it has been proposed, the more contact one has had with the victim, the less likely one will be to justify one's harmdoing. On the other hand, the more difficult and costly it is for harmdoers to provide not only compensation but the exact amount to set things right, the less likely they are to objectively restore justice (Berscheid & Walster, 1967; Berscheid, Walster, & Barclay, 1969). These findings suggest some rather counterintuitive propositions. Although the ease with which responsibility and guilt are elicited may work to the victim's advantage when adequate compensation is readily available, it may work to the victim's disadvantage when restitution is difficult. The greater the victim's deprivation, the more difficult it may be to set things right. Thus, those who suffer the greatest deprivations may be the least likely to be compensated. Rather than finding themselves the recipients of restorative acts, they may find themselves blamed and/or derogated by the very persons who have exploited them.

THE OBSERVER

Victims are derogated not only by those who exploit them, but also by those who merely observe their suffering (Lerner & Simmons, 1966). The more innocent the victim and the more severe the consequences, the greater the potential for rejection and blame (Burger, 1981; Jones & Aronson, 1973; MacDonald, 1972; Smith, Keating, Hester, & Mitchell, 1976; Stokols & Schopler, 1973).

Why do observers derogate those who suffer? And why are the most blameless the most rejected, especially when the consequences of their victimization are severe? Several explanations have been suggested and they all include a self-protective motive.

THE NEED FOR A JUST WORLD

Lerner (1970, 1977; Lerner & Miller, 1978; Lerner, Miller, & Holmes, 1976) purports that individuals have a need to believe that they live in a world where people generally get what they deserve. The need for this belief, according to Lerner, evolves from the personal and social contract developed in childhood as one advances from the pleasure to the reality principle. The essence of the contract is the acceptance of earning or deserving as the effective way to obtain what one wants. "The belief in a just world serves several functions. It enables the individual to confront his physical and social environment as though they were stable and orderly" (Lerner & Miller, 1978, p. 1030). It facilitates the pursuit of long-range goals and the socially regulated behavior of day-to-day life. Because the belief that the world is just serves such an important adaptive function, people are reluctant to relinquish it and can be greatly troubled if they encounter evidence that suggests that the world is not really just or orderly after all. Injustice to others as well as to oneself threatens the integrity of the personal and social contract and calls into question "one's own prior commitments, efforts, unfulfilled investments, and present beliefs and allows one's immediate impulses and desires to surface" (Lerner *et al.*, 1976, p. 137). "If others can suffer unjustly then the individual must admit to the unsettling prospect that he too could suffer unjustly" (Lerner & Miller, 1978, p. 1031). Injustice to others, Lerner asserts, has clear implications for the individual's own future fate. Consequently, a discrepancy between another's fate and that which he or she deserves elicits a motivation to restore justice.

> One way of accomplishing this is by acting to compensate the victim; another is by persuading oneself that the victim deserves to suffer. Victims can deserve

> their fate as a consequence of having a "bad" character or as a consequence of engaging in "bad" acts. Good people can deserve a "bad" fate if their acts are "careless" or "foolish." (Lerner & Miller, 1978, p. 1031)

Empirical tests of the just-world hypotheses have consistently demonstrated that observers will act to restore justice objectively when they can. Inability or uncertainty as to one's ability to objectively compensate leads to derogation and rejection of an innocent victim (Lerner & Simmons, 1966; Lincoln & Levinger, 1972; Mills & Egger, 1972; Simons & Piliavin, 1972; Sorrentino & Hardy, 1974).

> The sight of an innocent victim suffering, with no possibility of reward or compensation motivates people to devalue the attractiveness of the victim bringing about a more appropriate fit between her fate and her character. (Lerner & Miller, 1978, p. 1032)

In addition, and also consistent with the just-world hypothesis, victims who can be viewed as objectively responsible for their fate are less likely to be devalued (Lerner & Matthews, 1967).

THE NEED FOR A CONTROLLABLE WORLD

Whereas Lerner asserts a need to believe that the world is just, others assert a need to believe that the world is controllable. Furthermore, some have suggested that the need for a controllable world elicits a tendency to reject the blameless, particularly when the consequences of their victimization are severe. Accidents, natural disasters, acts or events with severe negative consequences provoke compelling attempts to identify who, often to the exclusion of what, was responsible (Drabeck & Quarantelli, 1967). The belief that severe negative outcomes can happen at random implies that they could happen to us.

> When we hear of a small loss, it is easy to feel sympathy for the sufferer, attributing his misfortune to chance and acknowledging that unpleasant things can happen to people through no fault of their own. As the magnitude of the misfortune increases, however, it becomes more and more unpleasant to acknowledge that "this is the kind of thing that could happen to anyone." (Walster, 1966, pp. 73–74)

The identification of someone as responsible establishes the event as predictable and controllable. It allows people to feel they can protect themselves by isolating or reforming the guilty party. If the person held responsible is the victim, people need only assure themselves that they are a different kind of person or that they would behave differently under similar circumstances.

The need for a controllable world has generated specific hypotheses. The worse the consequences of a seemingly random event, the greater the tendency to assign responsibility to someone possibly responsible. Furthermore, the more serious the negative consequences suffered by victims, the more they will be held responsible. Scripted studies of violent crimes have shown that perpetrators are sentenced to longer imprisonments for the rape of a nun, social worker, married woman, or virgin than for the rape of a divorcee. Furthermore, more responsibility is attributed to stereotypically "respectable" rape victims as compared to the "less respectable," and to the unacquainted ("innocent") victim than to the acquainted (Jones & Aronson, 1973; Smith *et al.*, 1976). The less responsible for his fate the victim of a violent crime is portrayed as being, the lower his rated attractiveness (MacDonald, 1972). The more severe the consequences, the more negatively the victim is evaluated (Stokols & Schopler, 1973).

The hypotheses have also been tested in a series of studies concerned with motivational distortions in the attribution of responsibility for an accident. Burger (1981), in a meta-analysis of 22 such studies, concludes that there is, in fact, a tendency to attribute more responsibility for a severe as compared to a mild accident. An important outcome of these and other studies, however, has been the identification of two of the factors that reduce or eliminate this tendency. When subjects receive empathy-inducing instructions the derogation effect is not only eliminated but reversed (Aderman, Brehm, & Katz, 1974; Regan & Totten, 1975). Subjects who anticipate a similar fate also assign less responsibility and are more lenient in their judgment (Chaiken & Darley, 1973; Shaver, 1970; Sorrentino & Boutilier, 1974). Personal similarity (i.e., similarity of attributes and attitudes) does not seem to affect the blaming and rejecting tendencies. Instead of altering their perceptions of responsibility, subjects simply deny the similarity (Shaver, 1970). Identification seems to require the possibility of a similar fate rather than the possibility of similar attributes (Lerner & Matthews, 1967). Consistent with these findings, Shaver (1970) has suggested that the hypothesis that severe negative outcomes will not be attributed to chance will be overridden by the desire to avoid being blamed. Under conditions of fate similarity, people will prefer to believe in a capricious world rather than a just or controllable one to avoid being the object of blame. It has been suggested that the relationship between observers' perceived risks and their judgments of victims is curvilinear. When there is very little chance of sharing the victim's fate, others may be normatively sympathetic or simply indifferent, depending on the extent to which the victim is perceived as part of their moral community. When the risk is more moderate, others may be inclined to avoid the thought of their own victimization by invoking derogation and blame. When the risk is very high, observers may begin to think what their own experiences as a victim would be like and so judge unfortunate others more leniently (Shaver, 1970).

INDIVIDUAL DIFFERENCES

Our review of the normative literature may be misleading unless we note that justice studies are sometimes characterized by self-selection factors (Lerner & Simmons, 1966; Steil, 1983). Some subjects in the Lerner studies, for example, condemn the experiment rather than the victim. In other studies, some subjects who benefit from a discriminatory allocation system seek change on behalf of the disadvantaged whereas others who are also benefited do not (Steil, 1983). These findings highlight the role of individual differences in the response to injustice. Investigations of these differences are limited but suggest a relationshp between several measures of ideological conservatism and the tendency to blame the victim. Although there is wide variability in the extent to which individuals believe the world is just, those who do are also likely to endorse attitudes that emphasize hard work as a value in its own right and as key to success, to be high in authoritarianism, and to be oriented to an expectancy for internal controls of reinforcement. High just-world believers are more likely than nonbelievers to admire fortunate people, political leaders, and existing institutions. They are also more likely to denigrate victims and endorse negative attitudes toward underprivileged groups (Rubin & Peplau, 1975). Similarly, individuals who score higher on measures that endorse sex-role stereotyping, interpersonal violence, and adversarial sexual beliefs are more likely than those who score low on such measures to accept rape myths that deny or reduce the perceived injury of the victim or blame victims for their own victimization (Burt, 1980).

THE VICTIM

As recently as 1978, Walster *et al.* in a review of the literature reported that "we know a great deal about exploiters; we know very little about the "exploited" (Walster, Walster, & Berscheid, 1978, p. 43). In another review article that same year, Lerner noted that

> a considerable amount of research has addressed the question of whether observers blame the victims of misfortune for their own fate, but virtually

> no research has examined the question of whether people who are the victims of misfortune tend to blame themselves for their fate. (Lerner & Miller, 1978, p. 1043)

Victims are more sensitive to and more aggrieved by an injustice than others are. Perhaps even more than the other parties, basic assumptions about their world and themselves are jeopardized. These include the belief in the world as just and meaningful, their view of themselves as positive and deserving, and the belief in their own invulnerability (cf. Janoff-Bulman & Frieze, 1983). How does the victim respond to the shattering of these assumptions? Theoretically, the victim is expected to respond by attempting to set things right, either by securing restitution or by retaliating against the harmdoer (Walster *et al.*, 1978). Successful restitution may allow a reaffirmation of the victim's primary belief system. But what if the victim feels unable to set things right?

Under these conditions the aversive aspects of victimization, particularly the prospect of devaluation by others, can lead the victim to seek means of self-enhancement. These strategies usually involve minimizing or denying their own victimization. Taylor *et al.* (1983) specify five self-protective mechanisms that victims use. These include comparing themselves to less fortunate others; focusing on attributes that make one appear advantaged ("I had a . . . small amount of surgery. How awful it must be for [those] who had . . . [more]," Taylor *et al.*, 1983, p. 29); creating hypothetical worse worlds, construing benefit from the victimizing event ("you take a long look at your life and realize that many things you thought were important are totally insignificant," Taylor *et al.*, 1983, p. 33); and manufacturing normative standards of adjustment that make one's own adjustment appear exceptional ("I think I did extremely well. I know there are some who aren't strong enough, who fall apart. . . ." p. 34).

Another, more costly perhaps, mechanism of psychological distortion is self-blame. Both field studies and clinical reports provide extensive evidence of this phenomenon. Indeed, studies of rape victims suggest that self-blame is not only prevalent (Burgess & Holmstrom, 1979; Libow & Doty, 1979; Martin, 1978; Notman & Nadelson, 1976; Schram, 1978) but is second only to fear in frequency of occurrence. Surprisingly, perhaps, self-blame is a far more common reaction than anger (Janoff-Bulman, 1979). Studies of battered women find that anywhere from one-quarter to more than half blame themselves for their husbands' violence (Frieze, 1978, 1979; Pfouts, 1978) as compared to only 20% of controls (Frieze, 1978). Victims, it seems, sometimes blame themselves more than others do. Although self-blame seems a costly response, some have asserted that it allows the victim to retain a sense of control (Bulman & Wortman,

1977; Janof-Bulman, 1979). Characterological self-blame which is dysfunctional, is distinguished from behavioral self-blame, which is adaptive. Characterological self-blame is believed to be esteem related because it involves attributions to a nonmodifiable source (one's character) and is associated with a belief in personal deserving for past negative outcomes. Characterological self-blame is associated with depressive symptoms (Janoff-Bulman, 1979; Peterson *et al.*, 1981) and may be an impediment to the amelioration of suffering when it is no longer inevitable (Comer & Laird, 1975). Behavioral self-blame, by contrast, is believed to be control related because it suggests a belief in the future avoidability of a negative outcome. Although blaming one's behavior is probably less costly when blaming one's character, it is, as yet, unclear whether it is as adaptive as some believe. Behavioral self-blame may allow victims to retain a sense of power, but at what cost? Assuming responsibility for provoking violence allows the battered woman, for example, to believe that by altering her behavior she can control the battering. But the self-deception involved stands in the way of constructive change.

Self-blame is related to victims' expectations as well as their sense of deserving. People expect to get what they deserve and to accept responsibility for it. More surprisingly, perhaps, people will alter their sense of deserving to fit that which they expect. Laboratory studies consistently show that subjects who have been led to expect a negative fate frequently alter their self-esteem to believe they deserve it, even when, by any objective assessment, that fate has been inflicted on them. Furthermore, those who blame themselves often choose to follow through with an expected negative event even when it subsequently becomes avoidable (Aronson, Carlsmith, & Darley, 1963; Comer & Laird, 1975; Foxman & Radtke, 1970).

Expectations derive from past experiences, societal ideologies, and, not unrelated to these, our perceptions of ourselves. Subjects who were led to believe they were less competent at a task than their peers expected to do less well and were more likely to assume personal responsibility for their lower outcomes than controls or "competent" subjects, even when the outcomes were clearly the result of a discriminatory allocation system (Steil, Levey, & Sanders, 1982). In one of the few investigations of the effect of ideologies, it was found that ideological position was associated with subjects' perceptions of personal deserving, their fairness judgments, and their desire for change. Orthodox Jewish women who endorsed fundamentalist views were less likely to feel entitled to participate in religious practices, were less desirous of participating, and had lower expectations regarding the possibility of change. In addition, they were less likely to view sex differences in access to participation as unfair and were less likely to see change as desirable. The more strongly the women endorsed feminist views, the more likely they were to feel entitled to participate in religious practices, whether they wanted to or not, and the more likely they were to perceive sex differences as unfair and change as desirable (Weinglass & Steil, 1981). Ideologies instruct people as to what they deserve. They also instruct as to the locus of blame. Crosby (1984) suggests that an understanding of the variables affecting the locus of blame is imperative to an understanding of victim self-evaluation. Citing Pettigrew (1978), she notes that the disadvantaged can interpret their inferior status as a statement about themselves or as a statement about the system with differing consequences. Steil (1983) demonstrated that subjects who label a system as fair are more likely to denigrate their own abilities. When subjects receive social support to validate their perception of the system as unfair they are less likely to blame themselves. Victims, it seems, are most likely to blame themselves when they feel powerless to alter their fate and their view of themselves as well as the dominant ideologies lead them to feel they deserve it. It seems a vicious and self-perpetuating cycle. The greater the injustice the more difficult the solution. The longer people are victimized, and the more they expect their outcomes to be negative, the more likely they will alter their conceptions of their own deserving. Under these conditions, outside help may be essential.

THE RESPONSE TO INJUSTICE: A PSYCHOANALYTIC PERSPECTIVE[3]

CLINICAL IMPLICATIONS

The social psychological research on injustice generally points to the likelihood that people who are victimized will be blamed for their fate rather than helped or compensated. The tendency to blame the victim characterizes the responses of apparently uninvolved observers, actual harmdoers, and victims themselves. The theoretical constructs used most frequently to account for these findings involve people's needs to believe in a just and controllable world.

Before these research findings can be considered in a clinical context, the psychoanalytic interpretation of the need for a just and controllable world, and the internal dynamics associated with the blaming

[3]I would like to thank Drs. Larry Epstein, Stanley Rosenman, and Felicia Rozek for their valuable suggestions.

observer, harmdoer, and victim need to be spelled out. It appears that there are characteristic defensive styles connected with this inappropriate blaming of the victim, and also that the cognitive distortions described by the just- and controllable-world hypotheses have developmental derivations.

The clinical implications of these findings most directly involve treatment situations in which a major aspect of the patient's difficulties centers on his or her actual role as victim or harmdoer (e.g., in the treatment of a rape victim or a child abuser). The patient's conscious and unconscious experience of this dimension of his or her life, as well as the therapist's objective and subjective countertransferential responses need to be considered. Most importantly, one must examine the extent to which the therapist–patient relationship recapitulates or alters the patient's history as the victim or the perpetrator of injustice, and what must be done in the treatment to help the patient integrate this life experience. Within this context, then, the role of distorted perceptions of the victim and the harmdoer described in the research literature must be reexamined in terms of their effects on the treatment. This will be our second focus, and will include the discussion of a clinical case.

On a broader level, the experiences of *all* patients may include self-perceptions as both victim and harmdoer. In this sense, an analysis of the individual's experiences of injustice can be viewed as an inevitable part of any psychoanalytic treatment, and these broader implications of the injustice research will be considered last.

THE NEED FOR A JUST AND CONTROLLABLE WORLD

Within the social psychological framework, the need for a just or a controllable world reflects people's need to see their environment as essentially predictable and orderly. Facing the presence of injustice, however, forces them to consider themselves potentially vulnerable to mistreatment. To the extent that there is a tendency in many people to explain or justify the misfortunes of others, there may be common personal experiences that motivate this need. Within a psychoanalytic framework, the need for a just and controllable world reflects an attempt to deny or distort reality, and thus stems from defensive processes. The question then, becomes: What is the purpose of this reality distortion, and what is its original source? The need for a just and controllable world seems to protect individuals from experiencing the ways in which they are, and have been, vulnerable to unjust or uncontrollable external forces. Although abstract cognitive concepts of justice evolve in adolescence or later, there are probably precursors to such experiences in childhood and even in infancy. In fact, it is in childhood that one is most vulnerable to the whims and variability of adults, who are "just" (fair) to the child only at times. Beginning in infancy, the baby is confronted with unpredictable (and thus uncontrollable) internal drive states (e.g., hunger). When the "good enough" mother meets these needs sufficiently well to allow the infant a brief experience of omnipotence (cf. Winnicott, 1965), the infant gets a sense of the world as a good, satisfying (just?) place. However, later development involves the mother's and the world's progressive failure to meet these needs in a perfect way. The baby and then the child are faced with both its own absolute, and then relative dependence, and also with the imperfections and vulnerabilities of its parents. For the child to fully acknowledge its own powerlessness, as well as the inevitable rage at parents who fail to be perfect and to satisfy perfectly, the child must fully abandon the fantasy of the parents' omnipotence and goodness. Later, especially during the oedipal phase, the child again comes up against the "injustice" of the family dynamics, which stand in the way of libidinal wishes. At this stage, the parents paradoxically represent both a moral resistance to the child's wishes and a potential aggressor against the child. When the child confronts the impossibility of oedipal wishes in the face of the parents' superior knowledge and power, a second layer is added to the belief in a just world. In bowing to the parents' "moral" stance, the child symbolically accepts the superior "justice" of the parents, and through the identification process a stronger (superego) basis for the development of values and therefore justice is formed. Paradoxically, however, the oedipal resolution also involves an unconscious identification with the parent as aggressor, and thus can represent an internal rejection of the victim vis-à-vis the aggressor. There are many clinical examples of adults who fail to abandon infantile and oedipal wishes. What may be true in a more general, less pathological way is that the need to believe in a just and controllable world reflects our unconscious wishes, as adults, to hold onto the infant and child's view of first the parents and then the self, as good, satisfying, powerful, and thus "just."

PSYCHOANALYTIC UNDERSTANDINGS OF THE BLAMING OBSERVER, HARMDOER, AND VICTIM

THE OBSERVER

Although there is a generalized tendency for people to believe in a just world, the literature has also pointed to individual differences that affect the observer's tendency to denigrate the victim. People who are strong just-world believers are those who hold rigid, conservative, and authoritarian beliefs. To the extent that a strong belief in a just world involves both a rejection of the innocent victim and a denial of the real (malevolent) qualities of the world, the just-world believer must employ massive defenses to maintain this distorted picture of reality. This denial of both the victim's innocence and the world's malevolence may be associated with a similar kind of internal dissociative process, which can fragment the ego leaving a portion of it split off and thus weakened. Individuals who resort to this distortion are those whose sense of self is especially vulnerable to narcissistic injury when confronted with a victim who is in some way similar to themselves. By denigrating the victim who belongs to their own group, they protect themselves from experiencing their own vulnerability, loss of self-esteem, and guilt (Rosenman, 1982). Through the use of denial, splitting, and projection, the observer can maintain a good and omnipotent (though fragile) self-image while viewing the victim as bad and weak. This allows the observer to avoid confronting internal anxieties about the self's competence, strength, and value. Thus, for example, paranoid individuals should be particularly drawn to the just-world hypotheses. In contrast, the ability to face and accept the presence of injustice, to tolerate one's own powerlessness to change things while offering whatever help one can, requires that the individual integrate a sense of helplessness with the wish for potency and justice. It requires the ability to experience fear (signal anxiety) without annihilative anxiety, and thus reflects a more developed ego structure. Presumably this develops out of early experiences in which the baby was both allowed an early experience of omnipotence, and in which these feelings were then disrupted gradually rather than abruptly, allowing for a better integration of this loss of absolute control.

The Harmdoer

In a similar way, once people have, intentionally or unintentionally, done harm to someone, individual difference factors may account for the harmdoer's response to the victim. (I am here excluding a consideration of intentional, repetitive, large-scale harmdoing, as in wartime where social and coercive factors play a large role.) An attempt to approach and compensate the victim of one's own destructiveness requires the capacity to experience and tolerate feelings of guilt and responsibility as well as the ability to experience one's potential goodness. The harmdoer who is unwilling or unable to attempt restitution may suffer from either the absence of a sense of guilt, or an inability to tolerate such feelings.

Additionally, such harmdoers may have no faith in their ability to actually offer anything of value to the victim, that is, may suffer from an overwhelming sense of internal badness. In developmental terms, the capacity to make restitution for one's wrongdoing probably derives from the infant's (and mother's) integration of loving and hateful feelings toward loved ones. This capacity to tolerate ambivalence eventually is expressed in attempts at reparation for feelings and expressions of destructiveness. These processes are generally believed to originate in infancy, (when the infant first integrates the maternal part objects) and to continue through the oedipal phase (when ambivalence primarily toward the same sex parent must be dealt with) (Klein & Riviere, 1937; Winnicott, 1965).

THE VICTIM

Finally, the dynamics of the self blaming victims need to be considered. What is it about being victimized that cuts off the individual's capacity for anger and for action? Any experience of victimization involves being overwhelmed by destructive external forces, and may call up whatever developmental precursors exist. All children are probably at times the object of undeserved blame and even unconscious hate on the parents' part, and often internalize this sense of blame, in part to protect the fantasied goodness of the parents (cf. Bloch, 1978). It may be that the adult victim's similarly passive responses to mistreatment occur in those repetitive situations that render one impotent and in which external forces seem insurmountable.[4] When the adult victim assumes responsibility and blame for this mistreatment, the victim may also be expressing a sense of hopelessness about making an impact on the object through protest or action, or, unconsciously, may be holding onto the fantasy that the harmdoer (parent) will ultimately set things right. Thus, for some victims, there may be an emotional investment in maintaining the victimized stance. This investment may be based on an unconscious attempt to repeat and master earlier experiences and thus, unconsciously, to gain some sense of control over them (cf. Wortman, 1983). For others who suffer from a sense of internal badness (or from internalized persecutory objects), an experience of actual victimization may be perceived as an external repetition of an ongoing internal experience.

[4]See Freud (1916) for a discussion of the "Exceptions," that is, people whose response to victimization is to thereafter demand special, compensatory treatment.

It may thus be either welcomed (because it relieves internal self-punishment) or submitted to as simply another such experience.[5]

In order for an individual to respond to victimization with action rather than self-blame, then, it would be necessary for the external experience to be one that does not mirror the internal one. Thus, the active victim would presumably be one who had had sufficient earlier experiences of satisfaction and reparation in transactions with powerful adults to be largely free of persecutory and depressive anxieties.

It is impossible to deal here with all the meanings or developmental processes related to the dynamics of victimization, since these are clearly related to individual life experiences. Additionally, a developmental understanding of these phenomena may be circular at times, since it is difficult to define precisely what, for example, constitutes a sufficient but not excessive experience of infantile omnipotence. The major point, however, is, that the cognitive social psychological explanations for injustice related behavior have parallels in psychoanalytic language, and may have developmental antecedents as well.

THE THERAPIST'S RESPONSES TO THE VICTIMIZED AND HARMDOING PATIENT

The psychoanalytic treatment of a patient who has been victimized or who has victimized others is, of course, subject to the same therapeutic issues and transference–countertransference dynamics that are evoked in any long-term treatment experience. These issues will be determined by the nature of the psychoanalytic process itself, the unique characteristics of patient and therapist, and by the developing patterns of interplay between the two. What is of interest here, however, is whether there are special characteristics involved in the treatment of the victim and victimizer that are related to the cognitive distortions described in the research literature. It is this latter area that will be the focus here, despite the obvious limitation built into any consideration of psychoanalytic treatment that omits a focus on the unique aspects of different cases.

Although the position of a therapist differs from that of the uninvolved observer in that therapists are trained (and paid) to listen and

[5]See Freud (1920) for a discussion of the repetition compulsion and Panken (1973) for a consideration of psychoanalytic theories of masochism.

accept without judgment a variety of feelings in the patient and in themselves, therapists are also vulnerable to experiencing countertransference distortions. These distortions may be based largely on the therapist's subjective, idiosyncratic responses to the patient's material (Freud, 1910); they may be induced by the patient's conscious or unconscious wish to be responded to in a particular way, (Epstein, 1979; Little, 1957; Racker, 1968; Spotnitz, 1969) or they may be an objective emotional response to the patient's material (Winnicott, 1958). In order to evaluate the therapist's response to the patient in these terms, however, it is first necessary to consider what is likely to be a therapeutic response to the victimized or victimizing patient. Although the research literature implies that the main inappropriate response to the victimized patient is one of blame, it is possible that a victim *is* actually responsible for his or her plight and is inappropriately seen as innocent. Similarly, it is possible to inappropriately distort the harmdoer's role by shifting the locus of blame. In the therapeutic context, all sorts of distortions can be harmful to the patient. What is most crucial to the therapeutic process is that the victim's distorted self-perceptions of *either* innocence or guilt be slowly analyzed and worked through. A balanced sense of the combined contribution of chance, malevolent forces, and one's own conscious or unconscious contribution to the victimizing experience is the ultimate goal. Thus, for example, self-blaming victims need help externalizing anger, whereas the defensively innocent victims would need to integrate their role in provoking the attack with a sense of being mistreated. Harmdoing patients need to acknowledge and integrate their destructiveness, and perhaps would also need to be helped to make reparation.[6] In order to work effectively with such patients (as with many severely disturbed patients), the therapist must understand the potential for these countertransference reactions. Active (but usually silent) work by the therapist is often necessary in order to sort out and understand the nature of the particular countertransference responses (subjective, induced, or objective) and their impact on the therapeutic process.

Although the research literature is limited to understanding people's distortions of the plight of the innocent victim (and thus should be expanded to consider the abovementioned possibilities), the available data suggest that therapists' subjective countertransferential distortions are most likely when the therapist is vulnerable to the particular issues involved. This is most common when the therapist feels helpless to help

[6]Here one must distinguish the harmdoer who is coerced (e.g., by the courts) into therapy from one who voluntarily seeks help.

the patient (cf. Lerner & Miller, 1978), especially in the case of a severe accident (Burger, 1981), and when the therapist is moderately vulnerable to the patient's own fate (cf. Walster, 1966). Thus, for example, a newly married woman therapist hoping to become pregnant might be most blaming of a woman patient who was the victim of a miscarriage. Similarly, an older male therapist with protective feelings toward young women might be unable to see how his young woman patient silently participated in her husband's mistreatment of their children. In general, of course, subjective countertransferential responses are less likely in the better integrated therapist. More particularly, these responses to victimized and harmdoing patients would seem least probable when the therapist is capable of tolerating feelings of destructiveness and injustice in himself or herself and in the world, and also when the therapist has the capacity for reparation.

In considering the therapist's response to the victimized or harmdoing patient, one must examine the possibility that the interplay between patient and therapist is what produces the therapist's countertransference response of blame or pity. That is, it is possible that the patient unconsciously induces feelings of blame or pity in the therapist, as a function of the patient's transference feelings. For example, a patient who was unconsciously invested in being masochistically victimized might elicit sadistic emotional responses from the therapist, who would not immediately be clear about their source. Although this kind of induced response can occur in any treatment situation, it is particularly likely to involve countertransference feelings of blame or pity when a salient feature of the patient's problems involves his or her role as a victim or harmdoer. Although such an induced countertransference is destructive when it is simply acted out in a repetitive way, the working through of such a repetition can make up the resolution of the patient's victimized or harmdoing stance. This working through may require that the therapist communicate to the patient something of the patient's emotional impact and the therapist's emotional response.

A blaming or pitying response to a victimized or harmdoing patient is occasionally an *objective* emotional reaction. This might be true, for example, in the treatment of a massively abused or abusing patient. Although the uncontrolled expression of such reactions in rarely appropriate within a treatment context, there are again times when the therapeutic response to such a patient may, in fact, be an expression of modulated objective hate or pity (Winnicott, 1958). This would be therapeutic when it provided the patient with a maturationally appropriate experience of making this impact on the object. This might be necessary late in the treatment of psychopathic patients, or with others whose histories were marked by the absence of genuine emotional interplay between parents and child.

In summary, the major point to be made about the therapist's response to victimized and harmdoing patients is that a neutral, sympathetic stance is not necessarily the therapeutically correct one. An example of the interplay of these transference–countertransference dynamics in the treatment of a young woman will be presented next.

Mrs. B.[7], a married woman with a young child, sought treatment for sexual frigidity at her husband's request. Despite some initial resistance, Mrs. B. quickly formed a positive connection to the therapist. During the initial phase of treatment, Mrs. B. focused on telling the therapist about her early life, which she felt was the cause of her current sexual problems. Mrs. B. is from a lower class family and grew up in a small midwestern city. Her early history was filled with experiences of physical abandonment and abuse, mostly by her mother, and of a nurturing relationship with the paternal grandmother. Then, beginning in early adolescence and lasting through her teens, Mrs. B. was sexually molested by her father on a nightly basis. She cooperated in this both out of fear and in order to extract favors in return. Mrs. B. was able to recall her shame and hatred of her father and, later, the physical pleasure she experienced.

The first phase of treatment (which lasted for many months) largely elaborated Mrs. B.'s memories of herself as the innocent victim of abuse. Mrs. B. experienced herself as a helpless victim of parental (both paternal and maternal) abuse, and the therapist responded with rather strong feelings of empathy, pity, and respect for her honesty in investigating this painful area. In light of the material that emerged, this phase of treatment seemed to serve the purpose of allowing the patient to form a positive reciprocal connection to the therapist in which the therapist became the caretaking grandmother and the patient the abused and helpless child.

The second phase of therapy began when Mrs. B. turned from a total focus on her past and began presenting present-day issues. Initially, this involved a description of her abusive, volatile husband and served to demonstrate to the therapist how she and, importantly, her daughter were being victimized in the present. However, when the therapist tied Mrs. B.'s choice of a husband to her earlier masochistic ties to her mother and father (and thus began investigating the ways in which Mrs. B. was

[7]Certain aspects of the patient's social and personal history have been disguised to preserve her anonymity. None of these facts was crucial to the matters under discussion here.

responsible for her present dilemma), the therapy's focus shifted. Mrs. B. slowly revealed that she and her husband (who were strikingly ignorant about child development) regularly abused their daughter both emotionally and physically. Mrs. B. described these incidents with feelings of rage at the husband and child and with some very vague discomfort at her own behavior. The therapist's response was one of intense anxiety and concern for the daughter, and she rapidly began confronting the patient with the fact of her abusiveness and interpreting what she understood about its transference implications (Mrs. B. seemed to be acting out her mother's hatred of her, vis-à-vis her daughter). The therapist's stance, then, became actively focused on the patient's responsibility for her current role as victimizer rather than as victim, and the therapist's feelings of empathy largely shifted from Mrs. B. to her daughter. Interestingly, Mrs. B. responded to these interventions with relief, by seeking and using concrete information about appropriate childrearing, accepted the reality of her abusiveness, and virtually stopped beating and mistreating her daughter. During the five years of treatment, this abuse has never recurred, and she has begun to develop a mothering stance vis-à-vis her daughter that is based on empathy.

The therapist's response to Mrs. B. as a harmdoer seemed multi-determined. The subjective countertransference factors mostly involved the therapist's own recent experience of motherhood, which intensified her strong feelings of protectiveness toward children and her horror of child abuse. Those responses probably made her intervene more rapidly and with more concrete help than would a therapist with less of an identification with child victims. Additionally, Mrs. B.'s very positive response to these interventions suggests that induced countertransference factors were also involved. Later investigations revealed that Mrs. B. was unconsciously trying to get the therapist to nurture and protect her daughter for her. In doing so, the therapist was both performing a reparative function (insofar as Mrs. B. was identified with her daughter) and also acting as a sort of auxiliary ego for Mrs. B., who was helplessly acting out an introjected sense of her hating mother when she abused her own child. Additionally, to the extent that the mistreatment of a child is an objectively horrifying act, aspects of the therapist's response may be related to objective countertransference factors. The confluence of the therapist's subjective countertransference with the patient's own transference needs probably minimized the degree to which the former interfered with the therapy. Nevertheless, later developments in the treatment suggest that a price was paid for this rapid resolution of the patient's abusive behavior. This was the postponement of the analysis and the working through of Mrs. B.'s own hate toward her parents and the therapist, and of her experience of herself as a victim in the therapeutic situation. This process did not begin until the patient (and therapist) were no longer focused on the patient as an active harmdoer.

In summary, then, the early phases of Mrs. B.'s therapy involved an analysis of her role as an innocent victim, a responsible victim, and as a harmdoer. It seemed necessary that each aspect of Mrs. B.'s experience be mirrored by the therapist, and thus that the therapist at times experience her with an empathic, objective, and blaming response. Interestingly, as Mrs. B.'s sense of merger with her daughter dissolved, she no longer had an emotional investment in the therapist's intervening on behalf of her daughter. The treatment then moved toward an investigation of her adult life (and her sexual life) with her husband (where she is also both a victim and a victimizer).

These clinical data raise questions about the effects of different external stances (e.g., blaming, empathic) on the victim and on the harmdoer. Empirical studies might investigate the nature of the external circumstances that determine whether a blaming, objective, or empathic response most facilitates action on the part of the victim, or reparation on the part of the harmdoer.

INJUSTICE AND PSYCHOANALYTIC TREATMENT: GENERAL IMPLICATIONS

Although this chapter is mainly concerned with the treatment of victims and harmdoers, an argument can be made for the application of these dynamics to virtually any treatment situation. To the extent that we all, as children, were helpless in the face of both parental responses and environmental impingements, we have all been the innocent victims of our own life circumstances. Additionally, we all have either directly or indirectly elicited the hurtful responses of our parents (and then, repetitively, of other adults), and in this sense have become responsible for these experiences. It is also probably true that the internal sense and expression of feelings of hate and destructiveness toward loved ones is a universal phenomenon. In this sense, any in-depth psychoanalytic treatment would require that the patient come to terms with those reality and fantasy experiences in which one was a victim and a harmdoer.

Although it is often necessary in a social context to free victims from any responsibility for their fate in order to actually rectify reality injustices that have been perpetrated, different issues emerge in a treatment context. Here, the therapeutic process often involves helping patients integrate a sense of themselves as responsible for the victimization with an identification of those external factors that were, in fact, harmful (or helpful). It is probably most common for the therapy to begin with an exploration of the patient's early environment and its impact on the patient. In this sense, the first phase of therapy often involves an analysis of the patient as innocent victim. Later, through an analysis of the reenactment of these experiences in the transference and in the current life situation, the ways in which the patient created (and creates) his or her own bad experiences, that is, is a responsible victim, become clarified.[8] It is in the transference reenactment that the patient can reexperience the self in the original situation and yet make and receive a different, curative emotional response.

> In psychoanalysis as we know it there is no trauma that is outside the individual's omnipotence. Everything eventually comes under ego-control, and thus becomes related to the secondary processes. The patient is not helped if the analyst says: "Your mother was not good enough . . .' Changes come in an analysis when the traumatic factors enter the psychoanalytic material in the patient's own way and within the patient's omnipotence. The interpretations that are alternative are those that can be made in terms of projection. . . . The paradox is that what is good and bad in the infant's environment is not in fact a projection. (Winnicott, 1958, pp. 37–38)

REFERENCES

Adams, J. S. Toward an understanding of inequity. *Journal of Abnormal and Social Psychology*, 1963, *67*, 422–436.

Adams. J. S. Inequity in social exchange. In L. Berkowitz (Ed.), *Advances in experimental social psychology* (Vol. 2). New York: Academic Press, 1965.

Aderman, D., Brehm, S. S., & Katz, L. B. Empathic observation of an innocent victim: The just world revisited. *Journal of Personality and Social Psychology*, 1974, *29*, 342–347.

Aronson, E., Carlsmith, J. M., & Darley, J. M. The effects of expectancy on volunteering for an unpleasant experience. *Journal of Abnormal and Social Psychology*, 1963, *66*, 220–224.

Austin, W., & Walster, E. Reactions to confirmations and disconfirmations of expectancies of equity and inequity. *Journal of Personality and Social Psychology*, 1974, *30*, 208–216.

Berscheid, E., & Walster, E. When does a harmdoer compensate a victim? *Journal of Personality and Social Psychology*, 1967, *6*, 435–441.

Berscheid, E., Walster, E., & Barclay, A. Effect of time on tendency to compensate a victim. *Psychological Reports*, 1969, *25*, 431–436.

[8]This sequence is certainly not linear, and involves a continual back-and-forth movement between understanding the patient as a responsible and an innocent victim, despite this overall pattern.

Bloch, D. *So the witch won't eat me.* Boston: Houghton Mifflin, 1978.

Brock, T. C., & Becker, L. A. "Debriefing and susceptibility to subsequent experimental manipulations. *Journal of Experimental Social Psychology*, 1966, *2*, 314–323.

Brock, T. C., & Buss, A. H. Dissonance, aggression, and evaluation of pain. *Journal of Abnormal and Social Psychology*, 1962, *65*, 197–202.

Brock, T. C., & Buss, A. H. Effects of justification for aggression and communication with the victim on postaggression dissonance. *Journal of Abnormal and Social Psychology*, 1964, *68*, 403–412.

Bulman, R. J., & Wortman, C. Attributions of blame and coping in the "real world": Severe accident victims react to their lot. *Journal of Personality and Social Psychology*, 1977, *35*, 351–363.

Burger, J. Motivational biases in the attribution of responsibility for an accident: A meta-analysis of the defensive attribution hypothesis. *Psychological Bulletin*, 1981, *90*, 496–512.

Burgess, A. W., & Holmstrom, L. L. Adaptive strategies and recovery from rape. *American Journal of Psychiatry*, 1979, *136*, 1278–1282.

Burt, M. R. Cultural myths and supports for rape. *Journal of Personality and Social Psychology*, 1980, *38*, 217–230.

Cahn, E. *The sense of injustice.* Bloomington: Indiana University Press, 1949.

Chaiken, A. L., & Darley, J. M. Victim or perpetrator? Defensive attribution of responsibility and the need for order and justice. *Journal of Personality and Social Psychology*, 1973, *25*, 268–275.

Cialdini, R. B., Darby, B. L., & Vincent, J. E. Transgression and altruism: A case for hedonism. *Journal of Experimental Social Psychology*, 1973, *9*, 502–516.

Comer, R., & Laird, J. D. Choosing to suffer as a consequence of expecting to suffer: Why do people do it? *Journal of Personality and Social Psychology*, 1975, *32*, 92–101.

Crosby, F. A model of egoistical relative deprivation. *Psychological Review*, 1976, *83*, 85–113.

Crosby, F. *Relative deprivation and working women.* New York: Oxford University Press, 1982.

Crosby, F. Relative deprivation in organization settings. In B. M. Staw & L. L. Cummings (Eds.), *Research in organizational behavior (Vol. 6).* Greenwich, Conn.: JAI Press, 1984.

Davis, K. E., & Jones, E. E. Changes in interpersonal perception as a means of reducing cognitive-dissonance. *Journal of Abnormal and Social Psychology*, 1960, *61*, 402–410.

Deutsch, M. Awakening the sense of injustice. In M. Ross, & M. Lerner (Eds.), *The quest for justice.* New York: Holt, Rinehart, & Winston, 1976.

Drabeck, T. E., & Quarantelli, E. L. Scapegoats, villains and disasters. *Trans-Action*, 1967, 4(4), 12–17.

Epstein, L. The therapeutic function of hate in the countertransference. In L. Epstein, & A. H. Feiner (Eds.), *Countertransference.* New York: Jason Aronson, 1979.

Foxman, J., & Radtke, R. C. Negative expectancy and the choice of an aversive task. *Journal of Personality and Social Psychology*, 1970, *15*, 253–257.

Freedman, J., Wallington, S., & Blass, E. Compliance without pressure: The effect of guilt. *Journal of Personality and Social Psychology*, 1967, *7*, 117–124.

Freud, S. The future prospects of psychoanalytic therapy. In J. Strachey (Ed. and Trans.), *The standard edition of the complete psychological works of Sigmund Freud* (Vol. 11). London: Hogarth Press, 1910.

Freud, S. Some character types met with in psychoanalytic work. In J. Strachey (Ed. and Trans.), *The standard edition of the complete psychological works of Sigmund Freud* (Vol. 14). London: Hogarth Press, 1916.

Freud, S. Beyond the pleasure principle. In J. Strachey (Ed. and Trans.), *The standard edition of the complete psychological works of Sigmund Freud* (Vol. 18). London: Hogarth Press, 1920.

Freud, S. *New introductory lectures on psychoanalysis*. New York: W. W. Norton, 1933.

Frieze, I. Perceptions of battered wives. In I. Frieze, D. Bar-tel, & J. Carroll (Eds.), *New approaches to social problems*. San Francisco: Jossey-Bass, 1979.

Frieze, I. *Self perceptions of the battered woman*. Paper presented at the meeting of the Association of Women in Psychology, Pittsburgh, March 1978.

Heider, F. *The psychology of interpersonal relations*. New York: Wiley, 1958.

Homans, G. C. *Social behavior: Its elementary forms*. New York: Harcourt, Brace & World, 1961.

Janoff-Bulman, R. Characterological versus behavior self-blame: Inquiries into depression and rape. *Journal of Personality and Social Psychology*, 1979, *37*, (10), 1798–1809.

Janoff-Bulman, R., & Frieze, I. A theoretical perspective for understanding reactions to victimization. *Journal of Social Issues*. 1983, *39*(2), 1–17.

Jones, C., & Aronson, E. Attribution of fault to a rape victim as a function of respectability of the victim. *Journal of Personality and Social Psychology*, 1973, *26*(3), 415–419.

Klein, M., & Riviere, J. *Love, hate and reparation*. London: Hogarth Press, 1937.

Lerner, M. J. The desire for justice and the reactions to victims. In J. Macaulay, & L. Berkowitz (Eds.), *Altruism and helping behavior*. New York: Academic Press, 1970.

Lerner, M. J. The justice motive: Some hypotheses as to its origins and forms. *Journal of Personality*, 1977, *45*, 1–52.

Lerner, M. J., & Matthews, G. Reactions to the suffering of others under conditions of indirect responsibility. *Journal of Personality and Social Psychology*, 1967, *5*, 319–325.

Lerner, M. J., & Miller, D. T. Just world research and the attribution process: Looking back and ahead. *Psychological Bulletin*, 1978, *85*, 1030–1151.

Lerner, M. J., & Simmons, C. H. Observer's reaction to the "innocent victim": Compassion or rejection? *Journal of Personality and Social Psychology*, 1966, *4*, 203–210.

Lerner, M. J., Miller, D. T., & Holmes, J. G. Deserving and the emergence of forms of justice. In L. Berkowitz, & E. Walster (Eds.), *Equity theory: Toward a general theory of social interaction*. New York: Academic Press, 1976.

Leventhal, G. S., Weiss, T., & Long, G. Equity, reciprocity and reallocating rewards in the dyad. *Journal of Personality and Social Psychology*, 1969, *13*, 300–305.

Libow, J. A., & Doty, D. An exploratory approach to self-blame and self-derogation by rape victims. *American Journal of Orthopsychiatry*, 1979, *49*, 670–679.

Lincoln, A., & Levinger, G. Observers' evaluations of the victim and the attacker in an aggressive incident. *Journal of Personality and Social Psychology*, 1972, *22*, 202–210.

Little, M. "R"—The analyst's total response to the patient's needs. *International Journal of Psychoanalysis*, 1957, *38*, 240–254.

MacDonald, A. P., Jr. More on the Protestant ethic. *Journal of Consulting and Clinical Psychology*, 1972, *39*, 116–122.

Martin, D. Battered women: Society's problem. In J. R. Chapman, & M. Gates (Eds.), *The victimization of women*. Beverly Hills, Calif.: Sage Publications, 1978.

Maslach, C. Burned-out. *Human Behavior*, September 1976, 16–22.

Mills, J., & Egger, R. Effect on derogation of a victim of choosing to reduce his distress. *Journal of Personality and Social Psychology*, 1972, *23*, 405–408.

Notman, M. T., & Nadelson, C. C. The rape victim: Psychodynamic considerations. *American Journal of Psychiatry*, 1976, *133*, 408–412.

Panken, S. *The joy of suffering*. New York: Jason Aronson, 1973.

Peterson, C., Schwartz, S. M., & Seligman, M. P. Self-blame and depressive symptoms. *Journal of Personality and Social Psychology*, 1981, *41*, 253–259.

Pettigrew, T. F. Three issues in ethnicity: Boundaries, deprivations, and perceptions. In J. M. Yinger & S. J. Cutler (Eds.), *Major social issues: A multidisciplinary view*. New York: Free Press, 1978.

Pfouts, J. H. Violent families: Coping responses of abused wives. *Child Welfare*, 1978, *57*, 101–111.

Racker, H. *Transference and countertransference*. New York: International Universities Press, 1968.

Rawls, J. *A theory of justice*. Cambridge: Harvard University Press, 1971.

Regan, D. T., & Totten, J. Empathy and attribution: Turning observers into actors. *Journal of Personality and Social Psychology*, 1975, *32*, 850–856.

Rosenman, S. Compassion versus contempt toward holocaust victims: Difficulties in attaining an adaptive identity in an annihilative world. *Israel Journal of Psychiatry*, 1982, *19*, 39–73.

Ross, M., & DiTecco, D. An attributional analysis of moral judgments. *Journal of Social Issues*, 1975, *31*, 91–109.

Rubin, Z., & Peplau, L. A. Who believes in a just world? *Journal of Social Issues*, 1975, *31*, 65–89.

Schram, D. D. Rape. In J. E. Chapman, & M. Gates (Eds.), *The victimization of women*. Beverly Hills, Calif.: Sage Publications, 1978.

Shaver, K. G. Defensive attribution: Effects of severity and relevance on the responsibility assigned for an accident. *Journal of Personality and Social Psychology*, 1970, *14*, 101–113.

Simons, C., & Piliavin, J. A. Effect of deception on reactions to a victim. *Journal of Personality and Social Psychology*, 1972, *21*, 56–60.

Smith, R. E., Keating, J. P., Hester, R. K., & Mitchell, H. E. Role and justice considerations in the attribution of responsibility to a rape victim. *Journal of Research in Personality*, 1976, *10*, 346–357.

Sorrentino, R. M., & Boutilier, R. G. Evaluation of a victim as a function of fate similarity/dissimilarity. *Journal of Experimental Social Psychology*, 1974, *10*, 84–93.

Sorrentino, R. M., & Hardy, J. E. Religiousness and derogation of an innocent victim. *Journal of Personality*, 1974, *42*, 372–382.

Spotnitz, H. *Modern psychoanalysis of the schizophrenic patient*. New York: Grune & Stratton, 1969.

Steil, J. The response to injustice: Effect of varying levels of social support and position of advantage and disadvantage. *Journal of Experimental Social Psychology*, 1983, *19*, 239–363.

Steil, J., Tuchman, B., & Deutsch, M. An exploratory study of the meanings of frustration and injustice. *Personality and Social Psychology Bulletin*, 1978, *4*, 393–398.

Steil, J., Levey, C., & Sanders, A. *The response to injustice: Effects of perceived self-competency and position of advantage or disadvantage*. Paper presented at the meetings of the Eastern Psychological Association, Baltimore, April 1982.

Stokols, D., & Schopler, J. Reactions to victims under conditions of situational detachment: The effects of responsibility, severity, and expected future interaction. *Journal of Personality and Social Psychology*, 1973, *25*, 199–209.

Sykes, G. M., & Matza, D. Techniques of neutralization: A theory of delinquency. *American Sociological Review*, 1957, *22*, 664–670.

Symonds, M. Victims of violence. *The American Journal of Psychoanalysis*, 1975, *35*, 19–26.

Taylor, S., Wood, J., & Lichtman, R. It could be worse: Selective evaluation as a response to victimization. *Journal of Social Issues*. 1983, *39*(2), 19–40.

Tillich, P. *Love, power, and justice*. Oxford University Press, 1954.

Walster, E. Assignment of responsibility for an accident. *Journal of Personality and Social Psychology*, 1966, *3*(1), 73–79.

Walster, E., Walster, W., & Berscheid, E. *Equity: Theory and research*. Boston: Allyn & Bacon, 1978.

Weinglass, J., & Steil, J. *When is unequal unfair: the role of ideology*. Paper presented at the meeting of the American Psychological Association, Los Angeles, 1981. (ERIC: Clearinghouse Document No. ED 218 530, Resources in Education, December 1982.)

Winnicott, D. W. Hate in the countertransference. In D. W. Winnicott, *Through pediatrics to psychoanalysis*. New York: Basic Books, 1958.

Winnicott, D. W. *The maturational processes and the facilitating environment: Studies in the theory of emotional development*. New York: International Universities Press, 1965.

Wortman, C. B. Coping with victimization: Conclusions and implications for future research. *Journal of Social Issues*, 1983, *39*(2), 195–221.

III

DEVELOPMENTAL PSYCHOLOGY

This section of the book focuses on four developmental periods and the complex issue of gender-role identity. Vickers, in Chapter 11, is concerned with several lines of research that focus on the development of infant competence during the first 18 months of life. After a careful examination of the literature on vision, hearing, and attachment, he concludes that competence in these areas depends upon the successful infant–caretaker negotiation of early tasks. Vickers presents a case vignette to illustrate how encountering a nonreciprocating preverbal environment thwarts the development of a sense of self.

Jonathan and Sally Bloom-Feshbach are concerned with the period from ages 3 to 12, a portion of life that, for the psychodynamic clinician, includes the crucial oedipal period and the critical developments of the latency stage. They show how a knowledge of research findings will lead to some specific technical approaches, and illustrate their theoretical speculations with clinical examples. Chapter 12, because of space limitations, focuses on a few areas to make the general point. These areas are the relationship between parental childrearing styles and children's social and emotional functioning, the developing metacognitive abilities of children, and the importance of fathers and peers in child development. Each topic is important in its own right, and taken together they demonstrate the value of the authors' perspective.

One of the primary tasks of childhood, bridging infancy and the early childhood years, is the development of appropriate gender role. This involves biological, psychological, and social factors, each of which categories is considered by Lee in Chapter 13. The biological circumstances are the primary determinant of the child's sex, but the social and psychological are the major influence on gender role, or the manner in which children view their own sex and behave in ways consistent or deviant from stereotypes for that sex. As perceptions and behavior become increasingly deviant from social expectations, it becomes more likely that the child will be presented as a patient. Lee describes the entangled factors that contribute to gender role, and gives examples of the clinical implications of the findings bearing on this issue.

Adolescence lies between childhood and adulthood, and is sufficiently different from each to warrant separate study. It is vastly overrepresented in the popular press and public concern, and correspondingly underrepresented in the research literature. More interestingly, even the limited research findings that have appeared are often misrepresented in clinical and public debate, so that we have an area that not only has failed to incorporate research findings, but in which research findings are pointedly contradicted. Adelson, in Chapter 14, attempts to correct this problem by bringing together a number of findings about such issues as normalcy in adolescence and relationships of adolescents to their parents. Clinical applications are interwoven with the presentation of empirical findings.

As the child moves along the developmental continuum, the likelihood of receiving research attention decreases. Just as we have seen a reduction in information about adolescents, the failure to attend to adult life is even more striking. It is only within the past decade that approaches to midlife development have drawn attention, and even this remains limited and selective. Lasky, in Chapter 15, brings this material together, presenting some clinical speculations about adult development and about the attempts to develop a body of empirical literature to support these speculations. Using clinical illustrations, she clearly asserts that development does not necessarily stagnate at any point in the life cycle, and that a knowledge of research can help to develop an appropriate treatment plan for patients.

11

Implications of Infant Development Research for Clinical Practice

RICHARD L. VICKERS

INTRODUCTION

The concept of continuity in human development has long been a provocative subject. The riddle of the Sphynx drew an ironic parallel between the inadequacy of infancy and the infirmity of old age. When Oedipus solved the riddle and freed the city of Thebes, the Sphynx killed herself in disgust, not having expected anyone to answer her question. It took the genius of Freud to discover that psychological causes operate from the distance of childhood in a particular way, through transformations of the psychosexual stages and especially the Oedipus complex. Many objected to this answer too, but it has come to represent the dominant view of psychodynamic theory.

During the last 20 years psychoanalytic theory has become increasingly focused on developmental issues preceding the oedipal phase, with concomitant disagreement over issues of developmental transformation and continuity. At the same time our view of infancy has been revised. Basic perspectives on infancy are being challenged by technological advances in at least three areas: (1) accumulation of more sophisticated research than was previously available, (2) application of life

RICHARD L. VICKERS • Institute of Advanced Psychological Studies, Adelphi University, Garden City, New York 11530.

support techniques making it possible to save infants as young as 27 weeks gestational age and as small as 1000 g, and (3) progress in neurochemical, genetic, and evolutionary theory defining a psychobiology of human life. This progress has extended the concept of extrauterine life backward by several months and defined the full-term infant as far more sentient, biologically complex, and psychologically active than was formerly imagined. The present view of a self-impelled "competent" infant contrasts with earlier paradigms depicting babies as programmed by maturational imperatives (Gessell), controlled by stimulus–response contingencies (behavioral learning theory), or held away by somatic pressures (psychoanalytic drive theory).

Conceptual revisions taking place today are fundamental and contain implications for clinical theory and practice. Some have been realized: The National Center for Clinical Infant Programs and the World Association for Infant Psychiatry represent new areas of clinical work. The third *Diagnostic and Statistical Manual of Mental Disorders* (American Psychiatric Association, 1980) contains the first listing of disorders "first evident in infancy, childhood, or adolescence," thereby extending psychodiagnostic practice to a new clinical population. Reviews of infant work appear with some regularity in psychoanalytic journals, and these discussions illustrate the complex extrapolation required to establish continuity between infant development and adult personality. Do meaningful connections exist? There is no lack of speculation about pre-oedipal personality development in psychoanalytic writing, but most of this theorizing is retrospective, that is, reconstruction based upon the patient's "archaic transference." Examples include statements about the origin of preambivalent defenses such as splitting and projective identification, topics receiving much attention in current literature on the narcissistic and borderline personality. Disagreement abounds over whether retrospective hypotheses elucidate early development or simply fantasies about early development, that is, current formations that are not really contiguous with early mental life.

Infant research contributes a prospective view of personality development to these discussions. Studies on security and resiliency of attachment, for example, are very much in line with clinical issues. Babies, because they cannot use language, force researchers to study actual events. Concretion has advantages in portraying phenomena that become indistinct to the practitioner because of excessive redefinition of terms in clinical theory. This approach also accommodates the urgings of Kohut (1971), Khan (1974), and others that psychoanalysis move away from its preoccupation with gross events in early life and become more attuned to the detailed and cumulative nature of trauma. A historical comparison may be drawn between present-day interest in pre-oedipal disorders, research on infancy, and the era in which Anna Freud (1973) first applied analytic understanding to direct observation of children. Nearly 40 years ago Hartmann and Kris (1945) urged that this work continue and suggested that direct observation would have particular success in defining conflict-free spheres of development. Research on infancy has become so productive, complex, and textured that a new era of more comprehensive understanding of early development is underway.

The task at hand is to survey findings and practice from a personal perspective that includes both analytic practice and infant intervention. A complete review of the literature is beyond the scope of this chapter and the reader is referred to Osofsky's (1979) state-of-the-art summary. My emphasis will be on several lines of research that establish infant "competence" during the first 16 months of life. I shall then discuss the implications of these findings for developmental and clinical theory. In the final section I shall present clinical applications of these concepts, first in adult work and then in the field of infant intervention.

REARCH ON INFANT DEVELOPMENT: BASIC COMPETENCE

Infant research is complex. It ranges across multiple, interrelated lines of development rather than the unitary sequences most familiar to clinicians. Included are physical and neurological maturation, cognitive development, social behavior, and various other adaptive capacities. Infants are difficult to study because momentary fluctuations of state are superimposed on various maturational developments, all in the context of intense exchange with the primary caregiver and the research scientist. Although methodological problems are formidable, adequate strategies have been created to maintain experimental clarity (see Escalona, 1976). Many findings are supported by convergent data obtained through different methods: Examples of correlated data include degree of nerve tract myelinization, cortical evoked potential, and sensory discrimination level. Natural experiments have also provided opportunity for observation outside the laboratory: Fraiberg's (1977) poignant work with congenitally blind infants demonstrates the resilience of adaptive capacities in face of major disability. These findings give dramatic emphasis to the concept of "competence," which has come to summarize infant studies from the period 1960 to 1980.

Infants are competent in the sense of the general definition "capacity equal to requirement," in this case having a repertoire of skills equal to developmental tasks under a range of conditions. All of the senses are functioning in some manner at birth. Early visual experience provides an important basis for social exchange: Ocular muscles are among the first small muscles to come under volitional control. Visual pathways are partially myelinated at birth; the fovea is underdeveloped; ciliary muscles are too weak to adjust the lens. A newborn's eyes, therefore, work like a camera using low-resolution film with focus fixed at about 19 cm, which approximates the face-to-face distance between mother and nursing baby. Estimates of visual acuity range from 20/150 (Dayton, Jones, Ain, Rawson, Steel, & Rose, 1964) to 20/800 (Fantz, Ordy, & Udelf, 1962), and this resolves to 20/20 at 4 months when studied by evoked potential methods (Sokol & Dobson, 1976). The neonate makes conjugate eye movements in order to track a moving stimulus (Wickelgren, 1967) but cannot maintain a stable retinal image of the moving object (Dayton, Jones, Steele, & Rose, 1964). Babies first show interest in high-contrast edges, curved lines, and simple patterns (Salapatek, 1975); by 8 to 10 weeks attention is held by more complex stimuli (Brennan, Ames, & Moore, 1966). Color vision is well documented by 10 weeks (Fagan, 1974; Oster, 1975) and evolves to categorical color perception by 16 weeks (Bornstein, Kessen, & Weiskopf, 1976). Simple discriminatory skills develop into more complex conceptual abilities toward the middle of the first year of life. These include some degree of size, shape, and even object constancy (Cohen, DeLoache, & Strauss, 1979; Day & McKenzie, 1971), as well as ability to perceive compound shapes as organized gestalts, for example, configurations of facial expression (Fagan, 1976). A full discussion of visual processes and perception of space, sound, and speech in infancy can be found in the two review volumes by Cohen and Salapatek (1975a, 1975b).

Infants clearly do not perceive their surroundings as the "dimly moving masses" imagined by Freud (1896/1966); instead they are avid seekers and processors of visual information. Nor do they hear the world as the "blooming, buzzing confusion" once postulated by William James (1913/1980). Auditory sensation is present before birth (Bernard & Sontag, 1947). Acuity is hampered by mesenchymal tissue trapped in the middle ear for a few days after delivery, and auditory thresholds are higher than for an adult. These decrease from 3 to 8 months. The infant can discriminate sounds by pitch (Engel & Young, 1969), intensity (Bartoshuk, 1964; Eisenberg, 1976), duration (Clifton, Graham, & Hatton, 1968), and location (Hammond, 1970) within the first few weeks. Infants prefer frequencies from 1000 to to 3000 Hz, in the range of the female voice (Keele & Neil, 1965). In fact, the elevated pitch, repetition, exaggerated intonation, and singsong cadence of adult "baby talk" match the infant's tonal preferences and information-processing abilities to a remarkable degree. Infants prefer complex sounds (S. Hutt, Hutt, Lenard, von Bernuth, & Muntjewerff, 1968), are distressed by high-frequency pure tones and low-frequency square waves (C. Hutt, von Bernuth, Lenard, Hutt, & Prechtl, 1968), startled by loud or sudden noises (Peiper, 1963), and soothed by various low-frequency sounds and continuous or rhythmic auditory stimulation (Brackbill, Adams, Crowell, & Gray, 1966). The research data indicate that infants possess well-developed auditory skill soon after birth.

Recent findings indicate that very young infants also show intermodel perception, that is, syncretic integration across sense modalities. Thus, 4-month-olds prefer films having synchronous movement and sound (Kuhl & Meltzoff, 1982). Meltzoff and Borton (1979) inserted two kinds of nipples—nubbed and smooth—into the mouths of 4-week-old infants wearings blindfolds and discovered visual-tactile fluency: With blindfold removed, 25 of 32 infants immediately directed their gaze to the type of nipple placed in their mouths. More startling are reports of synchronous responses of infants a few hours old to human speech (Condon & Sander, 1974), and even the ability to imitate complex facial expressions with 76% accuracy during the second day of life (Field, Woodson, Greenberg, & Cohen, 1982).

These findings introduce the area of social interaction and reveal the complex behavioral repertoire infants bring to bear upon prelinguistic communication. As summarized by Bruner (1974, 1975), Harris (1975), and Ryan (1974), and recently critiqued by Gratch (1979), research on early forms of communication and understanding has shifted focus from acquisition of grammar and syntax to problems of meaning and communication. Topics include nonverbal origins of communication (Stern, 1976), development of mutual reference (Lempers, Flavell, & Flavell, 1975), and beginnings of intersubjectivity (Trevarthen & Hubley, 1978). Gratch sees the focus on reciprocity (Brazelton, Koslowski, & Main, 1974) and dialectic (Riegel, 1975) as a search for the origins of thought and affect in terms of the infant's social matrix. He interprets this trend as a partial rejection of the Piagetian "metaphor" of the construction of reality and a reintroduction of the "game" metaphor of social theoreticians such as George Herbert Meade and Harry Stack Sullivan. Stern's (1976) approach is a case in point. Using frame-by-frame microanalysis of infant–caretaker interactions, Stern paid particular attention to episodes of eye contact and associated postural adjustments by both partners. He noted reliable difference between mother–infant pairs in terms of duration of contact and frequency with which each partner initiated and terminated episodes. He comments, as have others, on the remarkable synchrony of motion between mother and infant, and noted that approach or withdrawal behavior initiated by one partner receive complementary withdrawal or approach response by the other within 1.5 sec. He likens this to a waltz danced poorly or well according to each partner's ability to anticipate timing, rhythm, and sequence, and to learn from mistakes. In the best case a pattern becomes established in which the competent infant presents discernible cues to a mother whose responses are good (contingent) enough to establish a coherent exchange, that is, a balance of influence in which neither partner becomes overwhelmed or confused by the other. Deviations from an optimal range may be characterized by terms familiar to the clinician—"avoidant," "controlling," "overstimulating"—and are considered to affect the coping mechanisms of both partners. Success gained through mutual influence combined with intermodal communication helps to initiate mutual empathy and intersubjective awareness. Using this approach Stern has documented continuity of interactive patterns from 3 to 14 months, and suggests that infants' internal representations begin to coalesce around repeated experiences of this nature. Behavioral competencies are emphasized, especially the captivating quality of a infant's gaze which Stern labels a "cardinal attachment behavior."

These findings mesh well with Ainsworth's classic studies of attachment behavior (Ainsworth, 1979), Bowlby's (1969) major theoretical work in this area, and the follow-up studies of Sroufe (1979a, 1979b) and others demonstrating consequences of early attachment behaviors with through age 5. The key to predictability lies in tracing adaptive themes through a course of age-specific tasks. Thus infants who are "securely attached," that is, effective at seeking reunion with their caretakers following stressful separation, emerge as toddlers more able to engage in problem solving on their own, more flexible in impulsive controls, and more competent with their peers. Infants who are "anxiously attached," that is, mix contact seeking with angry resistance, later have difficulty using their caretakers as a secure base for exploration. Finally, infants whose early attachment is "avoidant," that is, whose reunion behavior is angry and rejecting toward the caretaker, may appear precociously independent as babies but later show signs of impoverished resources. These findings are striking in predictive reliability and make clear the point that continuity of adaption is epigenetic. In fact, a sort of developmental paradox exists in that infants who are effectively dependent become more autonomous and independently competent as preschoolers. Attachment research underscores the need for a coherent and satisfying signal system between infant and caretaker during the first half year of life, and findings bridge the data on neonatal competence and Mahler, Fine, and Bergman's (1975) work on separation-individuation. The point is not that personal resources are fatefully determined by early experience, but rather that there is a coherence of individual development based upon adaptive/maladaptive negotiation of early interpersonal tasks. Relative success or failure at each developmental step depends upon the assemblance of expressive and receptive skills, as well as the affective resources both partners bring to bear upon the issue at hand.

At a higher level of abstraction, infant researchers have discussed the ontogenesis and developmental transformation of affect for many years, beginning with Bridges's (1932) month-by-month analysis of emotional development. Genetic-dynamic theories are intimately related to clinical theory and practice, for example, in Kernberg's (1975) postulate that pre-oedipal defenses become organized "under the influence" of libidinal (good) and aggressive (bad) drive derivatives. I understand this to mean drives transformed into perceptual-affective dispositions, essentially affective organizations that become major problems for the borderline personality when they remain fixated in their early "nonmetabolized" state. Researchers have approached questions of emotional development from various perspectives: (1) searching for prototypes of mood and affect (Thomas, Chess, Birch, Hertzig, & Korn, 1963); (2) defining milestones of affective organization such as smiling, stranger anxiety, and negativism (Emde, Gaensbauer, & Harmon, 1976; Spitz, 1965); (3) tracking the origin of specific affect systems such as pleasure-joy, wariness-fear, and rage-anger (Sroufe, 1979b); and (4) examining the interface between affect and cognitive or ego development (Brody & Axelrod, 1970; Loevinger, 1976).

There is consensus that true emotions, as opposed to reflexes, require some degree of cognitive engagement. Traditional developmental perspectives view emotional and cognitive growth in parallel terms: Both are subject to epigenetic reorganization. Primitive states are evoked by enteroceptive stimulation. These are the obligatory reflexes and global states of the first few weeks of life. Second, affective prototypes help engage caretakers in reciprocal exchange and introduce some awareness of external reality. Third, true emotion is a hallmark of rudimentary self or ego formation: Temporal awareness involving memory and anticipation comes into play. Finally, full emotional life presupposes object constancy and evocative memory, leading to interplay among affect, symbolism, and language: These are the terms of oedipal and ultimately tripartite structures and conflicts.

Critical issues concern how linkages between affect and meaning take place before development of symbolic thought. Can subjective experience antedate self-differentiation? Can such experience be encoded? Does it have developmental continuity? The literature on competence suggests that continuity exists and organizing modes include motoric dispositions in addition to perceptual-affective self- and object representations. For example, infants' gaze aversion is temporally correlated with peaks of heart rate acceleration during approach by a stranger (Waters, Matas, & Sroufe, 1975). This may be seen as both a defensive operation and what might be called a motor affect. The concept of organ affect has long been recognized in psychoanalysis (Fenichel, 1945). Are motor phenomena also subject to developmental fixation, for example, in the idiosyncrasies of gaze demonstrated by adult paranoid personalities? Instead of conflict-related disturbances, disorders of internal regulation, emotional control, and self-regard are now commonplace in clinical work and notoriously difficult to treat (Kernberg, 1975; Kohut, 1971). Developmental research frames and articulates issues of pre-oedipal adaptation that have direct bearing on these clinical concerns. The general indication is a coherence of individual development based upon adaptive/maladaptive regulation of internal states. Relative success–failure depends partly on ability of the infant–caretaker unit to link these states with phase-appropriate reality, that is, to place them in meaningful contexts.

DISCUSSION

UPDATING DEVELOPMENTAL AND CLINICAL THEORY

Visual and auditory competence, attachment behavior, and emotional organization are current topics of interest in infant research. In this section I shall consider findings that are pertinent to developmental and clinical theory, noting initially that the need for theoretical revision is a matter of dispute.

Those who question such need (1) ask whether data obtained under laboratory conditions reflect typical life experiences, (2) point to immaturity of the cerebral cortex during the first months of life, (3) suggest that growth processes are too discontinuous for reductionist theoretical revision, (4) maintain that phase dominance of drives cannot be ignored, and (5) state that symbolic information of the type retrieved in treatment does not appear until well into the third year of life. Some clarification of issues can be attempted, first, by reviewing the general theories of pre-oedipal development shown in Figure 1. These schemata reflect issues of concern to each theorist, and they can be viewed as heuristic devices offering the clinician a range of interpretive frameworks. Examining the

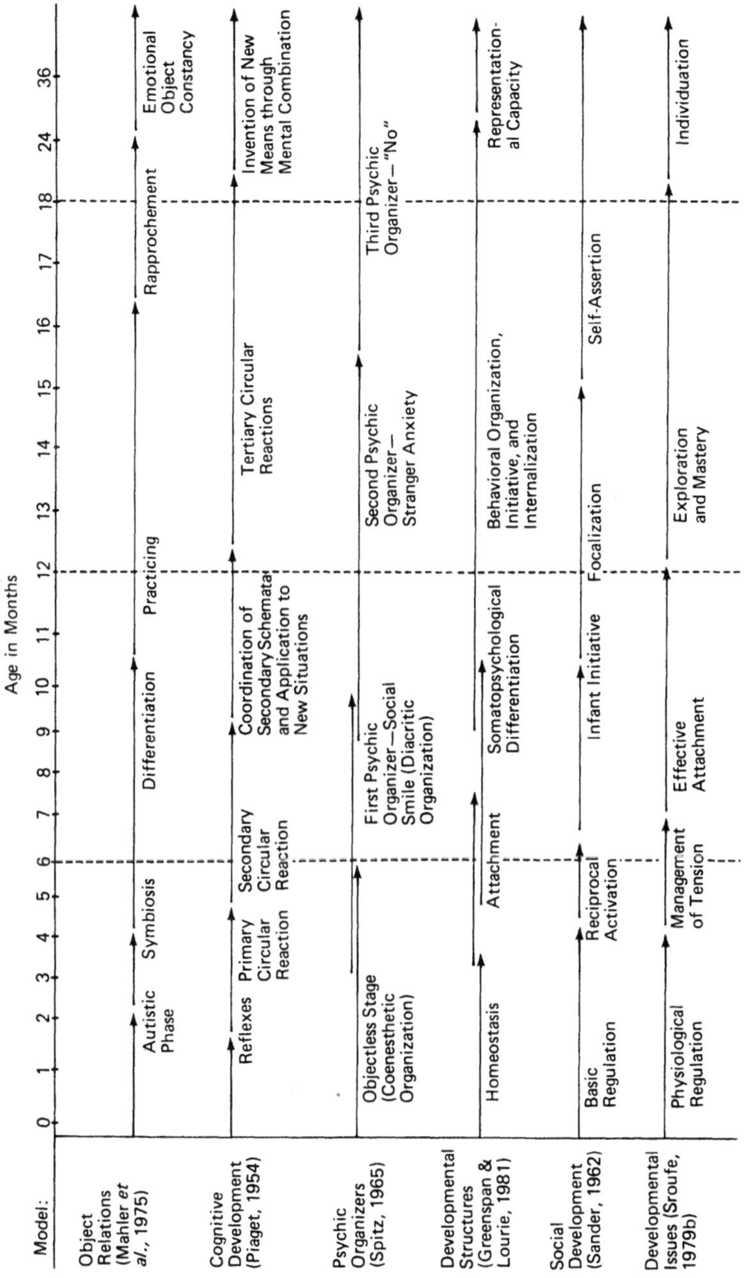

FIGURE 1. Schematic models of development in infancy.

first 6 months, a contrast emerges between theorists who emphasize unilinear differentiation (Mahler *et al.*, 1975; Piaget, 1954; Spitz, 1965) and theorists who emphasize interactional and regulatory process (Greenspan & Lourie, 1981; Sander, 1962; Sroufe, 1979b). During the second 6 months a contrast emerges between further differentiation and individuation (Greenspan & Lourie, 1981; Mahler *et al.*, 1975; Piaget, 1954; Spitz, 1965) and attachment and initiative (Sander, 1962; Sroufe, 1979b). Finally, during months 12 to 18 a more consensual focus appears on related themes involving exploration, assertion, and mastery, all associated with locomotion. Recent findings in infant development emphasize the latter categories listed during the first year of life: that is, regulatory process, interaction, and initiative. This seems especially clear during the first 6 months because of data manifestly incompatible with traditional points of view. Neonates who imitate facial expression are accomplishing a Stage 4 achievement in Piagetian theory. Moreover, findings regarding intermodal sensory organization from 1 to 4 months suggest that innate abilities operate to a degree not formerly recognized in such young infants.

Such complex abilities seem unlikely to be subcortical or vestigial in nature like the grasp or Moro reflex. Although functional appearance of these capacities is limited to relatively infrequent periods of alert inactivity, that they exist at all contradicts a number of clinical and metapsychological assumptions (see Wolff, 1960). Among these are the "bird-in-egg" model of a normal autistic phase (Mahler *et al.*, 1975), in which any rupture of the stimulus barrier is met by a global, diffuse, syncretic response, as well as the belief that hallucinatory wish fulfillment is the highest mental achievement reached during this period. Peterfreund (1979) has gone so far as to argue against "any conception of an undifferentiated phase," including normal autism, the symbiotic stage, primary narcissism, stimulus barrier, and so on. A more formidable, and perhaps more reasonable, proposition would be to integrate recent findings with developmental ego theory. The data suggest that inborn biological precursors of ego function embrace a wider, more integrated, and even more volitional range of activities than envisioned by the ego theorists. A biological function cannot be construed as volitional unless already under some degree of ego control; hence the data may argue for an earlier appearance of autonomous ego function than assumed. Theoretical consequences then follow, including a further "broadening of the concept of the reality principle" (Hartmann, 1958) and reintroduction of the issues raised by White's (1963) argument for independent ego energies. Finally, phase and mode dominance are concepts firmly rooted in psychoanalytic experience and literature. These positions are not challenged by new findings in infancy research. Rather, new data supplement and enlarge established models: Phase tensions are correlated with developmental competencies, for example, oral-visual, anal-motoric, in sets that may predispose affective organizations. The question of continuity of preverbal constructions is relevant to implications that new findings hold for clinical theory.

Developmental orientations in clinical theory may be classed as general comprehensive approaches to diagnosis and treatment of pre-oedipal disorders (Kernberg, 1975; Kohut, 1971; Masterson, 1976) or specific concerns regarding such technical matters as setting, use of reconstruction, and so on. (Giovacchini, 1975). General theories serve multiple purposes: ordering data, facilitating communication, prescribing technique. Efficacy of a particular approach therefore depends upon numerous considerations; the issue of accuracy of interpretation gathers importance with respect to an individual case. Guided by well-founded theory, clinicians make better judgments and more deeply moving interpretations to the patient.

Regarding general theory, we can follow Hartmann's (1958) dictum not to "make assumptions which clash with observations of behavior." The suggestion that cross-modal communication may affect intersubjective awareness between caretaker and infant attests to the complexity of early interaction. Newborns do not "know" their caretakers in the sense that parents would like to believe, but precursors of emphathic regard are present nearly from the beginning of extrauterine life. In this sense psychological birth precedes separation-individuation, and the caretaker's "primary maternal preoccupation" (Winnicott, 1965), involves more than ministration of need. Circular and reciprocal transactions come into play very early. One might compare the parent's role to that of a tourist who must negotiate the language, time schedule, foods, and so on of a new culture. Much has been assumed regarding the residue of early experience in adult patients; vicissitudes of development are too complex, even at this early time, to expect isomorphic recapitulation at preverbal structures in adult patients. The question is open as to how coherently specific states are translated into verbal-symbolic form. Constellations of nonverbal interaction (gesticulation, voice pattern, gaze, synchronous movement, etc.) are embedded in early contexts of meaning. Do these evolve into psychic structures? Certainly patients do reveal symbiotic confusion and the gamut of pre-oedipal transference states described in the literature, but it becomes difficult to distinguish true archaic transference from fantasies of primitive involvement—not unlike

Freud's dilemma regarding the theory of seduction. I would consider it more likely for many people to experience residues of early states in somatic-perceptual-motor terms including psychotic symptoms, addictive behaviors, momentary regressive states such as the Isakower phenomenon (Isakower, 1938; Spitz, 1965), autoerotic behaviors, and disturbances of body image. These affect patients' transference involvements in highly individualistic ways, almost always combined with later—for example, oedipal—contexts of interaction. Clarification of meaning is difficult because early residues have too little ideational content and later ones too much to be guided by manifest state.

Infant research also provides metaphors for the clinician to transcend familiar avenues of understanding patients trapped in early binds. Bruch (1982) has discussed how contact with new models of child development helped revise the treatment of primary anorexia nervosa. She contrasts views of this condition as a symoblic expression of oral conflicts with a model of developmental etiology based upon deficient mother–infant interaction. The latter view leads the therapist to analyze functional interaction rather than explore symbolic content, and "find entry into the rigid denial" of pervasive feelings of ineffectiveness and impaired control. The approach is aimed at establishing interactional coherence by putting into words what was previously expressed through the body. Rather than helping the patient resolve conflict buildup on top of regulatory problems, the therapist helps the patient code behaviors never before transformed into symbolic representation. Technical recommendations followed in this approach involve greater activity and special attention to the ongoing communicative stream: silence, speech, kinetics, and gaze, as well as verbalization.

Stone (1961) and Kohut have commented on activity/inactivity, use of the couch, and other issues related to the empathic tone in psychoanalytic treatment. The etiological model of pre-oedipal disorders suggested by infant research is in accordance with the view that a more active, flexible, face-to-face setting provides an effective holding environment for many patients with pre-oedipal difficulties. I believe that salient therapeutic issues for many such patients revolve around momentary surges of disorganizing affect more than organized states of transferential distortion. Finally, I suggest that judicious use of reconstruction with these patients is helpful. General reconstructions are of uncertain historical accuracy, but their therapeutic value lies in communicating empathic awareness of fundamental influences and compromises in the patient's life. Blum (1977) and Greenacre (1981) offer helpful discussions of this topic.

IMPLICATIONS FOR PRACTICE

Adult Treatment: Case Example

Therapeutic attention to motor affect and use of reconstruction are exemplified in the following vignette, taken from an early phase of twice-weekly intensive psychotherapy with a woman patient in her mid-thirties.

The woman's presenting difficulties were inability to confront her husband regarding neglect of her 10-month-old daughter, chronic depression, and recurrent binges of alcohol abuse in which she would dissipate her tension through wild nights on the town. She was raised in a morally upright professional family, and well educated. The mother suffered from chronic heart disease dating from the patient's early childhood. In fact, the mother was bedridden in the patient's earliest memory, in which her father was angry because she did not like the egg he had cooked her for breakfast. She remembered that her mother could only join in one dance with her father, and then had to sit out the evening. Both parents died during her early twenties. The only nourishment, care, and protection she received in adult life was from her first husband, but this relationship was one of cloying dependency and no sexual gratification. She sought excitement with the immature man who became her second husband. She blamed herself for lack of fortitude and expected little from others. A previous attempt to obtain help in psychotherapy was unsuccessful. Her initial presentation was remote, and the therapist's notes after the second session read: "She floods herself with alternate 'maybe' hypotheses in an intellectualized manner without integrating affects."

My intention here is to suggest her multiple concerns, unfocused thought, and pervasive attachment problems. For current purposes I would like to focus on her interest in dancing, the role it played in her life, and the meaning it expressed during the early phase of treatment. Her first dream was as follows: "I'm dancing in a ballet, taking a curtain call. I see that I have strong legs, dancer's legs, as I bow, and I am proud. Someone else, an older woman, is being recognized, and then I see that everyone is clapping for her. She is in charge of something, and I notice that she is my former principal." The patient's associations concerned her aspirations in dancing and choreography, which she felt were thwarted by her parents' wishes that she do something "practical." The dream seemed to her "optimistic" in that she discovered some previously unrecognized strength, and she mentioned that she did not want therapy to become a "crutch." The therapist remembered that the first time he saw her for consultation she wore dancer's warmups on her lower legs, but made no further comment or inquiry regarding dancing. Several months later she had reported that nightly episodes of rocking violently from side to side began as far back as she could remember and lasted into her first marriage. Like many of her "insights" it seemed to have no meaningful connection with anything else, past or present. At this point the therapist commented that perhaps she had been rocked as a small child because of her mother's illness and depression, and that she had invented a way of rocking herself. She then recalled that as a small child she had created the notion that her own leg was a doll, which she named, rocked, and kept warm in a blanket. She demonstrated this with her calf and lower leg, the same body parts wrapped in woolen warmups during the first session. This part of her body was used to maintain regulatory control and represented a precursor to transitional phenomena (Gaddini, 1978). Symbolic expressions of similar affects were created in the dream image of her legs transformed into dancer's legs, and her worry that therapy could become a crutch if she came to rely on it. However, these metaphorical statements carried less emotional weight for the patient than did increasing awareness of escape to concrete forms of expression such as drinking and dancing.

Familiarity with early developmental issues also emphasizes the salience of body expression. This brings to mind Freud's 1923/1961 statement that "the ego first and foremost is a body ego." Although resistance, transference, and conflict were expressed in verbal-symbolic terms, in this example the patient did not trust her ability to negotiate important matters through such means—either in therapy or in her marriage. At a later point she reported a dream in which an older female figure attempted to communicate with her in five different languages, and was unsuccessful in each. This marked the beginning of more focused awareness on her own inaccessibility as the key problem she needed to face in treatment. She half humorously alluded to her problem as "learning how to put my foot down and take a stand."

What can infant research contribute to understanding this patient's presentation and dynamics? If one assumes that she developed the normal complement of receptive and expressive skills, precursors to attachment, and affect systems reviewed in the first part of this chapter, these autonomous functions encountered a nonreciprocating environment because of the mother's illness and depression. (In this sense she was correct in saying that she had been "thwarted.") Profound attachment problems ensued, in which autoerotic behavior rather than empathic exchange continued to regulate internal states. Rather than pre-oedipal, core elements of her personality remained *preverbal* and *presymbolic*, depriving her of the representational capacities with which she could establish a sense of self. Precursors to self-expression were lucid and expressed concretely. They were not embedded in psychotic processes. She sought external validation (mirroring) through exhibitionistic pursuits, that is, dancing, but could not look into herself. She once dreamed that she looked into a mirror while dancing, only to see an infinite regress of other dancers in other rooms rather than her own reflection. On another occasion, she impulsively cut herself with a razor while looking into a mirror. Resort to such primitive modulating behaviors in a person who is otherwise functional suggests very early fixation. The primary therapeutic task with preverbal patients is to link internal states to the therapeutic relationship. Discussion of technique is beyond my present scope, but I suggest that parallels exist between this therapeutic task and the kinds of caretaker–infant exchange now being defined in developmental research.

Infant Intervention: Treatment and Prevention

Infant stimulation programs of the early 1970s (Beller, 1979), preventive education programs of the mid-1970s (Bromwich, 1981; Lief, 1979), and recently described clinical programs (Fraiberg, 1980; Greenspan, 1981) have established the rapidly growing field of parent–infant intervention. Huntington (1979) lists research findings she considers most significant for mental health professionals dealing with infants. Some of these data are discussed in the first section of this chapter, and they provide a much broader knowledge base for the clinician than was available 10 years ago.

Clinical work has moved beyond developmental guidance to include interactional assessment, crisis intervention, and intensive parent–infant psychotherapy. Efficacy of work in the last category has been made dramatically clear with such severe problems as failure-to-thrive (Fraiberg). Greenspan and Lourie's (1981) ordering of developmental structures (Figure 1) and corresponding adaptive or maladaptive caretaking environments from birth to 40 months is a major step forward. He describes observable landmarks for early development that take into account the complexity of social, emotional, and cognitive well-being. This kind of detailed awareness should be taught to those in most direct contact with infants, namely, pediatricians, in order to provide parents with access to services ranging from preventive education to psychotherapy. Greenspan (1981) outlines a spectrum of intervention strategies meant to address problems at each specific stage of development. These skills should be taught to clinicians.

Our program at the Center for Parents and Children at Glen Cove offers preventive education, developmental guidance, and infant assessment in a general hospital setting. Mother–infant and father–infant groups meet weekly for 3 years in a semistructured format involving lecture, discussion, and concrete demonstration of infants' growing capacities. When babies become toddlers they are able to move from our group room to the playroom, where a skilled play teacher helps them negotiate separation issues in this supportive setting. Families with particular problems are provided with the hospital's pediatric staff. Our curriculum presents basic information regarding infant development in a timely and salient manner. Parents who have not received a college course in child development are in many cases misinformed about basic facts. One mother expressed relief when she learned that her son's stranger anxiety marked a developmental milestone; she had been worried that he would become fearful of people and not "grow up to be a man." In other cases parents are unaware of how their own behavior affects a particular issue; they demonstrate *in vivo* the difficulty they are facing. During a discussion concerning how much vacation parents should take away from a young child, one mother was discussing her child's distress over separation when she suddenly realized that she had left the child's bottle out in her car. She bolted from the group to retrieve it, offering no explanation of her departure to her son. He became quite upset, and the leader was compelled to hold him up to the window and reassure him that "mommy would return in a moment." She had not realized that the separation issues just discussed in the context of vacations also applied to her present behavior. Some parents face the difficulty of giving appropriate stimulation to a slow infant, or one whose temperament evokes strong reactions. Others face multiple and conflicting role demands that complicate the logistics of childrearing. We have been struck by parents' eagerness to learn and willingness to attend our support groups for an extended period of time, and see this as *prima facie* evidence of the need for this kind of service. Their enthusiasm and commitment suggest the appeal and usefulness of new knowledge in infant development.

REFERENCES

Ainsworth, M. D. Infant-mother attachment. *American Psychologist*, 1979, *34*, 932–937.

American Psychiatric Association. *Diagnostic and statistical manual of mental disorders*. (3rd ed.). Washington, D.C.: Author, 1980.

Bartoshuk, A. K. Human neonatal cardiac responses to sound: A powerful function. *Psychonomic Science*, 1964, *1*, 151–152.

Beller, E. R. Early intervention programs. In J. D. Osofsky (Ed.), *Handbook of infant development*. New York: Wiley, 1979.

Bernard, S., & Sontag, L.W. Fetal reactivity to tonal stimulation: A preliminary report. *Journal of Genetic Psychology*, 1947, *70*, 205–210.

Blum, H. P. The prototype of preoedipal reconstruction. *Journal of the American Psychoanalytic Association*, 1977, *25*, 757–786.

Bornstein, M. H., Kessen, W., & Weiskopf, S. The categories of hue in infancy. *Science*, 1976, *191*, 201–202.

Bowlby, J. *Attachment and loss: Vol. 1. Attachment*. New York: Basic Books, 1969.

Brackbill, Y., Adams, G., Crowell, D. H., & Gray, M. L. Arousal level in neonates and preschool children under continuous auditory stimulation. *Journal of Experimental Child Psychology*, 1966, *4*, 178–188.

Brazelton, T. B., Koslowski, B., & Main, M. The origin of reciprocity in the mother-infant interaction. In M. Lewis & L. Rosenblum (Eds.), *The effect of the infant on its caregiver*. New York: Wiley, 1974.

Brennan, W. M., Ames, E. W., & Moore, R. W. Age differences in infants' attention to patterns of different complexities. *Science*, 1966, *151*, 1354–1356.

Bridges, K. Emotional development in early infancy. *Child Development*, 1932, *3*, 324–341.

Brody, S, & Axelrod, S. *Anxiety and ego formation in infancy*. New York: International Universities Press, 1970.

Bromwich, R. *Working with parents and infants: An interactional approach*. Baltimore: University Park, 1981.

Bruch, H. Anorexia nervosa: Therapy and theory. *American Journal of Psychiatry*, 1982, *139*, 1531–1538.

Bruner, J. S. *From communication to language: A psychological perspective*. (Mimeographed manuscript). Oxford University, 1974.

Bruner, J. S. The ontogenesis of speech arts. *Journal of Child Language*, 1975, *2*, 1–19.

Clifton, R. K., Graham, F. K., & Hatton, H. M. Newborn heart rate response and response habituation as a function of stimulus duration. *Journal of Experimental Child Psychology*, 1968, *6*, 266–278.

Cohen, L. B., & Salapatek, P. (Eds.). *Infant perception: From sensation to cognition: Basic visual processes*. New York: Academic Press, 1975(a).

Cohen, L. B., & Salapatek, P. (Eds.). *Infant perception: From sensation to perception: Vol. 2. Perception of space, speech, and sound*. New York: Academic Press, 1975(b).

Cohen, L. B., DeLoache, J. S., & Strauss, M. S. Infant visual perception. In J. D. Osofsky (Ed.), *Handbook of infant development*. New York: Wiley, 1979.

Condon, W. S., & Sander, L. Neonate movement is synchronized with adult speech: International participation and language acquisition. *Science*, 1974, *183*, 99–101.

Day, R. H., & McKenzie, B. E. Constancies in the perceptual world of the infant. In W. Epstein (Ed.), *Stability and constancy in visual perception: Mechanisms and processes*. New York: Wiley, 1977.

Dayton, G., Jones, M., Ain, P., Rawson, R., Steele, B., & Rose, M. Developmental study of coordinated eye movements in the human infant. I. Visual acuity in the newborn human: A study based on induced optokinetic nystagmus recorded by electrooculography. *Archives of Ophthalmology*, 1964, *71*, 865–870.

Dayton, G., Jones, M., Steele, B., & Rose, M. Developmental Study of conditioned eye movements in the human infant. II. An electrooculographic study of the fixation reflex in the newborn. *Archives of Opthalmology*, 1964, *71*, 871–875.

Eisenberg, R. B. *Auditory competence in early life*. Baltimore: University Park, 1976.

Emde, R., Gaensbauer, T., & Harmon, R. Emotional expression in infancy: A biobehavioral story. *Psychological Issues Monograph Series*, 1976, *10* (No. 37).

Engel, R., & Young, N. B. Calibrated pure tone audiograms in normal neonates based on evoked electroencephalographic responses. *Neuropädiatrie*, 1969, *1*, 149–160.

Escalona, S. K. The study of individual differences and the problem of state. In E. N. Rexford, L. W. Sander, & T. Shapiro (Eds.), *Infant Psychiatry: A New Synthesis*. New Haven: Yale University, 1976.

Fagan, J. F. Infant color perception. *Science*, 1974, *183*, 973–975.

Fagan, J. F. Infants' recognition of invariant features of faces. *Child Development*, 1976, *47*, 627–638.

Fantz, R. L., Ordy, J. M., & Udelf, M. S. Maturation of pattern vision in infants during the first six months. *Journal of Comparative and Physiological Psychology*, 1962, *55*, 907–917.

Fenichel, O. *The psychoanalytic theory of neuroses*. New York: W. W. Norton, 1945.

Field, T., Woodson, R., Greenberg, R., & Cohen, D. Discrimination and imitation of facial expressions by neonates. *Science*, 1982, *218*, 179–181.

Fraiberg, S. *Insights from the blind. Comparative studies of blind and sighted infants*. New York: Basic Books, 1977.

Fraiberg, S. (Ed.). *Clinical studies in infant mental health: The first year of life*. New York: Basic Books, 1980.

Freud, A. Infants without families: Reports of the Hampstead nurseries 1939–1945. In *The writings of Anna Freud* (Vol. 3). New York: International Universities Press, 1973.

Freud, S. The ego and the id. In J. Strachey (Ed. and Trans.), *Standard edition of the complete psychological works of Sigmund Freud* (Vol. 19). London: Hogarth Press, 1961. (Originally published, 1923)

Freud, S. Project for a scientific psychology. *The standard edition* (Vol. 1). London: Hogarth Press, 1966. (Originally published, 1895).

Gaddini, R. Transitional object origins and the psychosomatic symptom. In S. A. Grolnick & L. Baricin (Eds.), *Between reality and fantasy: Transitional objects and phenomena*. New York: Jason Aronson, 1978.

Giovacchini P. (Ed.) *Psychoanalysis of character disorders*. New York: Jason Aronson, 1975.

Gratch, G. The development of thought and language in infancy. In J. D. Osofsky (Ed.), *Handbook of infant development*. New York: Wiley, 1979.

Greenacre, P. Reconstruction: Its nature and therapeutic value. *Journal of the American Psychoanalytic Association*, 1981, *29*, 27–46.

Greenspan, S. *Psychopathology and adaptation in infancy and early childhood: Principles of clinical diagnosis and preventive intervention*. New York: International Universities Press, 1981.

Greenspan, S., & Lourie, R. S. Developmental structuralist approach to the classification of adaptive and pathologic personality organizations: Infancy and early childhood. *American Journal of Psychiatry*, 1981, *138*, 725–735.

Hammond, J. Hearing and response in the newborn. *Developmental Medicine and Child Neurology*, 1970, *12*, 3–5.

Harris, A. E. Social dialectics and language: Mother and child construct the discourse. *Human development*, 1975, *18*, 80–96.

Hartmann, H. *Ego psychology and the problem of adaptation*. New York: International Universities Press, 1958.

Hartmann, H., & Kris, E. The genetic approach in psychoanalysis. *Psychoanalytic Study of the Child*, 1945, *1*, 11–30.

Huntington, D. S. Supportive programs for infants and parents. In J. D. Osofsky (Ed.), *Handbook of infant development*. New York: Wiley, 1979.

Hutt, C., von Bernuth, H., Lenard, H. G., Hutt, S. S., & Prechtl, H. F. Habituation in relation to state in the human neonate. *Nature*, 1968, *220*, 618–620.

Hutt, S., Hutt, C., Lenard, H., von Bernuth, H., & Muntjewerff, W. Auditory responsivity in the human neonate. *Nature*, 1968, *218*, 888–890.

Isakower, D. A contribution to the pathopsychology of phenomena associated with falling asleep. *International Journal of Psychoanalysis*, 1938, *29*, 331–345.

James, W. *The principles of psychology*. New York: Dover, 1980. (Originally published, 1913).

Kahn, M. M. R. *The privacy of the self: Papers on psychoanalytic theory and technique*. New York: International Universities Press, 1974.

Keele, C., & Neil, E. (Eds.) *Samson Wright's applied physiology*. Oxford: Oxford University Press, 1965.

Kernberg, O. *Borderline conditions and pathological narcissism*. New York: Jason Aronson, 1975.

Kohut, H. *The analysis of self*. New York: International Universities, 1971.

Kuhl, P. K., & Meltzoff, A. N. The bimodel perception of speech in infancy. *Science*, 1982, *218*, 1138–1140.

Lempers, J. D., Flavell, E. R., & Flavell, J. H. *The development in very young children of tacit knowledge concerning visual perception* (Mimeographed manuscript). Institute of Child Development, University of Minnesota, 1975.

Lief, N. R. *The first year of life: A curriculum for parenting education*. New York: Keyway, 1979.

Loevinger, J. *Ego development*. San Francisco: Jossey-Bass, 1976.

Mahler, M. S., Pine, F., & Bergman, A. *The psychological birth of the human infant*. New York: Basic Books, 1975.

Masterson, J. F. *Psychotherapy of the borderline adult: A developmental approach*. New York: Brunner/Mazel, 1976.

Meltzoff, A. N., & Borton, R. W. Intermodal matching by human neonates. *Nature*, 1979, *282*, 403–404.

Osofsky, J. D. (Ed.). *Handbook of infant development*. New York: Wiley, 1979.

Oster, H. S. *Color perception in ten-week-old infants*. Paper presented at the biennial meeting of the Society for Research in Child Development, Denver, 1975.

Peiper, N. *Cerebral function in infancy and childhood*. New York: Consultations Bureau, 1963.

Peterfreund, E. Some critical comments on psychoanalytic conceptualizations of infancy. *International Journal of Psychoanalysis*, 1979, *6*, 427–441.

Piaget, J. *The construction of reality in the child*. New York: Basic Books, 1954.

Riegel, K. F. (Ed.). *The development of dialectical operations*. Basel: Karger, 1975.

Ryan, J. Early language development: Toward a communication analysis. In M. P. M. Richards (Ed.), *The integration of the child into a social world*. Cambridge: University of Cambridge, 1974.

Salapatek, P. Pattern perception in early infancy. In L. B. Cohen & P. Salapatek (Eds.), *Infant perception: From sensation to cognition: Vol. 1. Basic visual processes*. New York: Academic Press, 1975.

Sander, L. Issues in early mother-child interaction. *Journal of the American Academy of Child Psychiatry*, 1962, *1*, 141–166.

Sokol, S., & Dobson, V. Pattern reversal visually evoked potentials in infants. *Investigative Ophthalmology*, 1976, *15*, 58–62.

Spitz, R. A. *The first year of life: A psychoanalytic study of normal and deviant development of object relations*. New York: International Universities Press, 1965.

Sroufe, L. A. The coherence of individual development: Early care, attachment, and subsequent developmental issues. *American Psychologist*, 1979, *34*, 834–841.(a)

Sroufe, L. A. Socioemotional development. In J. D. Osofsky (Ed.), *Handbook of infant development*. New York: Wiley, 1979.(b)

Stern, D. N. A microanalysis of mother-infant interaction: Behavior regulating social contact between a mother and her 3½-month-old twins. In E. N. Rexford, L. W. Sander, & T. Shapiro (Eds.), *Infant psychiatry: A new synthesis*. New Haven: Yale University, 1976.

Stone, L. *The psychoanalytic situation*. New York: International Universities Press, 1961.

Thomas, A., Chess, S., Birch, H. G., Hertzig, M. E., & Korn, S. *Behavior individuality in early childhood*. New York: New York University Press, 1963.

Trevarthen, C., & Hubley P. Secondary intersubjectivity: Confidence, confiding, and acts of meaning in the first year. In A. Lock (Ed.), *Action, gesture, and symbol: The emergence of language*. New York: Academic Press, 1978.

Waters, E., Matas, L., & Sroufe, L. Infants' reactions to an approaching stranger: Description, validation and functional significance of wariness. *Child Development*, 1975, *46*, 348–356.

White, R. W. Ego and reality in psychoanalytic theory. *Psychological Issues*, 1963, *3* (3, Whole No. 11).

Wickelgren, L. Convergence in the human newborn. *Journal of Experimental Child Psychology*, 1967, *5*, 74–85.

Winnicott, D. W. *The maturational processes and the facilitating environment: Studies in the theory of emotional development*. New York: International Universities Press, 1965.

Wolff, P. The developmental psychologies of Jean Piaget and psychoanalysis. *Psychological Issues*, Monograph 5, New York: International Universities, 1960.

12

Psychodynamic Psychotherapy and Research on Early Childhood Development

JONATHAN BLOOM-FESHBACH AND SALLY BLOOM-FESHBACH

INTRODUCTION

Developmental experiences in years 3 to 12 have considerable implications for how children will negotiate adolescence and adulthood. The clinician, and especially the psychodynamic psychotherapist who believes that adult functioning depends upon the quality of earlier development, is keenly interested in the nature of the patient's life during this "early childhood" period that spans both the oedipal and the latency years. However, it is quite a different question how the therapist's understanding or *technique* might be affected by knowledge about the regularities or nomothetic characteristics of development during these life stages. Apart from the perspectives psychoanalytic developmental theory might gain from systematic examination of, and integration with, basic developmental research, the task of this chapter is to consider how clinical *practice* might benefit from such research knowledge.

JONATHAN BLOOM-FESHBACH • Georgetown University Counseling Center and Departments of Psychiatry and Behavioral Sciences and Child Health and Development, George Washington University, Washington, D.C. 20052. SALLY BLOOM-FESHBACH • American University Center for Psychological and Learning Services and National Research Council Committee on Child Development Research and Public Policy, Washington, D.C. 20016.

One problem that arises in attempting to refine clinical technique is that good clinicians tend to have a broader, more complicated, and more flexible technical repertoire than may be reflected in their theoretical positions, or even in their self-perceptions. As a result, when provided with a conceptual or research landscape from which technical implications may be drawn, the experienced clinican may well say "I do that already" or "Now I see the rationale for what I do." However, we do believe that a more detailed examination of basic research findings can communicate something new and useful to the practicing clinician, and perhaps shed further light upon the underpinnings of psychotherapy technique. From the vast array of research areas within the field of early childhood development, we have selected three research areas for discussion. These issues are intrinsically interesting, and also illustrate how such research findings may be used to inform clinical practice. The three areas are: (1) the relationship between parental childrearing styles and children's social and emotional functioning; (2) children's developing capacities to monitor their own cognitive processes—what is termed "metacognition"; and (3) the importance of multiple attachment or identification figures in child development—that is, the psychological impact of such nonmaternal figures as fathers and peers.

Our discussion of each of these research topics will interweave research findings with the technical perspectives suggested by the research literature. Where appropriate, clinical examples will illustrate conceptual points. Finally, we shall offer some hypotheses for future research that emerge from this consideration of the links between basic research on early childhood development and the practice of psychoanalytically oriented psychotherapy.

RESEARCH ON CHILDREARING STYLES

There are a number of parallels between the parent–child and the therapist–patient relationships. The familiar psychoanalytic focus on transference emphasizes the ways in which the therapeutic context evokes the patient's parental images and internalized relationship patterns. From a somewhat different vantage point, we can perhaps view psychotherapy itself as an analogous developmental process. By this we do not mean to imply that psychotherapy is "reparenting." Rather, psychotherapy can be conceptualized as a formative relationship that fosters development. From this perspective, we wish to consider how the therapist's stance or style of interaction can be informed by research on parental childrearing. In particular, what can the psychotherapist learn from research investigations on how different parental behaviors promote or interfere with children's healthy development?

One fundamental dimension of parent behavior that fosters the child's social and emotional development is parental warmth and nurturance. The earliest functions of parental nurturance are of course apparent before the child reaches 3. It is interesting to note, however, that despite long-standing disputes among empiricists about the importance of early experience, the evidence is beginning to mount that a secure parent–child bond during infancy predicts later social and cognitive competence (Matas, Arend, & Sroufe, 1978; Sroufe, 1978; Sroufe & Waters, 1977; Wachs & Gruen, 1982; Waters, 1978). Securely attached infants are those who seek their mother when she is present but do not cling and whine in her absence, and are explorative of their surroundings (Ainsworth & Wittig, 1969). Mothers of securely attached infants have been found to be sensitive to their infants' needs, whereas mothers of "avoidant" or "anxiously attached" infants are either emotionally unavailable or fluctuate in their emotional availability. Thus, the mother's sensitivity and warmth are crucial to the child's social development (Blehar, Lieberman, & Ainsworth, 1977; Bloom-Feshbach & Blatt, 1982; Clarke-Stewart, 1973).

The significance of paternal warmth and involvement is similarly evident in the research on sex-role development. Studies of sex-role acquisition do not yield direct correlations between fathers' masculinity or sex-role preferences and those of sons (Angrilli, 1960; Mussen & Rutherford, 1963; Payne & Mussen, 1956); rather, it is more nurturant fathers who have more masculine sons (Biller & Borstelmann, 1967; Hetherington, 1967; Reuter & Biller, 1973). In fact, regardless of the degree of parental masculinity, fathers who are warm and accessible to their children have sons who are more masculine (e.g., Bandura & Walters, 1969; Mussen & Rutherford, 1963) and daughters who are more feminine (Johnson, 1963). Conversely, Bronson (1959) found that tense father–son relationships produced a negative correlation between the father's and son's degree of masculinity. Reviewing this literature, Lamb (1982) comments that "these findings strongly imply that characteristics of the father-child relationship have greater formative significance than the fathers' avowed masculinity or punitiveness" (p. 19). Thus, the nurturant quality of the father–child relationship, rather than the father himself as an object of identification, is central to development.

Given that most clinicians would readily accept this evidence of the positive effects on children of parental warmth and nurturance, what implications does this literature have for therapeutic practice? First of all, it was only several decades ago that Carl Rogers felt compelled to emphasize the importance of the therapist's nurturance—of "unconditional positive regard"—in the therapy relationship. In psychoanalytic work today considerable debate has focused on the technical implications of Kohut's (1971) conceptualization of narcissistic disorders. In response to the work of Kohut and his colleagues, some analysts and psychoanalytically oriented psychotherapists have more recently shifted their therapeutic stance toward being more empathic, responsive, and warm. Better therapeutic progress is considered to result within the context of a warmer "real relationship," to use Greenson's (1967) term. Without a warm therapeutic relationship, even reflected in such simple cues as the subtlety of a smile or an empathic vocal agreement, the psychoanalytic goal of therapeutic neutrality may be construed as an atmosphere of emotional coldness and unresponsiveness. In misinterpreting neutrality, even healthier patients may experience the relationship with the therapist as cold and narcissistically wounding, making it difficult for them to make maximum use of psychotherapy. Thus, technical neutrality need not, indeed should not, imply emotional aloofness. As Kernberg (1977) states, "the empathic attitude of the therapist . . . has elements in common with the empathy of the 'good-enough mother' with her infant" (pp. 302–303).

Since one goal of psychodynamic treatment is to enable patients to identify with the therapeutic stance of reflecting upon experience from different vantage points, a nurturant interpersonal atmosphere promotes identification with this and other important dimensions of the therapist's analytic perspective. This view of the function of identification in the therapy process is developed in Loewald's (1960) classic article on the curative ingredients in psychoanalytic work, in which he discusses how the new object relationship between patient and analyst fosters the patient's ego development. Our principal point is that a warm, nurturant interpersonal environment may be a crucial variable in establishing the psychotherapy relationship and in facilitating identificatory and ego developmental processes.

Basic research on parent-child relationships in early childhood, like the psychoanalytic literature, points to the importance of setting clear and firm limits. The weight of developmental research evidence indicates that the consistent but nonpunitive exercise of parental authority fosters cognitive, moral, and social development in young children (e.g., Becker, 1964; Kagan & Moss, 1962). A widely acclaimed series of investigations by Baumrind and her colleagues (Baumrind, 1967, 1971, 1977; Baumrind & Black, 1967) has linked different patterns of childrearing to the personality characteristics of preschool children. Baumrind began by identifying three clusters of child behavior and then exploring associated patterns of parental behavior. In subsequent research she took the opposite approach—classifying childrearing styles and then exploring children's behavior patterns. With either methodology, the results demonstrated the beneficial effects of combining consistent nurturance with clear discipline.

Children who were best adjusted, who were self-reliant, self-controlled, explorative, and content, were reared by parents who were warm, rational, and receptive to the child's communications. At the same time, these parents were also moderately controlling and demanding. Baumrind labeled this pattern *authoritative*. In contrast, *authoritarian* parents, who exerted authority but were less warm in their childrearing style, had children who were discontent, withdrawn, and distrustful. The last childrearing category identified in the early research was the *permissive* style, that is, parents who were noncontrolling, nondemanding, and relatively warm. Children reared by permissive parents were the least self-reliant, explorative, and self-controlled. Thus, either too much authority and too little nurturance (the authoritarian style) or sufficient nurturance paired with an inadequate degree of discipline (the permissive style) promoted less adaptive patterns of personality development.

These research findings suggest that psychotherapists need to maintain a balance between providing a warm, empathic atmosphere and establishing firm, frustrating therapeutic boundaries. Although the hazards of overgratification by the therapist are generally recognized by psychoanalytically oriented clinicians, the potentially iatrogenic effects of too much frustration and too little nurturance are less widely acknowledged. Just as parents must adjust their balance of nurturance and limit-setting to the changing needs of the growing child, clinicians might consider whether and how therapeutic technique could be beneficially adjusted with certain patients and at various stages of the therapy process. Even with neurotic patients, it would be interesting to identify what aspects of the clinical picture suggest modifications in a "traditional" analytic stance. For example, might a slightly less frustrating affective atmosphere accelerate ego development for withdrawn patients? Or, can more varied emotional responsiveness at certain points work in concert with technical neutrality to facilitate the working through of preoedipal issues in relation to oedipal concerns? Furthermore, to what extent do aspects of the therapist's style serve as a model for the patient's identificatory processes?

During the childhood years, parents ideally must respond flexibly to their children's changing needs for nurturance and control. Patients too may benefit from a flexible balance of therapeutic warmth and frustration as the therapy process develops. The stage of treatment, the type of patient, the issue under consideration, and the therapist's personal style must all contribute to an assessment of what kinds of interventions will be most productive.

METACOGNITION

The last 10 years have seen the rise of a new area of cognitive developmental psychology, called "metacognition." A study by Flavell, Friedrichs, and Hoyt (1970) illustrates this process of cognitive monitoring of cognitive processes. Preschool and elementary school children were asked to study a set of items until they were certain they could remember them perfectly. The older children completed the task successfully, that is, when they said they were ready, their recall of the items was perfect. The younger children, however, indicated that they were ready but could not accurately remember the items. This study and others like it (Brown, 1978; Flavell, 1978; Flavell & Wellman, 1977; Markman, 1977) have led Flavell, a Piagetian theorist, to suggest that

> young children are quite limited in their cognition about cognitive phenomena, or in their *metacognition*, and do relatively little monitoring of their own memory, comprehension and other cognitive enterprises. (1981, p. 165)

Cognitive and *metacognitive* goals and processes are different, as the following example indicates. If schoolchildren sense (a metacognitive experience) that they do not know course material well enough to pass a test, they might reread their lessons (a cognitive process aimed at accomplishing the cognitive goal of increased knowledge). In order to assess whether rereading the material had prepared them adequately, a student might decide to wonder about possible questions the teacher might ask. The plan of asking oneself questions is a metacognitive strategy, and the process of asking these questions a metacognitive task. As Flavell (1981) states, "cognitive strategies are invoked to *make* cognitive progress, metacognitive strategies to *monitor* it" (p. 168).

There is a striking emphasis on metacognitive processes in psychotherapy (or, more precisely, on metacognitive-affective processes). In psychotherapy, patients learn to monitor their inner experience and to make more systematic assessments of their psychological processes. For example, a suicidal patient realizes at some point during therapy that unconscious, denied, rageful impulses precede suicidal thoughts. The patient may come to this insight by the therapist's noting the sequence of suicidal ideas being expressed soon after the announcement of a therapist's vacation. In this familiar example, such a patient is initially resistant to even entertaining the notion that suicidality emerges from a transformation of unconscious murderous rage. It is only as patients learn to observe their cognitive and affective processes that self-discovery and hence a modification of defense becomes a possibility. What facilitates such self-observation, or at least collaborative analysis with the therapist, are metacognitive strategies. The patient needs to invoke the principle, at least as a working hypothesis, that "when I feel suicidal, it is possible that I may be angry at someone and do not want to acknowledge this feeling." Such metacognitive principles—concepts that permit observational monitoring of cognitive-affective experience—can be specific as in the suicide/rage example. Or, metacognitive strategies can be more general, such as the notion that "whenever I notice myself feeling very pressured about something or someone, when I seem to be reacting out of proportion to the realistic situation, it is an occasion to reflect on the underlying (perhaps unconscious) factors motivating my feelings." As patients become increasingly psychologically minded and sophisticated, they employ more and more complex metacognitive principles.

An important feature of metacognitive thinking is that it develops through childhood. Metacognition involves some degree, however simple or uncomplicated, of reflecting about one's own internal processes, and the full ability to "think about thinking" only emerges during adolescence. However, research as well as clinical practice suggests that not all individuals achieve Piaget's final cognitive stage of formal operational thought. Because this advanced representational level may be a prerequisite for sophisticated kinds of metacognitive processes, some patients may not be able to utilize some complicated metacognitive perspectives.

However, even simple metacognitive capabilities developed in early childhood are essential to the conduct of psychoanalytically oriented psychotherapy with adults. One metacognitive process basic to psychotherapy—an awareness of inconsistencies—is illustrated in Markman's (1979) study of elementary school children. Markman investigated the process of how children recognize whether they understand something, whether something makes sense or is illogical. The findings indicate that young children in grades 3 to 6 are able to encode and retrieve large amounts of information, but have difficulty drawing inferences or retrieving and comparing propositions. Recognizing inconsistencies depends upon inferential processs, thus making it difficult for grade school children to see illogic in patterns of information. Psychotherapy calls for precisely this ability to go beyond a mere description of experience. Patients must ponder the meaning of and interrelationships among the separate elements of their lives, learning to make inferences about life events, to develop propositions about behavior, and to see patterns of inconsistency. As Markman (1979) found, although children in grades 3 to 6 do not spontaneously carry out the metacognitive tasks involved in knowing whether or not they understand something, older children and adults are able to perform these tasks. Consequently, therapists should maintain a developmental view toward metacognitive capacities. This perspective entails recognizing that metacognitive strategies will develop gradually throughout the process of psychotherapy, requiring time to be learned and utilized.

Therapists can facilitate their patients' ability to monitor key psychological processes by being more explicit about metacognitive strategies. For example, the patient's productive use of transference reactions depends upon observing and analyzing feelings about the therapist rather than acting on them. For the purpose of this discussion, we shall consider the principle of transference analysis a cognitive strategy of psychotherapy. What initiates this cognitive process is the patient's metacognitive recognition that a strong reaction to the therapist is a signal that analysis or self-reflection is called for. The patient's capacity for this metacognitive process (of recognizing the cues that indicate the need for further analysis of an issue) can more readily become an accepted principle within the therapist–patient relationship when the therapist makes the process explicit. Thus, analysis of specific transference reactions may be facilitated by the therapist's decision, at some appropriate point, to comment on the general nature of the therapeutic process. Such general principles can then be more easily understood and internalized by the patient, for application to new issues as they arise.

The patient's use of metacognitive strategies of self-observation facilitates both the working alliance and the transference relationship. The working alliance is promoted by the emphasis on the ego's observational capacity; the transference is promoted by providing some metacognitive understanding of the necessity and purposes of transference reactions. The patient who has trouble experiencing and expressing anger may be better able to analyze this difficulty if there is a conceptual context in which to place such reactions, and an established psychological basis within the psychotherapy relationship for exploration of new feelings and reactions.

Although child development research has heretofore studied metacognition without consideration of affective factors, any clinical discussion of metacognition is incomplete without the recognition that cognition and emotion are intertwined. Emotional disturbance can interfere with cognitive development during childhood and adolescence, and hence may be reflected in diminished cognitive competence as well as emotional maladjustment in adulthood. The psychological process of splitting in borderline patients is an example of an affectively based inability to perform the cognitive task of seeing the complexity in people and in emotional responses. For the borderline patient, there are points in psychotherapy when the most important metacognitive principle is that an overly extreme positive or negative reaction provides a clue to search for the "missing" side of his or her emotional experience. Paranoid patients may use complex but inaccurate metacognitive strategies because of emotional interference. When paranoid individuals feel psychological pain, they rigidly seek to externalize the conflict, often employing complicated metacognitive and cognitive rationales. Some obsessive patients may also defensively rely upon metacognition. At some point in the therapy process, obsessives come to recognize and accept the metacognitive principle that when they see themselves treading repeatedly over the same cognitive turf, it is likely a sign that they are avoiding feelings. Obviously, the significance of this recognition for the patient rests in its metacognitive-affective meaning; a metacognition that is not linked with affect (like an affectless cognition) remains an obsessive defense. These simplified examples illustrate that the therapeutic use of metacognition must recognize the developmental, dynamic, and characterological differences between patients.

Metacognition is important not only for the patient, but also for the psychotherapist and the therapist in training. Explicit articulation of metacognitive principles may improve the therapist's capacity to establish priorities, identify countertransferences, and make interpretations. Therapeutic understanding is achieved by exploring the connections between therapy content, therapy process, the patient's history, and the patient's current life experiences. Metacognitive processes are needed in *reminding* oneself to search for these interrelationships, in *remembering* to engage in certain therapeutic tasks or to use particular technical strategies, and in *instigating* the process of observing the internal responses that differentiate subjective versus patient-induced countertransferences.

MULTIPLE ATTACHMENT FIGURES

The last research area we shall discuss concerns nonmaternal attachment figures for the developing child. The most central of these figures, the father, was until recently considered to have a more peripheral role in child development. Now it is recognized that the father has a significant impact on many areas of the child's functioning, and even exerts substantial influence on the child's affective and interpersonal experiences as early as infancy (Cohen & Campos, 1974; Kotelchuck, 1976; Lamb, 1982; Parke & Sawin, 1976; Schaffer & Emerson, 1964).

Empirical studies of attachment during infancy and early childhood have used different indexes of attachment, and hence have yielded different findings. When protest at separation is used as the dependent measure reflecting strength of attachment, no parental preference is found; that is, infants protest the departure of father and mother equally (e.g., Cohen & Campos, 1974; Lamb, 1979; Spelke, Zelazo, Kagan, & Kotelchuck, 1973). Similarly, greeting behavior does not differentiate parental attachment; one study using this measure actually found an infant preference for fathers (Lamb, 1979). In stress-free situations, again children display no particular parental preference (Lamb, 1977a, 1977b). In situations of distress, infants respond similarly to whichever parent is available (e.g., Feldman & Ingram, 1975; Lamb, 1976a). When both parents are available, however, infants from age 10 to 18 months preferentially select their mothers (Cohen & Campos, 1974; Lamb, 1976a). Yet young infants (8 months) and toddlers (2 years) do not manifest this maternal preference in the distressing context (Lamb, 1976c). Fathers appear to have a special significance in evoking more affiliative behaviors from their children, such as smiling and vocalizing (Belsky, 1979; Clarke-Stewart, 1978; Lamb, 1977a, 1977b). Thus, both father and mother are viewed as highly significant attachment figures for the very young child. In fact, the nature of these attachments, in some respects and for some children, may be difficult to discriminate.

Freud (1924/1961, 1925/1961, 1909/1955) was the first observer of human behavior to postulate a role for the father in child development, emphasizing the father's significance during the oedipal phase (ages 4 to 6). Since Freud's initial conceptualization, psychoanalytic theorists (Meerloo, 1968; von der Heydt, 1964) have come to recognize the father's pre-oedipal relationship to the child, mainly viewing fathers as vehicles for differentiation from the potentially engulfing mother–child union. From this perspective, the father serves to introduce the child to the larger world of social interaction beyond the protective, though limiting, mother–child sphere (Loewald, 1962). Enlarging upon these ideas, Abelin (1971) and other psychoanalytic theorists (Burlingham & Freud, 1944; Greenacre, 1960; Mahler, Pine, & Bergman, 1975) have noted the specific attachment that develops between the father and the pre-oedipal child. However, these thinkers tend to view this attachment either as compensating for an inadequate mother–child relationship, or as a vehicle to spur separation from the mother. Similarly, the ethological scientific tradition developed by Bowlby (1969) deemphasized the father's role as a primary attachment figure.

Basic research in developmental psychology lends some support to the distinction in parental roles proposed by psychoanalytic thinkers.

The research indicates that the child's bond with each parent has distinctive features. Fathers interact more with their children in play situations, and in general are more vigorous, unpredictable, and physically active; in comparison, mothers interact more with their children in caregiving situations, and serve a relatively stronger security function (Kotelchuck, 1976; Lamb, 1976b). Because these distinctions are far from absolute, research finds that both mothers and fathers can share considerably in meeting children's affiliative and security needs. It may be useful, then, to consider the parents in terms of the quality of the many functions they serve, rather than simply labeling roles as maternal or paternal (Bloom-Feshbach, Bloom-Feshbach, & Gaughran, 1980).

Thus far, we have reviewed the importance of the father–child relationship during the child's early years. The formative nature of this bond makes it understandable that the father is shown to be an important figure in so many areas of child development research. The literature finds that the father influences the child's sex-role development (e.g., Biller, 1971; Biller & Weiss, 1970), moral development (e.g., Hoffman, 1971; Rutherford & Mussen, 1968), cognitive development (e.g., Radin, 1982), and general social and psychological adjustment. From a clinical perspective, the father's role in social-emotional functioning is of greatest interest. Paternal warmth and nurturance is important for both sons and daughters. These qualities are linked to the security of the father–child bond as well as to the child's ability to form good relationships outside the family (Lamb, Frodi, Hwang, & Frodi, 1982). Boys whose fathers are nurturant have more positive views of themselves (Coopersmith, 1967; Rosenberg, 1965; Sears, 1970), are more socially competent (Howells, 1969; Mussen, Bouterline-Young, Gaddini, & Morante, 1963; Rutherford & Mussen, 1968), and are better adjusted (Mussen, 1961; Mussen *et al.*, 1963; Reuter & Biller, 1973; Slater, 1962). Similarly, girls with warm, sensitive fathers have better personality adjustment (Baumrind & Black, 1967; Fish & Biller, 1973) and are happier in the heterosexual relationships they form in later life (Fisher, 1973; Lozoff, 1974). Taken together, the research literature points to the father's early and broad-reaching effect on children's psychological adjustment.

Before suggesting how this expanded conception of paternal influence on early child development might affect clinical technique, we shall discuss a second kind of nonmaternal relationship in early childhood, namely, friendship. As Sullivan (1953) first suggested, peer relationships help shape a child's sense of self, attitudes, and ways of relating to others (Hartup, 1983). Furthermore, mounting research evidence suggests that the quality of friendships in early childhood affects social and emotional functioning later in life.

The notion of friendship is commonly associated with the latency years, when the child begins elementary school. However, recent research finds that much younger children form meaningful relationships with their peers. By age 2, children typically engage in social play with others of the same age (Bronson, 1975; Mueller & Lucas, 1975). Two-year-olds who have repeated contact with particular peers come to prefer these "friends" to other children. As one might expect, research observations show that young friends play closer together, imitate each other more, and express warmer feelings than children less familiar with one another (Doyle, Connolly, & Rivest, 1980; Lewis, Young, Brooks, & Michalson, 1975). By age 3, children actively seek out their peers and are better able to engage in cooperative play.

For many years, peer influences were popularly thought of in negative terms, with the peer group and "peer pressure" cited as contributing factors to delinquency and antisocial attitudes. A quite different perspective has emerged from recent research on peer relationships in early childhood (Berndt, 1983). The literature finds that friendships can enhance children's academic achievement, prosocial behavior, and educational aspirations (Berndt, 1979; Ide, Parkerson, Haertel, & Walberg, 1981; Kandel, 1978). In addition, studies of peer acceptance and social standing generally find that young children who are well liked by their agemates interact positively with both peers and adults (Berndt, 1981; Bigelow & LaGaipa, 1980; Bixenstine, de Corte, & Bixenstine, 1976; Masters & Furman, 1981).

The young child's position in the peer group appears both to reflect and contribute to emotional and social development. Popularity in the early childhood years has important correlates in adolescence and adulthood. Children who are unpopular with their peers are more likely to engage in delinquent behavior, to fail to complete high school, and to manifest emotional disturbance as adults (Asher & Renshaw, 1981; Roff, Sells, & Golden, 1972; Ullman, 1957). Problems in fitting into the peer group in elementary school predict behavior disorders and other psychiatric problems in late adolescence and early adulthood (Cowen, Pedersen, Babigan, Izzo & Trost, 1973). Cowen *et al.* found that peer ratings of 8- and 9-year-olds were the most reliable predictors of mental health difficulties observed 11 years later, when the individuals had reached the ages of 19 or 20. In other words, the children's unpopularity, as perceived by other children in grade school, was more likely to predict later maladjustment than a host of other relevant factors (including self-report data, ratings by teachers, school records, achievement scores, grades, absentee rates, and IQ).

Of particular interest to the clinician is the nature of the link between parent–child and peer relationships. Research supports the clinical wisdom that healthy parent–child relationships provide the foundation for forming good relationships with others. For example, 3½-year-old children who were more popular with their peers in preschool—who served leadership roles and were sensitive to the feelings of other children—were more securely attached to their mothers at the age of 15 months (Waters, Wippman, & Sroufe, 1979). Similarly, children with secure attachments at preschool age are more reciprocal and less negative in their play (Lieberman, 1977). Throughout the early childhood years, good parent–child relationships are associated with peer group popularity.

When the parent–child relationship is troubled, the role of peers in development is less clear. Lewis and Schaeffer (1980) have suggested that parents who have difficulty developing secure attachments with their children may also not allow them adequate opportunities for peer contact. The resultant impairment in social competence likely compounds the negative relationship patterns developed within the family. Parental failures to recognize and provide for children's *friendship* needs are not commonly recognized in the clinical literature.

Given a lack of social skills and a conflicted internal life, it is not surprising that children from families lacking in warmth and support are both less popular with their contemporaries and more susceptible to negative peer influence (Bixenstine, *et al.*, 1976; Condry & Siman, 1974; Hirschi, 1969; Kandel, 1978; West & Farrington, 1973). An interesting question still to be investigated is whether, and in what ways, peer relationships provide growth-promoting alternatives to disturbed parent–child relationships.

The research on the importance of nonmaternal relationships in child development bears upon clinical practice in several respects. First, the research suggests that in formulating diagnoses, and in reconstructive efforts to understand patients' psychopathology and adaptive strengths, the therapist needs to assess the role of nonmaternal attachment and identification figures. Furthermore, it is useful to recognize that the traditional maternal function of providing nurturance and security can be fulfilled by fathers, or by other "substitute" caregivers (grandparents, older siblings, housekeepers, etc.). Therapists frequently work with patients who report a history of seriously disturbed parental relationships, but who display a striking degree of psychological strength despite such apparently negative early parenting. With such individuals, it is important to inquire in greater detail about other, alternative parental figures and about important peer relationships.

A more differentiated understanding of the patient's early object relationships can help the therapist anticipate and interpret the mixture of oedipal and pre-oedipal transference elements. For example, consider the depressed male patient whose alcoholic, psychotic, or manic-depressive mother never supplied sufficient nurturance. If the patient's father served a compensatory maternal attachment function, it may be dynamically problematic for the patient to feel anger toward a male figure. In order to work through his oedipal rivalry with father and contact his resentment of paternal authority, this patient will necessarily experience the simultaneous pre-oedipal threat of loss of father as a nurturant object. Complex dynamics that result from complicated and unusual constellations of early interpersonal relationships are familiar to the clinician, who recognizes the need for clarity and care in the interpretation of patient history, current behavior, and transference.

There is an additional, perhaps less familiar, level of complexity suggested by these research results, especially when they are integrated with psychoanalytic ideas about the development of intrapsychic structure. Because psychological structure develops in relation to the child's interpersonal experiences, the presence of alternative sources of attachment and identification leads to multiple levels of ego organization, each reflective of a specific formative object relationship. Thus, for example, a patient with a relatively healthy mother and a borderline father might develop significantly disparate "pockets" of ego organization. Under different conditions, and especially in different relationships, the ego strengths (derived developmentally from the mother) and the ego weaknesses (derived developmentally from the father) would emerge. Hence, the multiplicitous nature of parenting figures in early childhood, and the inconsistent aspects of parent–child relationships, are reflected in the varied levels of ego functioning therapists observe in adult personality. Thus, character structure, or patterns of psychopathology and defense, will be overlapping rather than mutually exclusive.

From this perspective, it is more useful for the therapist to think about the nature and extent of a patient's borderline, narcissistic, obsessive, or hysterical levels of functioning than simply to consider the patient a borderline, a narcissist, an obsessive, or a hysteric (Sugarman, Bloom-Feshbach, & Bloom-Feshbach, 1980). Of course, the different character components will vary in degree, giving a predominant cast to the patient's personality. But it is likely that for each patient, various styles or levels of functioning will emerge in fluctuating patterns throughout the therapy process, reflected in both the transference and the working alliance. Although psychoanalytic theory clearly acknowledges this fact in its understanding of regression during the therapy process, clinicians sometimes fail to recognize the necessity of shifting one's technical approach through the course of treatment. Advocates of a strictly "analytically neutral" stance (e.g., Langs) appear to overlook or discount this factor.

In psychoanalytically oriented psychotherapy, the division between neurotic versus more disturbed, or oedipal versus pre-oedipal difficulties, often serves as a criterion for determining the appropriate degree of frustration in the therapeutic relationship. For example, Stolorow and Lachmann (1978) point out how the severity of patient's psychopathology indicates whether technique should be guided by a more Kohutian emphasis on empathic intervention, or a more Kernbergian emphasis on interpretive intervention. Stolorow and Lachmann's "fixated developmental disorders" require more therapeutic nurturance than do disorders reflected in defensively maintained maladaptive processes.

Because each patient manifests various levels of ego functioning, the therapist may need to modify and regulate the balance of empathic understanding (usually experienced as nurturance) and interpretation of defense (usually experienced as more frustrating), even within the same psychotherapy session. Such flexibility is called for because of the coexistence of disparate levels of ego functioning, which may be differentially activated by different transference reactions. Clinical observation supports the fact that many patients with pre-oedipal problems *also* struggle with higher-level defensive processes, and many neurotic patients *also* have some pre-oedipal issues. Thus, the technical suggestions emerging from the chapter's first focus (on dimensions of parenting) should be refined to include this emphasis on the extraordinary complexity created by the multiple nature of parenting relationships.

IMPLICATIONS FOR RESEARCH

The areas of child development research we have reviewed suggest several directions for further inquiry. Future clinical research might examine how therapeutic process and outcome are affected by the therapist's flexibility in providing warmth while maintaining work expectations and clear limits. It would be useful to better identify what stages in therapy, or what issues, call for a particular therapeutic stance, just as responsive parenting reflects a recognition of the child's changing, and sometimes regressive, developmental needs. As for the reverse direction—the implications of clinical practice for research on early childhood—it would be interesting to study, for example, transference processes for the growing child. Patterns of distortion in children's perceptions of their parents, teachers, friends, or siblings might yield interesting information about social and emotional development. To date, children's subjective experiences are an important, though underemphasized, focus of developmental research (Bloom-Feshbach, Bloom-Feshbach, & Heller, 1982).

In the second section of the chapter we focused on metacognition—the cognitive monitoring of cognitive processes. We expanded this concept as defined by early childhood researchers to include the self-monitoring of psychological processes. From this perspective, we suggested that the explicit identification and articulation of metacognitive thinking might be helpful in psychotherapy treatment and in the training of therapists. Psychotherapy researchers might begin to identify which metacognitive principles foster therapeutic change. As a start, even an exploratory descriptive study that interviewed therapists and patients might provide interesting findings about the natural use of metacognitive strategies. On the other hand, this clinical consideration of metacognitive processes suggests that the developmentalist might study the role of affect in the child's acquisition of metacognitive abilities. Thus, in studying the development of metacognitive processes in elementary school children, researchers might explore how emotional factors interact with these cognitive capacities and, moreover, how these cognitive processes function in tasks that have a social or emotional context (e.g., memory for people vs. objects, or metacognitive strategies for dealing with anger or sadness). An example of a clinically useful direction for future research is the Dollinger and McGuire (1981) study of the development of psychological-mindedness, in which children's understanding of defense mechanisms was examined.

Another research direction is suggested by the multidimensionality of psychological structure. Diagnostic scales, and instruments that measure psychological structure (such as Loevinger's ego development scale, or Kolhberg's moral development scale) could be utilized as indicators of *constellations* of levels rather than as indicators of one dominant level of functioning. Studies of psychotherapy might examine various levels of psychological change, exploring the patternings of patients' psychopathology and emotional strengths.

This multidimensional view of psychological structure also has implications for basic research in early childhood development. Investigators might examine the impact of psychologically disparate parenting figures on children's emotional and social development (Bloom-Feshbach *et al.*, 1980). Other research might focus on the child's development of psychological structures, considering the complexity created by salient nonparental figures such as grandmothers (Cohler & Grunebaum, 1981), siblings (Lamb & Sutton-Smith, 1982), and peers (Berndt, 1983). It might be interesting to study the effects of incomplete or inconsistent attainment of cognitive stages, such as Piagetian levels or levels of interpersonal understanding (Selman, 1980). For example, some children may exhibit a range of conceptual capacities depending on the emotional and social properties of the task and the task context.

CONCLUSION

We view this chapter as an initial effort to promote more scholarly interchange between researchers in the field of early childhood development and psychotherapists who work with adults. The body of research on early childhood development is so large that we have had to be selective, omitting some important topics (e.g., regression). Our emphasis on psychoanalytically oriented psychotherapy is obviously based on our own theoretical bias, but also on the fact that clinicians with this orientation place greater weight on developmental theory and on the psychotherapy relationship. Thus, psychodynamic therapists may have the most to gain, and the most to contribute, in conceptual interchange with basic researchers of development. Research and clinical practice each offer a unique vantage point for understanding human behavior. In our view, the interaction between the empirical and conceptual rigor of research and the openness and depth of clinical theory and observation can enrich both endeavors.

REFERENCES

Abelin, E. L. The role of the father in the separation-individuation process. In J. B. McDevitt & C. F. Settlage (Eds.), *Separation-individuation: Essays in honor of Margaret Mahler.* New York: International Universities Press, 1971.

Ainsworth, M. D. S., & Wittig, B. A. Attachment and exploratory behavior of one-year-olds in a strange situation. In B. M. Foss (Ed.), *Determinants of infant behavior* (Vol. 4). London: Methuen, 1969.

Angrilli, A. F. The psychosexual identification of pre-school boys. *Journal of Genetic Psychology*, 1960, *97*, 327–340.

Asher, S. R., & Renshaw, P.D. Children without friends: Social knowledge and social skills training. In S. R. Asher & J. M. Gottman (Eds.), *The development of children's friendships.* Cambridge: Cambridge University Press, 1981.

Bandura, A., & Walters, R. H. *Adolescent aggression: A study of the influence of child-rearing practices and family interrelationships.* New York: Ronald Press, 1959.

Baumrind, D. Child care practices anteceding three patterns of preschool behavior. *Genetic Psychology Monographs*, 1967, *75*, 43–88.

Baumrind, D. Current patterns of parental authority. *Developmental Psychology Monographs*, 1971, *4* (1), Part 2.

Baumrind, D. *Socialization determinants of personal agency*. Paper presented at the biennial meeting of the Society for Research in Child Development, New Orleans, April 1977.

Baumrind, D., & Black, A. E. Socialization practices associated with dimensions of competence in preschool boys and girls. *Child Development*, 1967, *38*, 291–327.

Becker, W. C. Consequences of different kinds of parental discipline. In M. L. Hoffman & L. W. Hoffman (Eds.), *Review of child development research* (Vol. 1). New York: Russell Sage Foundation, 1964.

Belsky, J. Mother-father-infant interaction: A naturalistic observational study. *Developmental Psychology*, 1979, *15*, 601–607.

Berndt, T. J. Developmental changes in conformity to peers and parents. *Developmental Psychology*, 1979, *15*, 608–616.

Berndt, T. J. Relations between social cognition, nonsocial cognition, and social behavior: The case of friendship. In J. H. Flavell & L. D. Ross (Eds.), *Social cognitive development: Frontiers and possible futures*. Cambridge: Cambridge University Press, 1981.

Berndt, T. J. Peer relationships in children of working parents. In C. D. Hayes & S. B. Kamerman (Eds.), *Children of working parents: Experiences and outcomes*. Washington, D.C.: National Academy Press, 1983.

Bigelow, B. J., & LaGaipa, J. J. The development of friendship values and choice. In H. C. Foot, A. J. Chapman, & J. R. Smith (Eds.), *Friendship and social relations in children*. New York: Wiley, 1980.

Biller, H. B. *Father, child and sex role*. Lexington, Mass.: D.C. Heath, 1971.

Biller, H. B., & Borstelmann, L. J. Masculine development: An integrative review. *Merrill-Palmer Quarterly*, 1967, *13*, 253–294.

Biller, H. D., & Weiss, S. D. The father-daughter relationship and the personality development of the female. *Journal of Genetic Psychology*, 1970, *116*, 79–93.

Bixenstine, V. E., de Corte, M. S., & Bixenstine, B. A. Conformity to peer-sponsored misconduct at four age levels. *Developmental Psychology*, 1976, *12*, 226–236.

Blehar, M. C., Lieberman, A. F., & Ainsworth, M. D. S. Early face-to-face interaction and its relation to later mother-infant attachment. *Child Development*, 1977, *48*, 182–194.

Bloom-Feshbach, S., & Blatt, S. Separation and nursery school adjustment. *Journal of the American Academy of Child Psychiatry*, 1982, *21*, 58–64.

Bloom-Feshbach, S., Bloom-Feshbach, J., & Gaughran, J. The child's tie to both parents: Separation patterns and nursery school adjustment. *American Journal of Orthopsychiatry*, 1980, *50*, 505–521.

Bloom-Feshbach, S., Bloom-Feshbach, J., & Heller, K. Work, family, and children's perceptions of the world. In S. B. Kamerman & C. D. Hayes (Eds.), *Families that work: Children in a changing world*. Washington, D.C.: National Academy Press, 1982.

Bowlby, J. *Attachment and loss. Vol. 1. Attachment*. New York: Basic Books, 1969.

Bronson, W. C. Dimensions of ego and infantile identification. *Journal of Personality*, 1959, *27*, 532–545.

Bronson, W. C. Development in behavior with age-mates during the second year of life. In M. Lewis & L. A. Rosenblum (Eds.), *Friendship and peer relations*. New York: Wiley, 1975.

Brown, A. L. Knowing when, where, and how to remember: A problem of metacognition. In R. Glaser (Ed.), *Advances in instructional psychology*. New York: Halsted Press, 1978.

Burlingham, D., & Freud, A. *Infants without families*. London: Allen & Unwin, 1944.

Clarke-Stewart, K. A. Interactions between mothers and their young children: Characteristics and consequences. *Monographs of the Society for Research in Child Development*, 1973, *38* (Serial No. 153).

Clarke-Stewart, K. A. And daddy makes three: The father's impact on mother and young child. *Child Development*, 1978, *49*, 466–478.

Cohen, L. J., & Campos, J. J. Father, mother, and stranger as elicitors of attachment behaviors in infancy. *Developmental Psychology*, 1974, *10*, 146–154.

Cohler, B. J., & Grunebaum, H. U. *Mothers, grandmothers and daughters: Personality and childcare in three-qeneration families*. New York: Wiley, 1981.

Condry, J., & Siman, M. L. Characteristics of peer- and adult-oriented children. *Journal of Marriage and the Family*, 1974, *36*, 543–554.

Coopersmith, S. *The antecedents of self-esteem*. San Francisco: W. H. Freeman, 1967.

Cowen, E. L., Pederson, A., Babigan, H., Izzo, L. D., & Trost, M. A., Long-term follow-up of early detected vulnerable children. *Journal of Consulting and Clinical Pathology*, 1973, *41*, 438–446.

Dollinger, S. J., & McGuire, B. The development of psychological-mindedness: Children's understanding of defense mechanisms. *Journal of Clinical Child Psychology*, 1981, *10*, 117–121.

Doyle, A., Connolly, J., & Rivest, L. The effect of playmate familiarity on the social interactions of young children. *Child Development*, 1980, *51*, 217–223.

Feldman, S. S., & Ingram, M. E. Attachment behavior: A validation study in two age groups. *Child Development*, 1975, *46*, 319–330.

Fish, K. D., & Biller, H. B. Perceived childhood paternal relationships and college females' personal adjustment. *Adolescence*, 1973, *8*, 415–420.

Fisher, S. F. *The female orgasm: Psychology, physiology, fantasy*. New York: Basic Books, 1973.

Flavell, J. H. Metacognitive development. In J. M. Scandura & C. J. Brainerd (Eds.), *Structural/process theories of complex human behavior.*, Alphen aan den Rijn, The Netherlands: Sijthoff en Noordhoff, 1978.

Flavell, J. H. Metacognition and cognitive monitoring: A new area of cognitive-developmental inquiry. In E. M. Hetherington & R. D. Parke (Eds.), *Contemporary readings in child psychology* (2nd ed.). New York: McGraw-Hill, 1981.

Flavell, J. H., & Wellman, H. M. Metamemory. In R. V. Kail, J. W. Hagen (Eds.), *Perspectives on the development of memory and cognition*. Hillsdale, N.J.: Lawrence Erlbaum, 1977.

Flavell, J. H., Friedrichs, A. G., & Hoyt, J. D. Developmental changes in memorization processes. *Cognitive Psychology*, 1970, *1*, 324–340.

Freud, S. Analysis of a phobia in a five-year-old boy. In J. Strachey (Ed. and Trans.), *Standard edition of the complete psychological works of Sigmund Freud*. London: Hogarth Press, 1955. (Originally published, 1909).

Freud, S. The dissolution of the Oedipus complex. *The standard edition* (Vol. 2). London: Hogarth Press, 1955. (Originally published, 1924).

Freud, S. Some psychological consequences of the anatomical distinction between the sexes. *The standard edition* (Vol. 5). London: Hogarth Press, 1961. (Originally published, 1925)

Greenacre, P. Considerations regarding the parent-infant relationship. *International Journal of Psychoanalysis*, 1960, *41*, 571–584.

Greenson, R. *The technique and practice of psychoanalysis*. New York: International Universities Press, 1967.

Hartup, W. W. The peer system. In P. H. Mussen (Ed.), *Handbook of child psychology* (Vol. 4, 4th ed.). New York: Wiley, 1983.

Hetherington, E. M. The effects of family variables on sex typing, on parent-child similarity, and on imitation in children. In J. P. Hill (Ed.), *Minnesota Symposium on Child Psychology* (Vol. 1). Minneapolis: University of Minnesota Press, 1967.

Hirschi, T. *Causes of delinquency*. Berkeley: University of California Press, 1969.

Hoffman, M. L. Father absence and conscience development. *Developmental Psychology*, 1971, *4*, 400–406.

Howells, J. G. *Modern perspectives in international child psychiatry*. Edinburgh: Oliver & Boyd, 1969.

Ide, J. K., Parkerson, J., Haertel, G. E., & Walberg, H. J. Peer group influence on educational outcomes: A quantitative synthesis. *Journal of Educational Psychology*, 1981, *73*, 472–484.

Johnson, M. M. Sex role learning in the nuclear family. *Child Development*, 1963, *34*, 315–337.

Kagan, J., & Moss, H. A. *Birth to maturity: The Fels study of psychological development*. New York: Wiley, 1962.

Kandel, D. B. Homophily, selection and socialization in adolescent friendships. *American Journal of Sociology*, 1978, *84*, 427–436.

Kernberg, O. The structural diagnosis of borderline personality organization. In P. Hartocollis (Ed.), *Borderline personality disorders*. New York: International Universities Press, 1977.

Kohut, M. *The analysis of the self*. New York: International Universities Press, 1971.

Kotelchuck, M. The infant's relationship to the father: Experimental evidence. In M. E. Lamb (Ed.), *The role of the father in child development*. New York: Wiley, 1976.

Lamb, M. E. Effects of stress and cohort on mother- and father-infant interaction. *Developmental Psychology*, 1976, *12*, 435–443. (a)

Lamb, M. E. Interactions between 8-month-old children and their fathers and mothers. In M. E. Lamb (Ed.), *The role of the father in child development*. New York: Wiley, 1976. (b)

Lamb, M. E. Interactions between two-year-olds and their mothers and fathers. *Psychological Reports*, 1976, *38*, 447–450. (c).

Lamb, M. E. The development of mother-infant and father-infant attachments in the second year of life. *Developmental Psychology*, 1977, *12*, 637–648. (a)

Lamb, M. E. Father-infant and mother-infant interaction in the first year of life. *Child Development*, 1977, *48*, 167–181. (b)

Lamb, M. E. Separation and reunion behaviors as criteria of attachment to mothers and fathers. *Early Human Development*, 1979, 3/4, 329–339.

Lamb, M. E. (Ed.). *The role of the father in child development* (2nd ed.). New York: Wiley, 1982.

Lamb, M. E., & Sutton-Smith, B. (Eds.). *Sibling relationships: Their nature and significance across the lifespan*. Hillsdale, N.J.: Lawrence Erlbaum, 1982.

Lamb, M. E., Frodi, A. M., Hwang, C. P., & Frodi, M. Varying degrees of paternal involvement in infant care: Attitudinal and behavioral correlates. In M. E. Lamb (Ed.), *Nontraditional families: Parenting and child development*. Hillsdale, N.J.: Lawrence Erlbaum, 1982.

Lewis, M., & Schaeffer, S. Peer behavior and mother-infant interaction in maltreated children. In M. Lewis & L. A. Rosenblum (Eds.), *The uncommon child*. New York: Plenum Press, 1980.

Lewis, M., Young, G., Brooks, J., & Michalson, L. The beginning of friendship. In M. Lewis & L. A. Rosenblum (Eds.), *Friendship and peer relations*. New York: Wiley, 1975.

Lieberman, A. F. Preschoolers' competence with a peer: Relations with attachment and peer experience. *Child Development*, 1977, *48*, 1277–1287.

Loewald, H. On the therapeutic action of psycho-analysis. *International Journal of Psycho-analysis*, 1960, *41*, 16–33.

Loewald, H. Internalization, separation, mourning and the superego. *Psychoanalytic Quarterly*, 1962, *31*, 483–504.

Lozoff, M. M. Fathers and autonomy in women. In R. B. Kundsin (Ed.), *Women and success*. New York: William Morrow, 1974.

Mahler, M. S., Pine, F., & Bergman, A. *The psychological birth of the human infant*. New York: Basic Books, 1975.

Markman, E. M. Realizing that you don't understand: A preliminary investigation. *Child Development*, 1977, *48*, 986–992.

Markman, E. M. Realizing that you don't understand: Elementary school children's awareness of inconsistencies. *Child Development*, 1979, *50*, 643–655.

Masters, J. C., & Furman, W. Popularity, individual friendship selection, and specific peer interaction among children. *Developmental Psychology*, 1981, *17*, 344–350.

Matas, L., Arend, R., & Sroufe, L. Continuity of adaptation in the second year: The relationship between quality of attachment and later competence. *Child Development*, 1978, *49*, 547–556.

Meerloo, J. A. M. The psychological role of the father: The father cuts the cord. *Child and Family*, 1968, *10*, 102–114.

Mueller, E., & Lucas, T. A. A developmental analysis of peer interaction among toddlers. In M. Lewis & L. A. Rosenblum (Eds.), *Friendship and peer relations*. New York: Wiley, 1975.

Mussen, P. H. Some antecedents and consequences of male sex-typing in adolescent boys. *Psychological Monographs*, 1961, *75*(Whole No. 506).

Mussen, P. H., & Rutherford, E. Parent-child relations and parental personality in relation to young children's sex-role preferences. *Child Development*, 1963, *34*, 589–607.

Mussen, P. H., Bouterline-Young, H., Gaddini, R., & Morante, L. The influence of father-son relationships on adolescent personality and attitudes. *Journal of Child Psychology and Psychiatry*, 1963, *4*, 3–16.

Parke, R. D., & Sawin, D. B. The father's role in infancy: A reevaluation. *Family Coordinator*, 1976, *25*, 365–371.

Payne, D. E., & Mussen, P. H. Parent-child relations and father identification among adolescent boys. *Journal of Abnormal and Social Psychology*, 1965, *52*, 358–362.

Radin, N. The role of the father in cognitive, academic and intellectual development. In M. E. Lamb (Ed.), *The role of the father in child development* (2nd ed.). New York: Wiley, 1982.

Reuter, M. W., & Biller, H. B. Perceived paternal nurturance-availability and personality adjustment among college males. *Journal of Consulting and Clinical Psychology*, 1973, *40*, 339–342.

Roff, M., Sells, B., & Golden, M. *Social adjustment and personality development in children*. Minneapolis: University of Minnesota Press, 1972.

Rosenberg, M. *Society and the adolescent self-image*. Princeton: Princeton University Press, 1965.

Rutherford, E. E., & Mussen, P. H. Generosity in nursery school boys. *Child Development*, 1968, *39*, 755–765.

Schaffer, H. R., & Emerson, P. E. The development of social attachments in infancy. *Monographs of the Society for Research in Child Development*, 1964, *29* (Serial No. 94).

Sears, R. R. Relation of early socialization experiences to self-concepts and gender role in middle childhood. *Child Development*, 1970, *41*, 267–289.

Selman, R. L. *The growth of interpersonal understanding: Developmental and clinical analyses.* New York: Academic Press, 1980.

Slater, P. E. Parental behavior and the personality of the child. *Journal of Genetic Psychology*, 1962, *101*, 53–68.

Spelke, E., Zelazo, P., Kagan, J., & Kotelchuck, M. Father interaction and separation protest. *Developmental Psychology*, 1973, *7*, 83–90.

Sroufe, L. A. Attachment and the roots of competence. *Human Nature*, October 1978, pp. 50–57.

Sroufe, L. A., & Waters, E. Attachment as an organizational construct. *Child Development*, 1977, *48*, 1184–1199.

Stolorow, R., & Lachmann, F. M. The developmental prestages of defenses: Diagnostic and therapeutic implications. *Psychoanalytic Quarterly*, 1978, *47*, 73–102.

Sugarman, A., Bloom-Feshbach, S., & Bloom-Feshbach, J. The psychological dimensions of borderline adolescents. In J. Kwawer, H. O. Lerner, P. M. Lerner, & A. Sugarman (Eds.), *Borderline phenomena and the Rorschach test.* New York: International Universities Press, 1980.

Sullivan, H. S. *The interpersonal theory of psychiatry.* New York: W. W. Norton, 1953.

Ullman, C. A. Teachers, peers and tests as predictors of adjustment. *Journal of Educational Psychology*, 1957, *48*, 257–267.

von der Heydt, V. The role of the father in early mental development. *British Journal of Medical Psychology*, 1964, *37*, 123–131.

Wachs, T. D., & Gruen, G. E. *Early experience and human development.* New York: Plenum Press, 1982.

Waters, E. The reliability and stability of individual differences in infant-mother attachment. *Child Development*, 1978, *49*, 483–494.

Waters, E., Wippman, J., & Sroufe, L. A. Attachment, positive affect, and competence in the peer group: Two studies in construct validation. *Child Development*, 1979, *50*, 821–829.

West, D. J., & Farrington, D. P. *Who becomes delinquent?* London: Heinemann, 1973.

13

Normal and Pathological Gender-Role Development in Children

ANNA C. LEE

INTRODUCTION

The study of sex- and gender-role identity development subsumes a multitude of psychological differences between the sexes. Although based on sexual dimorphism, these differences are critically influenced by factors more profound than biological givens. These primarily include normal versus abnormal development, social stereotypes, and sex-role behavior. This chapter will discuss the influence of biological and social factors on psychosexual development in childhood. It will also investigate ways in which children come to develop particular perceptions, attitudes, affects, and behaviors consistent with their gender-role identity, focusing on the interaction of biological substrates and constitutional endowment with environment. This will be followed by brief reviews of the three major theories of gender-role development, psychoanalytic, social learning, and cognitive development. Issues concerning the development of sex-role stereotypes in early childhood will also be surveyed, with reference to both normal and pathological gender-role development in childhood and to the consequences when disturbances go untreated.

ANNA C. LEE • Rehabilitation Services, Hospital for Special Surgery, New York, New York 10021.

Second, I shall look at some empirical findings on psychotherapy involving issues of referral, diagnosis, formulating treatment goals, establishing the therapeutic alliance, and transference and countertransference, using case illustrations to show how these bear on gender-role development. Finally, the direction of future research will be considered in relation to clinical practice.

Before proceeding further, explication of some terms is in order. This chapter takes its definitions from Stoller (1974), who distinguishes between *sex* and *gender*. The term *sex* has a strictly biological connotation and refers to the chromosomes external genitalia, internal genitalia, gonads, hormonal states, and secondary sex characteristics. One's sex is either male or female. In contrast, *gender* is a term with psychological or cultural rather than biological connotations and refers to an individual's relative qualtity of masculinity or femininity. Whereas the proper terms for sex are "male" and "female," the corresponding terms for gender are "masculine" and "feminine" and may be quite independent of biological sex.

> *Gender identity* starts with the knowledge and awareness, whether conscious or unconscious, that one belongs to one sex and not the other, though as one develops, gender identity becomes much more complicated, so that, for example one may sense himself not only a male but a masculine man or an effeminate man or even a man who fantasizes being a woman. *Gender role* is the overt behavior one displays in society, the role which he plays, especially with other people, to establish his position with them insofar as his and their evaluation of him is concerned. While gender, gender identity and gender role are almost synonymous in the usual person, in certain abnormal cases they are at variance. (Stoller, 1974, p. 10)

This chapter will address mostly issues related to *gender-identity* and *gender-role* development rather than *sexual identity* and *sex role*, though the terms can be and often are used interchangeably.

Biological Factors

The role of biological factors in the development of gender and gender-role identification has been widely studied as one aspect of the nature versus nurture argument. Although it is generally accepted that biology plays a dominant role in the sexual behavior of animals, its role in human behavior seems much less clear. There can be no doubt that the fundamental difference between the sexes, namely, anatomy, is biologically determined. Past that point the influence of social and cultural factors appears to take precedence. Even Freud (1937/1955), who emphasized physical sex differences, remained convinced that anatomical sex differences set the pattern for the development of sexual and gender identification. For him such differences derived from a constitutional bisexuality, a "bedrock," the core through which further attempts at analysis of resistance would prove futile.

During the past 30 years a vast accumulation of studies in brain physiology have provided unequivocal support for Freud's contention of sexual "bipotentiality." Controlled laboratory studies of the neuroanatomical and neuroendocrine mechanisms of sexual behavior in animals have discovered hierarchies of hormones, internal and external perceptual inputs, releasing mechanisms, and new psychological experience that critically influence later sexual development (Money, 1965; Stoller, 1972a). Although the genetic structure (XX or XY) determines the differentiation of fetal gonads into testes or ovaries, hormonal control subsequently determines biological sexual differentiation. The resting state for the central mechanisms of gender is essentially female. Masculine behavior occurs only when the fetal brain is infused with androgens. This sexual bipotentiality exists, moreover, across all species. There seem to exist critical periods for the introduction of androgens to the fetal brain. These occur when the brain is most sensitive to the influence of sex hormones that organize tendencies toward masculine or feminine (not only reproductive in nature) behavior. The period of greatest sensitivity varies from species to species, some just before and some just after birth, but is invariant for individuals of both sexes within the species.

Money and colleagues (Money, 1965; Money & Ehrhardt, 1972; Money, Hampson, & Hampson, 1955) offer considerable evidence of human analogs for the effects of hormones on behavior. Their studies of individuals whose assigned sex differed from their genetic sex or from their hormonal histories have illuminated the role of hormones in contributing to temperamental, attitudinal, and behavioral sex differences (M. Lewis & Weinraub, 1979). Primary among those studied are individuals with Turner's syndrome (X0 chromosomal abnormality), testicular feminization syndrome, constitutional male hypogonadism, temporal lobe disorder, masculinization through large doses of progesterone in females, and the androgen insensitivity syndrome. Their findings argue for the primacy of rearing as the determinant of gender identity and gender-role behavior despite the child's genetic sex. Generally, genetic males raised as females (i.e., children who were progestin-induced hermaphrodites or those with female fetal androgenitization syndrome) evidenced more masculine attitudes, preferences, and behaviors than did controls. Likely to be described as "tomboys," they showed increased activity levels, preferred cars, trucks, and building materials to dolls, and were more career-oriented and less interested in infants than controls.

From all evidence accumulated over 20 years of research, Money and Ehrhardt, (1972) conclude that no one factor, neither chromosomal sex nor prenatally or postnatally administtered hormones, is sufficient to determine gender identity. Rather, the consistency of being reared as one sex or another seems to be the deciding factor, in particular in the period from 18 months to 3 to 4 years old. Past this point, a forced change of sex is likely to result in iatrogenic psychopathology.

THEORIES

We turn now to the various theories of gender development and gender-role development, beginning with the psychoanalytic perspective. Derived exclusively from observations made during the analyses of patients, psychoanalytic formulations must also be assessed in light of the prevailing historical context from which they originated.

PSYCHOANALYTIC THEORY

For Freud (1923/1961), the development of gender identity does not manifest clear signs of differentiation between male and female until the child's discovery of the penis (or lack thereof) at about age 2 or 3. Only with genital awareness does gender identity begin.

With the development of the Oedipus complex, the appearance of castration anxiety forces a fundamental distinction between the two sexes. Whereas in boys the complex is destroyed by the fear of castration by the father, in girls it is made possible and led up to by the castration complex (Freud, 1925/1961). Castration anxiety, then, operates to inhibit and limit masculinity and encourages femininity. In consequence, the boy relinquishes his libidinal cathexis toward his mother and identifies with his father. For the girl, recognition of her already castrated state causes her to abandon her wish for a penis. Instead, she wishes for a child with her father (the penis = baby equation), thereby taking her father as a love object. Her mother thus becomes the object of her jealousy. Penis envy causes the girl to disavow her castrated state (the masculinity complex) and to experience the narcissistic wound with a sense of inferiority. Furthermore, it is likely to affect deleteriously her relationship with her mother because she may hold her mother responsible for the lack of a penis. Lastly, penis envy may stimulate a massive repression against masturbation that, in puberty, will do away with a large amount of the girl's masculinity to make room for the development of her femininity. Freud adamantly stresses a qualitiative difference between the sexes in the aftermath of the castration complex: Whereas the boy's oedipal yearnings are dissolved once and for all and incorporated into the superego, the incestuous wishes of the girl are more slowly abandoned or repressed, thereby exerting profound effects on a woman's mental life, especially in her superego formation. There also comes the identification with the same-sex parent, thereby fostering the child's incorporation and development of the thoughts, affects, and behaviors appropriate for his or her gender.

Vigorous objections to Freud's view of female psychosexual development as degrading have since arisen from many quarters. Within the psychoanalytic camp, Karen Horney (1926) pointed to the masculine orientation within which psychoanalytic formulations had been derived and boldly asserted the revolutionary idea of womb and motherhood envy in men. Melanie Klein (1928) reported her conviction of early vaginal awareness, placing the origin of the oedipal conflict within the first year of life. Discussing current psychoanalytic thinking about female sexuality, Stoller, Buxbaum, and Galenson (1976) posit that previous psychoanalytic theories of masculinity and femininity were derived without the benefit of recent biological and psychological findings.

Although Freud was unable to study the earliest stages of core gender identity, his discoveries of penis envy, castration reactions, and oedipal complications remain valid. Stoller believes that the newer and more accurate veiw of the psychology of women is that

> women have a primary, unquestioned acceptance of their femaleness. A woman's psychology includes her disappointment at not being a male, but as a distinctly different, not inherently psychologically inferior being. (Stoller *et al.*, 1976, p. 142)

Rather than theorizing a masculine gender orientation as the earliest for both sexes, Fast (1978) proposes the existence of an original, undifferentiated matrix of gender identity, unattuned to sex differences and unaware of limitations imposed by membership in one sex or the other. These limits are gradually learned: The boy recognizes his mother's role in the origin of babies, and the girl becomes aware that she has no penis.

Current psychoanalytic theories focus on the child's pre-oedipal relationship with the mother, which serves as a foundation of the development of gender identity (Mahler & Furer, 1968). From observations of infants and children and their fantasy lives as revealed in play and in interaction with their families, the role of the pre-oedipal mother seems crucial in the infant's identity development. Galenson and Roiphe

(1980) suggest the existence of an early genital phase in both sexes from age 15 to 24 months. During this period, normally increasing genital sensations, concomitant with the maturational capacity for anal and urinary sphincter control and increasing self–object differentiation, are thought to contribute to the establishment of genital schematization and sense of genital identity coextensive with a sense of self.

Social Learning Theory

Social learning theory agreed with the premise of psychoanalytic theory in postulating identification as the motivating principle for acquisition of appropriate gender-role behaviors. According to the learning theorists, however, the psychoanalytic position failed to explain the actual mechanisms by which the learning occurred. Thus, social learning theory developed inthe 1940s and 1950s and attempted to apply principles of learning theory to explain the acquisition of gender-role behaviors. According to Mischel (1966), its major assumption is that children learn through observing behaviors associated with both parents. They do so without direct reinforcement, viewing their parents as powerful figures with control over rewards. Through differential reinforcement they learn to value gender-"appropriate" behaviors and devalue gender-"inappropriate" behaviors. The reinforcement comes from many quarters including parents, teachers, and peers.

Unlike other theories of gender-identity development, social learning posits no specific age at which any of the processes of gender identity or gender-role behaviors develop. It assumes that although the acquisition of gender-typed behaviors may be regulated by cognitive processes, the actual performance of these behaviors depends on reinforcement histories alone.

Support for the social learning theory can be seen in a number of empirical studies. Heilbrun (1965), for instance, found that males and females who identified with instrumental (masculine) fathers showed the most extensive and appropriate sex-role personality differences. Hence, identification with the instrumental father is associated with enhanced masculinity in the son and femininity in the daughter. In contrast, identification with an expressive mother failed to elicit equally extensive and appropriate sex-role differentiation. Lastly, parental warmth appeared critical in enhancing the child's identification with the parent.

As for differential treatment and socialization of young children, Maccoby and Jacklin (1974) in their extensive review of sex differences found "surprisingly little differentiation in parent behavior according to the sex of the child" (p. 338). They note only four areas of demonstrated sex differences: (1) greater verbal ability in girls than in boys; (2) greater visual-spatial ability in boys; (3) better mathematical ability in boys; and (4) more aggressive behavior in males. Since then, however, a proliferation of infant studies have contributed to greater understanding of the role of biological endowment and environmental influences. In her review of sex differences in the earliest years, Birns (1976) notes that although most sex differences in behavior related to personality and cognitive development are not apparent at birth or during infancy, they do become apparent during the preschool period. Such differences include, first of all, more aggressive, toy-manipulating, and exploratory play in boys. Girls, in contrast, are more sedentary, imitative, and attentive. Although cognitive differences in coding and memory were not observed, differences in field independence (characteristically seen in males during adolescence and adulthood) favor females in the preschool years.

More pertinent here is Birns's finding of differential socialization of males and females from the moment of birth. The differences were consistent with cultural sex stereotypes and were sufficiently pervasive to explain differences in adult roles. These included, first of all, differential behavior on the part of adults with infants as a function of the infant's gender. In studies of parent–infant interactions, fathers were found to hold, attend to, and provide greater physical and auditory stimulation for neonates than mothers, and fathers were most attentive to newborn boys (Osofsky & Danzger, 1974). Even during infancy males were found to receive more attention (a function of their increased crying and irritability), whereas girls learned to laugh and smile and to be content even when ignored (Moss, 1974). Boys who receive more proximal (touching and holding) attention are gradually shifted to the distal mode. The reverse occurs for girls: Originally less attended to (distal), this gradually changes to greater encouragement of close contact with parents (Lewis & Weinraub, 1974).

The influence of fathers has received greater research attention in the past decade. Fathers were found to exhibit more physical and less conventional play with their infants, and infants showed greater pleasure in response to fathers' play (Lamb, 1975). Fathers have also been seen to manifest more concern with gender-appropriate play for their children than mothers, appearing intent on producing "masculine" sons and "feminine" daughters (Fagot, 1974).

In nursery school, teachers were found to respond more vigorously to boys' disruptive behavior than to that of girls. In addition, teachers were similar to parents in responding to boys' requests for help in ways that gave more information to boys (Servin, O'Leary, Kent, & Tonick, 1968).

Cognitive Developmental Theory

Basing the cognitive developmental theory of gender identity on Piaget's (1952) concept of conservation of physical properties, Kohlberg (1966) contends that the child himself plays an active role in structuring the world. As such, children perceive the world in discrete stages, their perception and view of reality differing qualitatively from that of adults. According to Kohlberg, children's attribution of gender depends on physical qualities. It is not until they have developed the concept of conservation (at about age 5 or 6) that they have any sense of gender constancy, that is, that one's gender is invariant and irreversible. Thus, it appears pointless to refer to their gender identity until conservation of property is mastered. Although it it possible for a child of 3 to label himself correctly ("I am a boy"), he cannot yet know that (1) a person's gender never changes, (2) everyone has a gender, and (3) gender differences are physical/anatomical in nature. Only when conservation develops does the young child understand that a person's gender is invariant. As gender identity stabilizes, children begin to prefer gender-typed activities and objects, mostly because they come to value and wish to be like things they perceive to be similar to themselves (i.e., other boys and men). Thus, they actively seek objects, activities, and behaviors consistent with their gender. Although this sequence seems straightforward for boys, Kohlberg postulates that girls are not as "gender typed" as boys. They know they are girls and want to be like their mothers, but, like boys, they also equate "male" with "big" and, synonymously, "powerful." Thus, both boys and to some extent girls are likely to identify with their fathers and with male things in general.

Although studies seem more scant for the cognitive developmental model, empirical support has been found in those undertaken (DeVries, 1969; Emmerick, Goldman, Kirsh, & Sharabany, 1977). Having investigated the relationship of gender constancy to cognitive level and gender-role preference, Marcus and Overton (1978) concur with the position that cognitive growth is a necessary feature for the establishment of a stable gender identity. Furthermore, gender constancy was greater in relation to the self and later generalized consistently to others. Unexpectedly, however, children showed difficulty in maintaining gender constancy when presented live models undergoing changes in hairstyle and clothing. Initially fascinated by the transformations by the live model, they became increasingly uncomfortable and perplexed as the changes took place. Contrary, to theory, however, Marcus and Overton did not find that children tend to increase their same-gender preferences as gender constancy becomes established. Instead, as children come to understand the immutability of their gender, they seem less threatened by objects and activities inconsistent with their gender role. Thus, the average 5-year-old, in addition to demonstrating a lack of gender constancy, perceives conforming to gender-role expectations as necessary to his gender identity. The 8-year-old, in contrast, who has achieved gender constancy, recognizes that masculine and feminine characteristics reflect individual choice rather than biological endowment. He can then be more accepting of variations in roles and behaviors than the younger child. Finally, the model suggests that young children's (4 to 7) reliance on specific physical attributes that they do not yet possess may motivate a necessary tendency among boys to seek out the friends, games, or activities that validate or confirm their emerging gender identity (Ullian, 1981). Girls, in contrast, apparently have an easier transition since they grow up to be women who, by nature of their childrearing function, are allowed simply "to be."

Criticism of the cognitive developmental theory comes from both psychoanalytic and social learning perspectives. From the former it may well be argued that children, by 5, are aware of genital differences between men and women even if they do not report this to be the reason for the distinction between the two sexes. Moreover, such learning theorists as Maccoby and Jacklin (1974) have pointed out that gender constancy is not necessary for socialization into gender roles. Children by 3 years old show clear gender-typed behaviors, long before gender constancy is established. Lastly, the theory fails to account for individual differences, especially among persons with atypical identity and gender-role disturbances in the face of unrelenting social pressure toward normalcy.

Finally, Katz's (1979) model of sex-role development reaches well into adulthood. Katz maintains that sex-role behavior undergoes a gradual process of change. She proposes, hence, a life span perspective of sex-role development, emphasizing three distinct developmental levels: (1) learning what is appropriate behavior for a male or female child, (2) acquiring concepts about what is appropriate as a potential male or female adult, and (3) behaving in ways that are deemed appropriate for male and female adults across the life span. Although the model has not been much investigated thus far, it offers intriguing possibilities for the understanding of gender-role development and the demands and tasks indigenous and appropriate for each phase of life.

Each theory contributes to our understanding of particular aspects of sex-role development. The evidence so far demonstrates, however, that no one theory can account appreciably for all aspects of gender-role development. What seems needed, therefore, is a greater awareness, consideration, and dialogue among the theories in order to inform and validate those hypotheses that are viable and discard those that are not.

DISTURBANCES OF GENDER-RELATED BEHAVIOR

Having examined the role of biological factors, developmental processes, and social stereotypes on normative gender-identity development, we turn now to an examination of deviant gender-related behaviors. Although rare, they represent the extreme of a general continuum of gender identity the examination of which contributes to our understanding of how gender identity, gender role, and sexual object choice develop. Variations along the continuum are largely attributable to prevailing cultural norms and standards.

Among children, atypical and disturbed gender-identity development appears relatively rare, occurring mostly in boys who are also more often the subjects of research and clinical studies in gender disturbance. Green (1974) suggests several reasons for this trend, emphasizing that although masculinity in girls (tomboyishness) is much more common than femininity (sissiness) in boys, the adult clinical picture shows more males than females experiencing sexual identity conflict. Although many girls show tomboyishness in earlier years, they typically become feminine during adolescence. Furthermore, societal tolerance of tomboyishness often precludes identification of this pattern as deviant. In contrast, femininity in boys causes considerable social stress and censure. Available follow-up data on feminine boys reveals that "such behavior more often portends sexual identity conflict during adulthood" (Green, 1974, p. 142).

Rosen, Rekers, and Friar (1977) differentiate two basic syndromes in male child gender disturbance: "cross-gender identification" and "gender-behavior disturbance." Gender-behavior disturbance is defined as "simple effeminacy" and may be frequent and intense or defined as "simple effeminacy" and may be frequent and intense of infrequent and subtle, with no evidence of cross-gender identification. In contrast, boys with cross-gender identity not only behave in a feminine way but truly believe that they are girls (Stoller, 1968). Characteristics of the syndrome include: cross-gender clothing preferences, actual or imaginary use of cosmetic articles, feminine gestures and behavioral mannerisms, and an aversion to masculine sex-typed activities, coupled with a preference for girl playmates and feminine activities, preference for assuming a female role, feminine voice inflection and predominantly feminine speech content, and verbal statements about the desire or preference to be a girl or a mother and to bear children and breastfeed infants (Rekers & Lovaas, 1974). Although base rates for incidence and prevalence are estimates at best, gender-identity disturbance occurs in approximately 1 out of 200,000 boys, and gender-behavior disturbance in about 1 out of 80,000 (Rekers, Rosen, Lovaas, & Bentler, 1978).

Considerable controversy surrounds the treatment of gender disturbances (Nordyke, Baer, Etzel, & LeBlanc, 1977; Winkler, 1977), mostly involving the rationale and ethical implications of treatment. Rekers (1977), a strong proponent of therapeutic intervention, points to the social ridicule and ostracism elicited from peers by extreme cross-gender behaviors, thereby creating depression, frustration and negative behavior in gender-disturbed boys. He notes that

> the most adaptive psychological state appears to be one in which the essential biologically mandated and socially defined distinctions between the male and female roles are mastered by the child. Beyond that point, there should be sex-role flexibility. (p. 561)

Without intervention, moreover, a gender-disturbed child is at high risk for even more serious adjustment difficulties in adulthood, with men more often than women being prone to symptoms of disturbed gender identity and sexual deviations (Bentler, 1976; Green, 1974; Green & Money, 1969; Kinsey, Pomeroy, Martin, & Gebhard, 1953; Lebovitz, 1972; Money & Ehrhardt, 1972; Rosen, Rekers, & Friar, 1977; Stoller, 1968, 1970; Zuger, 1966).

Although a review of the etiological factors contributing to gender-identity disturbance is beyond the scope of this chapter, several pathological features of the parent–child relationship should be mentioned. Primary among these is an unusually intimate physical and psychological contact with the mother evidenced by boys with cross-gender identity (Rosen *et al.*, 1977). A diffusion of boundaries exists between mother and son with both encouraging this type of relationship. The relationship with the father, however, tends to be psychologically distant in the majority of cross-gender boys studied, with the fathers spending little time with their children. The mothers, then, are seen as the primary model of behavior, whereas the fathers are perceived as either absent or ineffective. This pattern of family dynamics is clearly reflected in Stoller's (1979) description of transsexual children from 15 families. The children, typically seen by the mother as unusually beautiful, are used by her as a cure for lifelong regret at not being a male.

> All the boys were very feminine and had been so since earliest childhood. The mothers all had the position of power in the family; all but three had histories of successful competition with boys (almost always in sports) and conscious desires in childhood to be male. They had an excessively close symbiosis with this son, had married passive and distant men, and had stayed with the marriage despite unending incompatibility. . . . The more profound and longer the symbiosis persists, the more feminine the boy. . . . The femininity does not occur unless both mother and father do their part. If either acted differently, this extreme form would not result. (Stoller, 1979, pp. 840–841)

As for the etiology of cross-gender disturbance in girls, the few available data (Stoller, 1972b) support the theory that very masculine females, by dint of a retiring or inadequate mother, come to experience heightened protectiveness over this mother. The latter, usually a tired, depressed woman unfulfilled by her husband, encourages the overprotectiveness in her daughter. The clinical syndrome, then, is that of a dominant father, a retiring, unavailable mother, more closeness between father and daughter, and parental selective reinforcement of the daughter's masculine behavior.

Treatment for gender disturbances in children focuses primarily on early detection and intervention. Psychoanalytic and behavioral techniques for such children (Green, Newman, & Stoller, 1972; Greenson, 1966; Rekers *et al.*, 1978) stress, first of all, the sex of the therapist, noting the salience with which the same-sex therapist would (1) serve as an object of identification, and (2) fill a void created by the absent or psychologically unavailable same-sex parent. Second, parental attitudes toward the cross-gender disturbance must be explored, especially to highlight its inappropriateness and attendant peer rejection for the cross-gender behavior. This pattern is especially critical in the case of feminine boys, whose parents often approve of the behavior either implicitly or explicitly. Such reinforcement should then be interrupted. Providing such children alternative ways of behaving and increasing their comfort in gender-role flexibility appears a central principle. Applying a response–cost contingency to behaviors and to identifying for the children (boys, in this case) gender-appropriate interpersonal skills and coping patterns, and avoiding stereotyped self-defeating and socially punished ways of behavior often serves to encourage appropriate gender-identity development. In general, group intervention may be utilized as a way of offering the children initial socialization experience with peers. Its major drawback, however, is the dilution of the relationship with the therapist when group therapy is used exclusively.

THERAPY ISSUES

The foregoing review has both direct and indirect relevance for psychotherapy. It may also have specific implications for certain aspects of psychotherapy. I shall examine from a developmental perspective the importance of gender-identity and gender-role development and their interface with personality formation. A fundamental organizing principle of identity, gender-identity development is essentially an element within individual development. It is in this context that I shall examine the impact of variables highlighted in the preceding sections on psychotherapy, regardless of the primary or secondary role the variable plays in the individual psychopathology.

I shall consider first the relevance of the foregoing empirical evidence for the referral and diagnostic processes, the formulation of treatment goals, and the stages of the treatment, including formation of the therapeutic alliance as well as transference and countertransference issues, especially during particular phases of treatment. Finally, future directions for research and clinical practice will be considered.

A REFERRAL

Several issues involving the referral process are noteworthy, particularly the indications for treatment, and by whom and for whom the treatment is sought.

From the psychoanalytic perspective, treatment is usually indicated when internal conflicts between psychic structures are sufficiently debilitating to necessitate neurotic compromises and, at times, impair ego functions. Anxiety and depression often accompany the neurotic symptoms and may impede further development. Where there is a fundamental disturbance of gender-role and gender identity, the research suggests obvious indications for treatment.

The need for treatment becomes less clear when the disturbance is less dramatic and involves some degree of gender-role reversal or same-sex sexual object preference. In these cases it may not be the gender/gender-role disturbance itself that mandates treatment but the intrapsychic and social adjustment difficulties that accompany such behavior. Again, the individual's perception that his gender-related behavior is not a part of his adjustment difficulties may make it difficult for the therapist to confront this issue in treatment. Such resistance is found in children and adults alike.

The research seems more optimistic about engaging in psychotherapy those children experiencing gender disturbances, since their defenses are less firmly entrenched (Green, 1974; Hay, Barlow, & Hay, 1981; Rekers, 1977; Rekers & Lovaas, 1974; Stoller, 1979). Although subject to individual variation, the results of treatment are maintained upon follow-up.

For children, indications for psychotherapy frequently occur as early as the pre-oedipal phase. The onset of genital awareness and an awareness of anatomical differences by 18 months of age (Galenson & Roiphe, 1980) involve various periods during which fixations, regressions, and crises may disrupt psychosexual development. This is especially so during the subphases of the separation-individuation process (Mahler, Pine, & Bergman, 1975), and may be sufficient to create disturbances in gender-identity development with long-term implications. These evolve within the context of primary object relationships, especially within the mother–infant dyad. Stoller (1968, 1979) emphasizes the very clear indications for treatment of cross-gender-disturbed children, especially where there is a symbiotic relationship between the child and the opposite-sex parent (especially mother–son dyads) wherein paternal physical absence and/or emotional unavailability, and/or severe marital conflict, prevail. Abelin (1975) reminds us of the importance of the father in the rapprochement subphase since he fosters the development of the child's earliest exploratory and phallic attitudes, thus enabling the child to disentangle himself from the regressive pull of the symbiotic bond. This function allows for a nonsymbiotic relationship with the father, a figure who remains essentially "uncontaminated" during rapprochement. Atkins (1981) points compellingly to the mother's role in catalyzing or discoloring the child's affective attachment to the father. Given the evidence of the relatively little actual contact between father and child during the pre-oedipal period, it seems that father representations originate from sources outside the father–child dyad, specifically, the way in which the mother either facilitates or retards the child's attachment to the father.

The presence of symptoms related to disturbed gender identity is sufficient to warrant psychotherapy, given the prognostication for adult sexual deviancy. Early intervention is strongly suggested in view of the tendency for greater retrenching of disturbances in object relationships, ego functions, self-representations, and, most importantly, in identity formation. The prevalence of borderline and pathological, narcissistic symptoms in adult sexual deviants, many of whom experienced gender-identity disturbance in early childhood, also indicates the need for early diagnosis and treatment.

The question for whom and by whom treatment is generally sought for children is controversial because it is either the parent or an outside agency (usually the school) who initiates the referral. The designation of pathological functioning rests, hence, on the judgment of the child's caretakers and becomes a complex issue, especially for children and/or parents who may minimize or even deny the extent of pathology that is manifest. Nordyke *et al.*, (1977) question, in part, whether deviancy necessarily constitutes pathology, especially where the symptoms appear ego-syntonic. They also question whether "placating the parents" is a viable indication for treatment. Rekers (1977) argues cogently for treatment on the grounds of the anxiety and depression he found in these children as a result of the ostracism they experience from peers. Their parents' fears about the child's present and future adjustment also merited consideration.

Objections of the kind raised by Nordyke *et al.* highlight the confusion of social deviancy with psychopathology. It appears that given the current relaxation of sexual repression, there exists much reluctance to designate deviance from the norm as pathological. This trend is evidenced, for instance, by the decision of the American Psychiatric Association to delete homosexuality *per se* as a diagnostic entity. Increased social tolerance of homosexuality and, indeed, other forms of sexual deviancy does not eliminate the possibility that this behavior is indeed pathological. The crucial issue, regardless of societal norms, remains the satisfaction of individuals with their behavior and the degree to which the behavior interferes with their overall adjustment. Removing the pathological connotation of sexual deviancy creates a reluctance among professionals to view such behavior as problematic, to disregard or minimize the enormous negative impact it may have on other aspects of the child's life. Such a posture by professionals may only reinforce the tendency of parents to minimize the already socially sensitive area of gender-related behavior disturbance in their children.

From the social learning model we are reminded that socialization proceeds from the earliest years of life and that differential socialization along gender lines is no exception. Such development, moreover, is a lifelong process, in many ways longer for women than for men (Barnett & Baruch, 1977). In this context, the research demonstrates the powerful and enduring effects of imitation and modeling on sex- and gender-role development. Initially occurring with parents and parental surrogates, it is later reinforced by peers and media, both of which have often contributed toward perpetuating traditional sex roles. Although the direct effect of modeling on gender-role behaviors (Green, 1978; Maccoby & Jacklin, 1974) is obvious, one is cautioned about the effects of delayed learning (Katz, 1979) with several possible consequences. Specifically, the effect of the model may be present but not easily ascertained because (1) we remain uncertain whom the child is modeling, (2) some aspects of behavior are not observable to children and cannot be learned on the basis of modeling, and (3) a very long time may have elapsed between the original observational learning and its performance. Indeed, it is often not until individuals are themselves adults that the effects of modeling vis-à-vis gender-role development are observable. This seems particularly true for parenting behaviors and socialization of children along gender lines.

Indications for treatment based on this model include the consideration of maladaptive patterns of behavior involving cross-sex preferences in dressing, play activities, roles assumed in play, choice of companions, gender-identity statements, and wishes to assume the gender role of the opposite sex as an adult. Treatment is particularly indicated where rigidity and exclusiveness characterize the cross-sex preferences and behaviors, such that the child avoids any attempt to expand the repertoire of attitudes and behaviors to include those consistent with his own gender and gender role. In terms of prevention, therapy may also be considered where the child manifests a relative discomfort with the gender role in the absence of pathological cross-sex preference, since the symptom may point to particular ambivalence and conflict about one's gender role and gender identity.

The evidence from the cognitive developmental literature has direct implications for treatment because it identifies specifically the age at which the gender identity can be considered fully established and irreversible. Gender constancy is not considered achieved until approximately age 6, a fact therapists must consider when determining the extent to which the child can be expected to master and integrate the facets of gender role before the maturation of necessary cognitive structures. Because the theory predicts an active attempt by children to seek out those things appropriate for their gender when identification is established, we may yet predict a greater penchant for passive learning before the achievement of gender constancy. Thus, children's ability to regulate their behavior in categorizing what is gender appropriate may depend on changes in gender stage (Ruble, Balaban, & Cooper, 1981).

DIAGNOSTIC ISSUES

The literature raises various diagnostic issues. Although gender-identity and -role disturbances seem generally clear-cut, some confusion still exists concerning determination of pathology. That is, are there *degrees* of pathology in cross-gender-role behavior? Such questions address the controversy about gender roles between radical feminist ideology and traditional psychiatric thinking (Lerner, 1978). Whereas the feminists decry the pernicious effects of sexism and stereotyping, traditionalists regard the depolarization of sex-role behavior as dangerous for children, possibly causing pathological role confusion prodromal to identity disturbance. Although these are the extreme positions, one must consider the point at which cross-gender behavior goes beyond relaxation of stereotypes into the area of pathology. This appears increasingly more difficult to differentiate as gender-role stereotypes are held up to closer scrutiny. Thus, diagnostic confusion is likely to result as the cultural milieu in which gender roles are embedded begins to shift toward depolarization of sex roles. For children, the confusion of gender roles can critically affect their gender identity.

Another diagnostic issue concerns the propensity to overlook gender disturbance in girls. Pathology is generally discernible in boys because of the culture's heightened intolerance of feminine behavior in males; and the very opposite is true for girls. Although many tomboys eventually adopt the feminine role as adolescents, retrospective studies of masculine women with gender disturbances (such as transsexuals) reveal a significant number who report undetected, unaddressed, and unresolved cross-gender preferences and gender conflicts as children (Green, 1974; Stoller, 1968). The pathological nature of such gender disturbance is, moreover, no less real.

Finally, the stereotyping of the mental health of girls and boys along gender lines exists even in mental health practitioners. Although sex bias in adult psychotherapy has been the subject of much controversy, it has been less studied in child psychotherapy. Some studies involving child mental health practices have supported the thesis that boys and girls tend to be viewed and treated differently (Jackson, Farley, Zimet, & Waterman, 1978; Silvern, 1978). Jackson *et al.*'s review of five child guidance clinics around the country revealed, for example, bias in the sex and race of children treated by practitioners and differential parental involvement in the treatment according to sex of the child. Professional discipline, sex of the therapist, and sex of the child also functioned to yield differential responses within the therapy (Zimet, Waterman, Jackson, & Farley, 1980).

Efforts to engage patients with gender-related disturbances in a therapeutic alliance may be particularly problematic. Resistance to the therapeutic process may be strong, given the emotion-laden nature of the issues in question. The child patient's resistance may be related to the child's fear of the therapist as an agent of the parents, especially when intense conflicts have already occurred vis-à-vis the presenting symptoms. Transferential reactions, particularly negative-oedipal ones, work against the child's motivation to engage in the therapy. Although object hunger seems a powerful dynamic in the majority of gender-disturbed children, it seems superimposed by anxiety and rage toward the transferential parental figure, thereby intensifiying the resistance to therapy.

Goals of Treatment

The setting of realistic treatment goals for patients with gender disturbance seems critical. Given the complexity of factors affecting sex-role development, treatment goals must recognize the relative weight various factors will exert on the final outcome. For example, in children, recognition of the development of gender disturbance will, in all likelihood, dictate intervention with pathological family and/or parental-marital constellations. One must keep in mind the purpose served by the symptoms and the role they play within the family constellation. Consideration of the quality of the child's object relationships, external and internal, seems critical in facilitating the ultimate goal of treatment—resumption of derailed development. Unlike with adults, therapy with children aims directly at changing the gender-related disturbances. Specific goals may include: (1) educating the child as to the invariability of his sex; (2) stressing to him the advantages of participating in appropriate gender-role activities with same-sex peers, thereby increasing comfort with the gender role; (3) educating the parents in how they may be fostering sexual identity conflict in their child; (4) advising the parents of the need to disapprove consistently of very cross-sex behavior and to approve of appropriate gender-role behavior; and (5) enhancing the relationship between the child and the same-sex parent (Green, 1974) with the goal of promoting the identification process. Some nonanalytic education of the parents and children may also be necessary before the actual treatment.

Transference

Within the process of psychotherapy, the patient's reaction to the therapist as a figure of displacement is an inevitable phenomenon that forms the core of the treatment. Regardless of age, patients form transference reactions of a particular form and content that reflect their unique experience and development. In addition, the patient may use the "real" qualities of the therapist, those inherently related to the therapist, apart from those of the primary objects. These qualities affect the nontransferential relationship as well, highlighting the patient's need to use the therapist as a figure of identification. Thus, for therapists a heightened awareness of the role model they provide for the patient is vital.

This attains special prominence during developmental phases when the child patient becomes particularly vulnerable to embracing or repudiating those qualities of the therapist that are especially gender related. Tyson (1980) argues for the importance of the therapist's gender in the treatment. It may play a powerful role in both pre-oedipal and oedipal situations. In the pre-oedipal condition, the therapist's gender could function as a focus of resistance of the transference, as an enhancement or attenuation of the transference, and in the service of developmental needs. The child in the oedipal situation, in contrast, displaces conflicts onto the treatment situation from the parent of the same gender as the therapist, especially at the beginning of the treatment (Tyson, 1980, p. 330).

The following case illustrates, for instance, a young patient's preoedipal longings as reenacted within the transference. A feminine boy, his identification with a same-sex therapist enhanced his capacity to make significant changes in gender identity and gender-role preference.

Johnny, a 4½-year-old boy, was referred for treatment by his teachers for frequent temper tantrums and cross-dressing. Although he played often with same-sex agemates, he preferred doll play, taking exclusively the role of mother or other female figures. Johnny identified himself as a girl and wished solely to be a mommy when he grew up. In fantasy play he identified with Wonder Woman, creating themes of rescue and extraordinary powers. In school he sought to stroke his teacher's hair, often in a coquettish way. The younger of two sons, Johnny was reportedly well related to his 8-year-old brother, a child who also manifested feminine behavior at approximately the same age. Johnny's mother, an overworked woman who almost singlehandedly raised the children, generally ignored the symptoms and refused to participate in her son's therapy, confident that he would outgrow his symptoms eventually. His father, a man whose work made him frequently absent from home, seemed a quiet, distant man who relegated most of the childrearing to his wife.

Johnny attached himself quickly to his male therapist. He looked forward to sessions, was immensely involved in the play, and protested vehemently upon leaving each session. His anger with his parents for their frequent quarrels and limit setting surfaced early in the therapy. His doll play, for instance, expressed his sadness and anger about their felt neglect, followed often by retaliatory beatings of the dolls. His anger was reflected in the transference as he sought to control and dominate his therapist. When the latter refused to acquiesce, he lashed out with temper tantrums and running away, only to return tearfully to reconcile and make reparation.

Separation-individuation conflicts proved the primary issue of the therapy. Using his therapist as a figure with whom to play hide-and-seek, he appeared thrilled upon reunion. After 3 months of therapy, he began to get increasingly upset if his therapist was not constantly engaged with him. His need for attention took the form of clinging demandingness since he found it excruciating to verbalize his dependency needs. By the end of the fourth month of treatment, however, he began to express a wish to play with gender-appropriate toys (trucks) and to grow up to be a "daddy." He yet retained much of his identification with his mother, however, reacting furiously when his therapist reminded him that he could only grow up to be a boy. By the end of the fifth month he began to identify himself consistently as a boy, all the while relating the fears surrounding his parents' arguments and battles. By the end of the ninth month, Johnny was consistently identifying himself as a boy and dressing up in male costumes as well. Although he occasionally donned feminine clothes, his play with them was qualitatively different in that he no longer evidenced a conscious belief or desire to become a girl. In sum, it seemed that Johnny's feminine identification was intimately associated with the power he perceived in the maternal figure. Arrests experienced during separation-individuation also affected the extent to which he could negotiate phase-appropriate developmental demands and establish an appropriate gender identity and role. Within a therapeutic relationship with a nurturing, benign male, Johnny was able to make the transition with which he embraced this process, moreover, revealed the object hunger that appeared ambivalently satisfied by primary figures, sufficient to engender masochistic longings in the face of persistent rejection.

The gender of the therapist also affects the parents' transference to the therapist (Tyson, 1980). This is particularly true when female therapists have to deal with mothers of patients. The mother may experience difficulty sharing her child with another woman, or view the therapist as an idealized mother. The mother's unresolved conflicts with her own mother may be triggered by the alleviation of the child's symptoms. In such cases, simultaneous treatment of the mother seems essential in order to preserve the therapy. This treatment should be aimed directly at helping the mother separate her feelings about her mother from those toward the therapist.

Patients may also consider the therapist's gender to be an important factor in deciding whether or not to enter treatment with a particular individual. This holds especially true for women seeking therapists whose views on feminism and the feminine role are consistent with their own. Thus, one effect of the feminist movement has been to politicize the issue of therapy, that is, by whom it is conducted. Such being the case, it is hardly realistic to dismiss the issue of the therapist's gender as inconsequential, as is often done in more traditional circles. It appears that the gender of the therapist affects most often the initial phases of treatment by enhancing or impeding the establishment of the therapeutic alliance. As treatment proceeds, the patient's habitual mode of relating usually takes ascendance, thereby rendering less crucial the gender of the therapist.

COUNTERTRANSFERENCE

Research experience in psychotherapy with gender-related disturbances shows that no other aspect of psychotherapy plays as critical a role as that of countertransference (Lothstein, 1977). The key to the role countertransference plays lies in the interaction between the therapist's gender and the patient's. Although countertransferential reactions span the spectrum depending on the degree of the patient's (and therapist's) pathology, they typically include confusion, rage, unrealistic total dedication to the patient, third-party countertransference reactions, and discontinuation of the treatment. Intense hostility in the therapist toward the patient may be aroused by sex/gender envy, superegeo conflicts, homosexual concerns, defenses against passivity and emotional surrender, and wishes to or rejection of the need to provide a role model for the patient. These reactions differ in nature according to the composition of the patient–therapist dyad, and may take subtle forms according to the self-awareness and experience of the therapist. Lothstein argues for an immediate acknowledgment and interpretation of the countertransference as a way of both understanding the patient and the clinical process and safeguarding against victimizing the patient.

Awareness of the role of socialization and acculturation in sex-role development is a *sine qua non* in child psychotherapy. Therapists, therefore, should be cautious about the way their own views about "sissiness" and "tomboyishness" can affect clinical judgment. Some typical countertransference reactions can be expected in treating children including the therapist's identification with the child, competition with his parents as a "better" model of identification or a source of nurturance, the need to control the child's heightened motoric discharge, and fears of regression when confronted with the child's characteristically regressive, immature capacity to deal with his affects and impulses. In addition to these, still other reactions may occur. The child, reacting to the therapist both as a figure of displacement and as a new object, may evidence an intensification of affect related to the therapist, depending both on the latter's gender and the particular phase of treatment. Tyson (1980) warns, in particular, of the female child's difficulties in relating to a female therapist when pre-oedipal ambivalent struggles with the mother are at issue. Countertransference conflicts are often revived in the treatment of such girls, especially in adolescents. These may be expressed in action before the therapist is aware of the countertransference, often leading the female therapist to accept the role of mother which these patients are attempting to force on them. The rapidity of this process poses particular hazards since it threatens to embroil the patient and therapist in a struggle for control that may be self-reinforcing and ultimately derail the treatment.

CONCLUSIONS

Research and therapeutic implications for the future study of sex-role development are manifold. They include the need to consider the confluence and interaction of biological, social, and environmental factors that affect individual development. A line of development may be suggested, thus, from the perception of oneself as genderless, to belonging to one sex, and toward embracing characteristics beyond one's appropriate gender role once gender identity is solidly established. Since this development seems, indeed, a lifelong process, the roles played by cognitive, intrapsychic, interpersonal, and social variables are important to recognize insofar as they influence either a normative or idiosyncratic resolution of the individual's gender-role development. This chapter has highlighted some of the more salient variables, especially those pertinent to psychotherapy, but is by no means exhaustive.

Several conclusions, however, bear reiteration. The deleterious effects of sex-role stereotyping are roundly acknowledged, and there has been an attempt in our culture to dismantle sex differences on a variety of fronts. However, attempts to rear children without an obvious emphasis on sex differences should keep in mind the time frame for the establishment of gender constancy. Along the way toward this goal, gender-role rigidity may be expected as the child learns what he is before learning what he is not. As this ability is acquired, greater flexibility may be expected in gender-role behavior without threatening the integrity of the gender identity. Where ridigity is maintained long past the age for establishment of gender constancy, other factors may be operative and may or may not be an indication for treatment depending on the overall functioning of the child. Sensitive and sensible therapists should also consider the intrapsychic and interpersonal needs served by rigid adherence to a sex-role stereotype by their adult as well as child patients. Where it serves as a further aid to maintaining a tenuously established individual identity, one should be careful about questioning its viability or advocating its abolition. The costs of rejecting conventional sex-role attitudes and behavior, despite the gains made by feminism, continue to exist and must be considered for patients who seek to deny or usurp them.

Lastly, there may be shifts in the definition of appropriate gender-role behavior within the next generation. Even now, as fathers come to exert a more obvious role on children's development, especially in the earliest years, the effects on children's sense of self and of others, both in internal representations and in external reality, may be profound. There may be changes, for instance, in the extent to which the father serves as the primary figure of attachment. The father's role in providing a more neutral figure of attachment, thereby interrupting the symbiotic pull of the mother–child dyad, deserves further investigation. Finally, the role fathers have played in fostering sex-role differentiation and socialization in their children may or may not shift as they assume greater parental involvement. This may be particularly true if this involvement occurs early in the child's development or as fathers assume more of the nurturing functions. These matters remain to be explored in future investigations and may lead to valuable insights about the nature of sex-role development. They point, moreover, to the need for longitudinal and cross-sectional studies of important facets already addressed, especially those related to pathological gender-role development.

It is clear from the literature that sex-role development is highly complex. There is a need to clarify what is indeed inherently pathological or merely socially deviant, especially with regard to psychotherapeutic and parenting practices. Also, there is much to be resolved in the conflict between traditional and alternative views of gender-role development. No doubt gender-role development has undergone some remarkable transitions in the past decade, the effects of which are not yet fully recognized. Future inquiry in this area will need to consider the richly complex and unique interaction between the individual and the social context.

ACKNOWLEDGMENTS

I wish to express grateful appreciation to Dr. Kevin O'Connor for his very helpful comments on the chapter during its preparation.

REFERENCES

Abelin, E. L. Some further observations and comments on the earliest role of the father. *International Journal of Psychoanalysis*, 1975, *56*, 293–302.

Atkins, R. N. Finding one's father: The mother's contribution to early father representations. *Journal of the American Academy of Psychoanalysis*, 1981, *9*, 539–559.

Barnett, R. C. & Baruch, G. K. *Women in the middle years: An overview*. Paper presented at the meeting of the Society for Research in Child Development, New Orleans, April 1977.

Bentler, P. M. A typology of transsexualism: Gender identity theory and data. *Archives of Sexual Behavior*, 1976, *5*, 567–584.

Birns, B. The emergence and socialization of sex differences in the earliest years. *Merrill-Palmer Quarterly*, 1976, *22*, 229–254.

DeVries, R. Constancy of gender identity in the years three to six. *Monographs of the Society for Research in Child Development*, 1969, *34* (3, Serial No. 127).

Emmerick, W., Goldman, K. S., Kirsh, B., & Sharabany, R. Evidence for a transitional phase in gender constancy. *Child Development*, 1977, *48*, 930–936.

Fagot, B. I. Some differences in toddlers' behavior and parental relations. *Developmental Psychology*, 1974, *10*, 554–558.

Fast, I. Developments in gender identity: The original matrix. *International Review of Psychoanalysis*, 1978, *5*, 265–273.

Freud, S. Psychoanalysis terminable and interminable. In J. Strachey (Ed. and Trans.), *Standard edition of the complete psychological works of Sigmund Freud* (Vol. 23). London: Hogarth Press, 1955. (Originally published, 1937)

Freud, S. Infantile genital organization. In J. Strachey (Ed. and Trans.), *Standard edition of the complete psychological works of Sigmund Freud* (Vol. 19). London: Hogarth Press, 1961. (Originally published, 1923)

Freud, S. Some psychical consequences of the anatomical distinction between the sexes. In J. Strachey (Ed. and Trans.), *Standard edition of the complete psychological works of Sigmund Freud* (Vol. 19). London: Hogarth Press, 1961. (Originally published, 1925)

Galenson, E. & Roiphe, H. The preoedipal development of the boy. *Journal of the American Psychoanalytic Association*, 1980, *28*, 805–827.

Green, R. *Sexual identity conflict in children and adults*. New York: Basic Books, 1974.

Green, R. Sexual identity of 37 children raised by homosexual or transsexual parents. *American Journal of Psychiatry*, 1978, *135*, 692–697.

Green, R. & Money, J. *Transsexualism and sex reassignment*. Baltimore: Johns Hopkins, University Press, 1969.

Green, R., Newman, L., & Stoller, R. Treatment of boyhood transexualism: An interim report of four years' duration. *Archives of General Psychiatry*, 1972, *26*, 213–217.

Greenson, R. A transvestite boy and a hypothesis. *International Journal of Psychoanalysis*, 1966, *47*, 396–403.

Hay, W. M., Barlow, D. H., & Hay, L. R. Treatment of stereotypic cross-gender motor behavior using covert modeling in a boy with gender identity confusion. *Journal of Consulting and Clinical Psychology*, 1981, *49*, 388–394.

Heilbrun, A. B., Jr. An empirical test of the modeling theory of sex-role learning. *Child Development*, 1965, *36*, 789–799.
Horney, K. The flight from womanhood: The masculinity-complex in women, as viewed by men and women. *International Journal of Psychoanalysis*, 1926, *7*, 324–339.
Jackson, A. M., Farley, G. K., Zimet, S. G., & Waterman, J. M. Race and sex as variables for children involved in treatment. *Psychological Reports*, 1978, *43*, 883–886.
Katz, P. The development of female identity. *Sex Roles*, 1979, *5*, 155–178.
Kinsey, A. C., Pomeroy, W. B., Martin, C. E., & Gebhard, P. H. *Sexual behavior in the human female*. Philadelphia: W. B. Saunders, 1953.
Klein, M. Early stages of oedipal conflict. *International Journal of Psychoanalysis*, 1928, *9*, 167–180.
Kohlberg, L. A cognitive-developmental analysis of children's sex-role concepts and attitudes. In E. Maccoby (Ed.), *The development of sex differences*. Stanford, Calif.: Stanford University Press, 1966.
Lamb, M. E. *Infants, fathers and mothers: Interaction at 8 months of age in the home and in the laboratory*. Paper presented at the meetingof the Eastern Psychological Association, New York, April 1975.
Lebovitz, P. S. Feminine behavior in boys: Aspects of its outcome. *American Journal of Psychiatry*, 1972, *128*, 1283–1289.
Lerner, H. E. Adaptive and pathogenic aspects of sex-role stereotypes: Implications for parenting and psychotherapy. *American Journal of Psychiatry*, 1978, *135*, 48–52.
Lewis, H. B. Gender identity. *Bulletin of the Menninger Clinic*, 1979, *43*, 145–160.
Lewis, M. & Weinraub, M. Sex of parent X sex of child: Socioemotional development. In R. C. Friedman, R. M. Richard, & R. L. van de Wiele (Eds.), *Sex differences in behavior*. New York: Wiley, 1974.
Lewis, M., & Weinraub, M. Origins of early sex-role development. *Sex Roles*, 1979, *5*, 135–153.
Lothstein, L. M. Countertransference reactions to gender dysphoric patients: Implications for psychotherapy. *Psychotherapy: Theory, Research and Practice*, 1977, *14*, 21–31.
Maccoby, E., & Jacklin, C. *The psychology of sex differences*. Stanford, Calif.: Stanford University Press, 1974.
Mahler, M. S., & Furer, M. *On human symbiosis and the vicissitudes of individuation*. New York: International Universities Press, 1968.
Mahler, M. S., Pine, F., & Bergman, A. *The psychological birth of the human infant: Symbiosis and individuation*. New York: Basic Books, 1975.
Marcus, D. E., & Overton, W. F. The development of cognitive gender constancy and sex role preferences. *Child Development*, 1978, *49*, 434–444.
Mischel, W. A social learning view of sex differences in behavior. In E. Maccoby (Ed.), *The development of sex differences*. Stanford, Calif.: Stanford University Press, 1966.
Money, J. Psychosexual differentiation. In J. Money (Ed.), *Sex research: New developments*. New York: Holt, Rinehart & Winston, 1965.
Money, J., & Ehrhardt, A. A. *Man and woman, boys and girls*. Baltimore: Johns Hopkins University Press, 1972.
Money, J., Hampson, J. G., & Hampson, J. L. An examination of some basic sexual concepts: The evidence of human hermaphroditism. *Bulletin of Johns Hopkins Hospital*, 1955, *97*, 301–319.
Moss, H. A. Early sex differences and mother-infant interaction. In R. C. Friedman, R. M. Richard, & R. L. van de Wiele (Eds.), *Sex differences in behavior*. New York: Wiley, 1974.

Nordyke, N. S., Baer, D. M., Etzel, B. C., & LeBlanc, J. M. Implications of the stereotyping and modification of sex role. *Journal of Applied Behavior Analysis*, 1977, *10*, 553–557.

Osofsky, J. D., & Danzger, B. Relationship between neonatal characteristics and mother-child interaction. *Developmental Psychology*, 1974, *10*, 124–130.

Piaget, J. *The origins of intelligence in children*. New York: International Universities Press, 1952.

Rekers, G. A. Atypical gender development and psychosocial adjustment. *Journal of Applied Behavior Analysis*, 1977, *10*, 559–571.

Rekers, G. A., & Lovaas, I. O. Behavioral treatment of deviant sex-role behaviors in a male child. *Journal of Applied Behavior Analysis*, 1974, *7*, 173–190.

Rekers, G. A., Rosen, A., Lovaas, I. O., & Bentler, P. M. Sex-role stereotypy and professional intervention for childhood gender disturbance. *Professional Psychology*, 1978, *9*, 127–136.

Rosen, A. C., Rekers, G. A., & Friar, N. R. Theoretical and diagnostic issues in child gender disturbances. *Journal of Sex Research*, 1977, *13*, 89–103.

Ruble, D. N., Balaban, T., & Cooper, J. Gender constancy and the effects of sex-typed televised toy commercials. *Child Development*, 1981, *52*, 667–673.

Serbin, L. A., O'Leary, K. D., Kent, R. N., & Tonick, I. J. A comparison of teacher response to the preacademic and problem behavior of boys and girls. *Child Development*, 1968, *39*, 103–112.

Silvern, L. E. Masculinity-femininity in children's self-concepts: The relationship to teachers' judgments of social adjustment and academic ability, classroom behavior, and popularity. *Sex Roles*, 1978, *4*, 929–949.

Stoller, R. J. *Sex and gender*. New York: Science House, 1968.

Stoller, R. J. Psychotherapy of extremely feminine boys. *International Journal of Psychiatry*, 1970, *9*, 278–280.

Stoller, R. J. The bedrock of masculinity and femininity: Bisexuality. *Archives of General Psychiatry*, 1972, *26*, 207–212. (a)

Stoller, R. J. Etiological factors in female transsexualism: A first approximation. *Archives of Sexual Behavior*, 1972, *2*, 47–64. (b)

Stoller, R. J. *Sex and gender* (Vol. 1). New York: Jason Aronson, 1974.

Stoller, R. J. Fathers of transsexual children. *Journal of the American Psychoanalytic Association*, 1979, *27*, 837–866.

Stoller, R. J., Buxbaum, E., & Galenson, E. Scientific proceedings—panel reports: Psychology of women. *Journal of the American Psychoanalytic Association*, 1976, *24*, 141–160.

Tyson, P. The gender of the analyst. *Psychoanalytic Study of the Child*, 1980, *35*, 321–338.

Ullian, D. Z. Why boys will be boys: A structural perspective. *American Journal of Orthopsychiatry*, 1981, *51*, 493–501.

Winkler, R. C. What types of sex-role behavior should behavior modifiers promote? *Journal of Applied Behavior Analysis*, 1977, *10*, 549–552.

Zimet, S. G., Waterman, J. M., Jackson, A. M., & Farley, G. K. Do child psychotherapists treat boys and girls differently? *Journal of the American Academy of Child Psychiatry*, 1980, *9*, 270–280.

Zuger, B. Effeminate behavior present in boys from early childhood: 1. The clinical syndrome and follow-up studies. *Journal of Pediatrics*, 1966, *69*, 1098–1107.

14

Adolescence for Clinicians

JOSEPH ADELSON

Three major problems confront us in reporting research on adolescence to practicing clinicians.

1. The first of these is that empirical work in most parts of the field is thin, yet widely spread. A vast number of topics have been studied, so much so that a first glance will offer the impression that we occupy a lively, bustling arena of research. But when we look more closely, we may discover that there are only a handful of studies on a given topic, that one or two of them are seriously out of date, another suffers from grievous flaws in methodology, another is simpleminded, and so on. It is not at all uncommon for the specialist in adolescent psychology to be approached by a colleague, or a journalist, with the observation that he or she has made a preliminary study of this or that topic in adolescence and has been unable to find much. Surely there is more, and where can one find it? More often than not, one will reply that there really is no more than that handful.

Here are some examples: We know very little about the effect of the biological changes of puberty on the child's behavior and psychological experience; as Petersen and Taylor (1980) point out in an excellent survey of the biological approach to adolescence, research

> is in a preliminary stage and has not kept pace with biological studies. . . . Consequently, despite a proliferation of theoretical positions

---

JOSEPH ADELSON • Department of Psychology, University of Michigan, Ann Arbor, Michigan 48109.

> hypothesizing causal connections between physiological and psychological changes in early adolescence, much of our thinking in this area remains in the realm of assertion and belief. (p. 132)

We also know surprisingly little about developmental changes in psychological testing during the course of adolescence, particularly in relation to "depth" and "personality variables." Aside from the seminal work of Greenberger and Steinberg (1981; Greenberger, Steinberg, & Vaux, 1981; Steinberg, 1982; Steinberg, Greenberger, Garduque, Ruggiero, & Vaux, 1982), we know little about the effects of the work experience upon adolescent psychological development. As we shall see later in this chapter, our knowledge of the adolescent's family is so underdeveloped as to be unreliable. Despite generations of work on the roots of adolescent delinquency, almost all we know is sociological in nature; work on the psychological sources and functions of crime is just getting under way. And so it goes; the examples are chosen almost randomly, and the list is easily extended.

With so few studies available, we find ourselves deprived of the clamor and dissonance we expect in a thriving scientific enterprise. It is rare in social science for studies to support each other fully, since even slight variations in method and population can produce significant variations in outcome. What emerges as scientific "truth" develops only gradually, from the juxtaposition and collision of findings. Deprived of that thickness of texture, the writer on adolescence soon finds himself leaning too heavily on the few empirical studies, and beefing up that knowledge by references to secondary sources (reviews, or reviews of reviews), to social commentary, to stray clinical observations, and the like. Although each of these genres is important to any intellectual endeavor, in the case of adolescent psychology we find a relative feebleness of the empirical mode, one that puts stringent limits on the confidence and scope of a chapter such as this one.

2. A second difficulty is in the systematic underrepresentation of important populations of the young, such that general statements about adolescence or "the adolescent" are, if not entirely false, then surely misleading. The most important example of this has been the failure, until recently, to study adolescent girls, or to represent them equitably in studies of adolescent problems. That failure has been evident for many years—it was brought to scholarly attention in the national survey of adolescents by Douvan and Adelson (1966), which pointed out the significant disparities between their male and female samples and suggested that the then prevailing theories of adolescence were tacitly and no doubt unconsciously theories of male adolescence, and that such important topics as the psychology of achievement involved an unwitting

imputation to girls of dynamics derived from and applicable to boys alone. In one form or another, that discovery keeps being made: Gilligan's recent studies of moral development (1982) suggest that the Kohlberg scheme does not sufficiently take into account differences between the sexes in the perception of moral issues. That bias is now beginning to change in that there are concerted efforts to study both sexes and the differences between them—one recent example is that the important longitudinal study of boys by the Offer group (Offer, 1969; Offer & Offer, 1975) is now being replicated on a sample of normal adolescent girls by Petersen. Nevertheless, there is an androcentric bias built into most general propositions about adolescence.

Another source of bias, much harder to document, is to be found in the overrepresentation of certain social strata. An informal survey of my own suggests that "normal" samples probably overrepresent the upper middle class, probably because those groups are more available for study. In one recent volume of a journal in developmental psychology, the samples in adolescent studies were these: college students; children from a suburban school; the children of professional parents; and those attending a private camp. There is also a contrary tendency: Many studies draw upon atypical populations from the other end of the status spectrum, especially youngsters in trouble with the law. One senses that there is a silent majority of teens, blue-collar and lower-middle-class, not commonly represented in proportion to their numbers.

3. These two problems—the thin development of empirical work, and the biases in research samples—are linked to a third, perhaps the most important of all: the presence and persistence of a number of fixed, largely false ideas about adolescence, shared by much of the laity, by the media, by many social scientists, and by most clinicians. These ideas involve a profound misunderstanding of what is and what is not typical or modal or "normal" in adolescence. There is a tendency to see adolescence as a more dramatic and tempestuous period than it is, and to see intensified inner and outer conflict as normative. Consider how the fact of peer relations is understood in most accounts of adolescence, as a revolutionary event wherein the parents' authority withers, to be replaced by the enthronement of the peer group. The underlying assumptions are false—in seeing parents and peers as intrinsically oppositional, involved in a zero-sum game wherein the achievement of influence by one produces a devaluation of the other. That way of framing the question simply does not capture a far more complex state of affairs.

The presence of these false ideas is not nearly as intriguing as their persistence. The idea of a sanguinary struggle between the adolescent and his parents has been known to be flawed for about 30 years. Yet it seems to have made no impact on the common awareness. As chance would have it, on the very day that I write these words the *New York Times* provides a perfect example. In a feature story headlined "Youth: A Change in Values," we learn that the Gallup Youth Survey has just found that American youngsters ages 13 to 18 report amicable ties to their parents: 60% say they get along very well; 82% think parental discipline is "about right"; 77% would consult parents (as against 18% for friends) if faced with a serious life decision. Yet the story is written entirely from the perspective of change. It assumes that the findings convey a remarkable shift in the attitudes of youngsters since the halcyon days of the youth revolution. The reporter asked a number of psychiatrists to play pundit, which they were glad to do, averring solemnly that there was indeed a return to traditionalism among the young. What no one seemed to grasp, neither the *Times* nor the experts they called upon, was that the results obtained were essentially the same as in all previous research.

It is not an isolated case. Much of what clinicians believe to be true about adolescents is unsupported by research findings; what is worse, much of it is flatly contradicted by those findings. When pressed upon the clinical consciousness, the data are resisted by means ingenious and otherwise: I can remember, at a convention of psychiatrists, one panelist saying that he did not care about findings, since he was a therapist and knew better. One might accept such dismissals if the data in question came from hostile theoretical positions; but as we shall see, most studies have been carried out by psychodynamic psychologists and psychiatrists. One might accept such dismissals if the issues involved were trivial or peripheral; but they involve issues central to an understanding of the adolescent experience. Clearly we are dealing with something of a collective illusion, stubbornly held, and for that reason it seems best to eschew any effort at a conspectus of adolescence research, to concentrate instead upon what is at issue between common clinical belief on the one hand and the empirical canon on the other.

How Normal Is the Normal Adolescent?

Most of us, I suspect, will shy away from any precipitous diagnosis of an adolescent patient. Not at the extremes, perhaps—that is, if the youngster is in serious trouble with the police, or is severely addicted to hard drugs, or is beginning to show signs of cognitive slippage, we are prepared to see him or her as psychologically troubled. Or if he or she is entirely free of symptoms, doing well in school, dating with normal frequency, and so on, we are then prepared to judge the youngster to be well adjusted. But between these extremes we find ourselves holding back, not only because we want to avoid stigmatizing someone so young, but also because we believe that adolescence is customarily marked by emotional turmoil. We may feel that psychoneurotic symptoms, let us say, may be misleading in adolescence, since they may reflect only transient or evanesecent conflicts. It is an era, we believe, that is "normally" marked by emotional distress, the sources of which are such stimuli as the dramatic biological changes of the period, psychosexual marginality, rapid cognitive growth, striking and at times volatile shifts in identity, the savage pressures of peer group homogeneity and exclusion, the anxiety one feels about an uncertain future, and so on.

The belief that a normal adolescence is "abnormal" can be traced as far back as G. Stanley Hall's *Adolescence* (1904). Yet the most important current expression stems from psychoanalysis, initially in the work of Anna Freud, later taken up and elaborated by Blos. This view of adolescence initially focused upon a revitalized conflict between the drives and the defensive and superego forces. Indeed, Anna Freud (1936) used what she took to be characteristic adolescent behaviors—asceticism or intense intellectualization—as a way of illustrating the operations of the ego under stress. In her later writing (1958), which is both succinct and marvelously lucid, she moves away from the ego–id conflict to the adolescent's struggle with revived infantile object attachments. It is this emphasis that is developed in Blos's many writings on adolescence (1962, 1967, 1974). Blos's version of the vicissitudes of adolescence, too complex to report here, sees the youngster's progress through the period as dominated by the need to overcome a continuing series of regressive dangers involving earlier introjects. The picture we derive from these writings is of a youngster beleaguered by the recrudescence of infantile drives, primitive fears, and threatening images, and holding on valiantly despite an affective volatility and a general stretching and impoverishment of ego functions. The child's relations to the "outer world"—peers and parents—are dominated by their status as threats and tempters or as defensive anchors.

Most of the work in this tradition—that is, adolescence-as-pathology—derives directly from or is heavily influenced by psychoanalytic writing. The data base, so to speak, seems for the most part to be youngsters suffering from severe psychoneurotic or character-neurotic problems, or—more recently—addictive, narcissistic, and borderline states. It is assumed in these writings as in most writings of the psychoanalytic tradition, that there is no fundamental difference in kind, though perhaps in intensity, between the states of mind uncovered in deep analytic work with these youngsters and what might be uncovered in therapy with those less disturbed.

Those working in the empirical tradition (adolescence-as-normality) have been drawn from a large number of disciplines, though primarily from the clinical fields, and have used diverse methodologies ranging from questionnaire studies to intensive clinical interviews and projective tests. On the whole, they share a psychodynamic outlook, a great many of them having been trained in clinical psychology or psychoanalysis. Most continue to practice. Commonly they began their research as believers and became heretics in the course of doing it—that is, they began with the assumption of adolescence as a period of turmoil, but were unable to find that confirmed in work with normal subjects. There is a remarkable uniformity to the findings. Every systematic study that has been published has been unable to discover a significant degree of psychopathology in the experience or history of the ordinary adolescent. Since the number of such studies is now too large to permit a complete report here, we shall limit ourselves to several of the most widely known, reporting them chronologically.

Douvan and Adelson's *The Adolescent Experience* (1966) was the first national survey of adolescent boys and girls, involving interviews with about 3000 of them; it was also one of the first to employ projective questions as part of the survey interview. A large number of topics were covered—family, friends, attitudes toward achievement, work and family, the future, and many more. The typical youngster was found to have a fairly realistic assessment of himself, to be more or less at peace with his family on such issues as discipline and values, to have conventional and essentially realistic work and family ambitions.

Masterson's well-known study (1967) made a direct comparison between a group of adolescents seen at an outpatient psychiatric clinic and a control sample of "normal" youngsters, that is, normal in that they had not been referred for psychological help. If adolescence is a period marked by regressiveness, a weakened ego, and the like, one would expect to see a high rate of symptomatology among the presumed normals, perhaps even matching the level found in those labeled "disturbed." In fact, Masterson found substantial differences between the two samples: The disturbed adolescents had an extremely high rate of symptoms of a portentous type—disturbances in behavior, such as antisocial conduct, and disturbances in the quality of thinking. Nearly 60% of the adolescent patients showed these symptoms, as against fewer than 20% of the nonpatients.

The studies by Offer and his associates (1969; Offer & Offer, 1975) were of particular importance in filling a methodological gap: They were longitudinal, following a group of adolescent boys from age 14 to 21. Beyond that, the researchers used intensive clinical methods—depth interviews, projective tests—upon a nonclinical sample. Whereas social psychological studies might be faulted for their use of survey-style, hence insufficiently "deep" questioning, and whereas Masterson and similar studies could be said to focus too narrowly on clinical data, the Offer research used depth measures on a normal sample, thus finessing both objects. The findings confirm the direction of all previous (and following) work: Adolescence in a nonclinical population of boys is marked by a general absence of symptoms, or other signs of intense inner conflict; by an absence of rebelliousness toward the parents; by a general mood of reason and amity in the family; and by an acceptance of the values of the larger society.

We might be willing to agree that adolescence is not quite the quagmire of psychopathology it is sometimes taken to be; yet we would still want to ask whether the level of inner turbulence is or is not equivalent to what we find among adults. In the course of a superb scholarly review of this material, Weiner (1982) had the happy idea of collating those extant studies of normal adolescent populations that attempted to assess the proportion of disturbed youngsters. These researches were carried out at different times, on rather different populations, by different investigators using different measures; nevertheless, one finds a strong central tendency in the data, a clustering of "disturbance" at the 20% range with the range itself extending from 16% to 25% (see Table 1). As Weiner points out, this 20% figure is about the same level we find in the more reliable studies of disturbed behavior in the population at large. One can only be suggestive here, but we may be willing to infer that about one in five persons in the general population shows symptoms of emotional disturbance, as adolescents do, and that the latter group is neither more nor less disturbed than others.

If that is the case, we may go on to speculate that clinicians are assuming a false base rate of pathology in their tacit internal portrait of "the adolescent." Some evidence that this is in fact so is provided through an ingenious study by Offer, Ostrov, and Howard (1981), which asked a group of 62 mental health professionals (psychologists, psychiatrists, etc.) to fill out a self-report inventory, designed for adolescents, as they believed a "mentally healthy/well-adjusted adolescent" would. The "scores" given by the professionals are considerably more "pathological" than those achieved by actual adolescents. Indeed, the imagined scores were more pathological than those obtained by seriously delinquent adolescents, and by deeply disturbed (hospitalized) youngsters. The normal adolescent receives a standard score of 50; the hospitalized a

TABLE 1. Incidence of Psychological Disorder among Samples of "Normal" (Nonpsychiatric) Adolescents[a]

Study	Subjects	Percentage with psychological disorder
Hudgens, 1974	110 12 to 19-year-olds	23
Kysar *et al.*, 1969	77 college freshmen	22.1
Leslie, 1974	150 13 to 14-year-olds	17.2
Masterson, 1967	101 12 to 18-year-olds	20
Offer & Offer, 1975	73 male high school students	21
Rimmer *et al.*, 1978	153 college sophomores	15.3
Rimmer *et al.*, 1976	200 14 to 15-year-olds	16.3
Smith *et al.*, 1963	86 18 to 20-year old college students	12

[a]Adapted from Weiner, 1982.

score of 43; the delinquent a score of 42. The mental health professional imagined the "well adjusted" as scoring at 40.

It is not hard to see how and why, given that mind-set, clinicians are so often led to err in their work with adolescents. They will tend much of the time to underdiagnose adolescent problems, that is, to minimize their severity; and at times to misdiagnose them completely. Let me illustrate from one of my own misjudgments. Some years ago a colleague and I organized an analytic group for adolescents. A 16-year-old boy found his way to our office, complaining of depression. We discovered that it began when he was thrown over by his girlfriend, who complained that he was becoming "too intense." We could see what she meant, for he is a brilliant youngster and very much in command of the language of passion. As we listened to his vivid account of the tempestuous love affair—his initial exhilaration, his slavish devotion, the shock of the loss, his current inconsolability—in one part of our minds we were able to recognize his suffering, yet in another we were busily discounting, drawing on a stock of familiar literary prototypes—Romeo, Werther, Moravia's adolescents—along with the familiar psychoanalytic observations we spoke of earlier, of the defensive and regressive aspects of adolescent love attachments. In these sources, both literary and psychodynamic, the emphasis is placed upon the *expectable* disruptions of adolescence, these being seen as transient, as characteristic of the period, or as necessary to later growth. In our diagnostic interviews with this boy, we were able to construe the defensive nature of the passionate falling-in-love, as a retreat from a revival of incestuous feeling.

Now that formula, as it happens, was quite correct, for we were able to see quite clearly during the therapy, especially through the intense transference reactions to the female group therapist, that this boy was indeed struggling against the reappearance of archaic oedipal attachments and trying to handle them by displacements and by various modes of splitting. Yet at the same time that formula was deeply misleading. In the tacit assumption that these are commonplace intrapsychic conflicts, one is led to the belief that the crisis will work itself out in the ordinary course of events, as the drive diminishes in intensity, as the regressive forces ebb, as the ego is strengthened, and the more primitive defenses are discarded. None of that took place, neither in this case nor—as I would argue—in most like instances. The boy was at the beginning of what was to be a tragic emotional career—one not yet completed—that involved several psychotic episodes, several serious suicide attempts, and a life so troubled as to inhibit completely his extraordinary intellectual gifts.

More often than not, the appearance of symptoms in adolescence is the precursor, or the first visible expression, of an established and continuing pattern of psychopathology. The prudent clinician ought not to assume that the youngster, or those about him—the family, or the school—are overreacting to what will be a transient phase in development. On these matters the laity seem to have a shrewd sense of what is and is not normative, and the decision to refer or self-refer can be taken as fairly good testimony that the youngster's distress should be addressed and treated.

THE MILIEU OF THE ADOLESCENT

As we have seen, the "turmoil" theory of adolescence imputes—I believe mistakenly—a high degree of internal discord to the period. In addition it assumes, as a normal state of affairs, the youngster to be in conflict with his social setting. The child's relation to the family is said to be marked by discord, as he seeks to emancipate himself from their expectations and values and to find new ones more appropriate to his own generation. The peer group in particular is viewed as a central instrument of emancipation, and at the same time the source of new values. Two phrases—now catchphrases—have been employed to capture these putative shifts in orientation: from parental to peer influence; and the generation gap.

Let us begin by looking at the adolescent's relation to the family, noting first the truly astonishing paucity of information on the topic. If one were to sample a series of the leading textbooks on adolescence, one would discover that most of them devote a chapter to the topic, and that these contain discussions, generally thoughtful, of such matters as the dissolution of the traditional family, the invention of alternative forms, the effects of the rising divorce rate, the impact of mothers working or of changes in sex roles, and in general the nature and effect of recent social, structural, and ideological changes as these impinge on the adolescent's family. What one would not find in these chapters is a systematic report of the actual distribution and range of family forms, or an empirical account of the processes of reward, punishment, indoctrination, and the like, nor any material based on extensive *in situ* observation, of the sort commonplace in anthropological field studies.

In the absence of such direct observation, the best empirical evidence we have is from the reports of adolescents themselves. Here the testimony is quite clear, and entirely at odds with the supposition of intense conflict. We saw earlier that the recent Gallup statistics show American adolescents speaking of their parents with approval, reporting good overall relations, satisfactory discipline, and a sense of moral reliance. These recent findings are not in any way unique, as we said, and confirm what all other studies have shown. Some of this research is based on questionnaires, some on interviews, and some on fairly probing clinical methods; in all cases an essentially positive picture emerges. The most consistent trend is that showing the stability and strength of parental as against peer influence. Curiously enough, that topic was initiated as a means of discovering the tempo at which the adolescent escapes parental influence, it being assumed by the early investigators that their job was to measure carefully what was otherwise palpable. What they discovered instead was the persistence of parental values into late adolescence and the onset of adulthood, the influence of the peer group being limited much of the time to such issues as hairstyle and other modes of consumption. It is of particular interest that on such major issues as whether to attend college, parental opinion is considered to be far more significant than that of best friends (Floyd & South, 1972; Kandel & Lesser, 1969; Lerner & Knapp, 1975).

Of course such studies can be looked at skeptically in that they rely on self-report, and may be suspect of striving to give a good—that is, sanctimonious—impression to the inquirer. But in those instances where we can make a direct comparison between adolescents and parents, much the same impression emerges, of continuity, agreement, and harmony as against the reverse. There is a substantial body of work stemming from political science—work not generally known among psychologists—on the transmission of political attitudes, where comparisons are made among the responses of adolescents and young adults, their parents, and their friends. These data have shown that parents were "more influential in shaping adolescent [political] orientations than were peers" (Jennings & Niemi, 1981). To be sure, the degree of parent–child concordance is modest—the child is by no means the parents' clone—yet the findings are consistent over a wide range of studies using a wide range of methods and measures and samples and countries.

To write in this vein is to expose oneself to the suspicion of an advanced Pollyannaism. Surely adolescents as a group are not the Eagle Scouts these studies seem to say. Surely they are not, and I hope it will be understood that the intention here has not been to turn the "turmoil" theory on its head simply for the sake of the argument. Our view is not that adolescents are paragons of mental health, only that they are fundamentally ordinary people, in a somewhat younger version, and show that range of strengths and weaknesses we would find in the general population.

Why do we believe otherwise? No doubt it is partly because of an occupational bias, which leads one to extrapolate one's experiences with disturbed young to the age group at large. A second reason may be a subtle class bias, in that we tend to make our judgments about adolescents as a group from those most visible to us, the children of academics and professionals, who are as a cadre "normal" psychologically, yet atypically articulate in speech and febrile in temperament.

There is a third reason that deserves particular attention, since it may seem to confute the position taken here. There has been a rise in all forms of social and psychological pathology among adolescents in recent years. The increases have been substantial. A survey by Shapiro and Wynne (1982), based on U.S. government statistics, shows increases ranging from two to four times on indexes of suicide, homicide, and illegitimate births. Roughly equivalent trends can be found in surveys of teen-age arrests, drug use, and the like—though in many of these instances we do not yet have genuinely long-range data. We may add to these such specific syndromes as the anorexia–bulimia pattern, once exotic, now apparently epidemic among upper-middle-class adolescent girls.

What we have had during the post–World War II era is an enormous rise in the size of age cohorts, such that the decade of the 1970s in particular was marked by a "youth bulge," meaning a substantial increase in the number of adolescents. There was in addition an extraordinary increase in the rate of certain pathologies. The rise in numbers compounded by the rise in rate has meant an explosive increase in the number of cases of destructive and self-destructive behavior. In turn, this has tended to sustain the idea of adolescence as a particularly vulnerable period of life. Yet we do well to keep the issues separate. Consider that the "pathological" view of adolescence was at the height of intellectual acceptance in the 1950s, just at the time when the true rate of adolescent pathology was at the lowest point in recorded American history for many indexes. That the rates are high now tells us nothing about the normative vulnerability of the adolescent period; it only tells us that there are processes at work in the culture that are stressful to those populations exposed to them.

The most exciting new hypothesis we have on this matter is that the source of these stresses is demographic. The position has been stated elegantly in a persuasive book by Easterlin (1980) that points out the many ways in which a crowded age cohort works to the extreme disadvantage of its members. In a crowded generation there is more competition for a relatively smaller number of places, producing a variety of effects—on the psychological side a lowered self-esteem and heightened feelings of alienation, and with respect to the hard facts of social pathology, such phenomena as increases in crime, suicide, divorce, and drug use. Some of the relationships teased out by the demographers are quite unexpected, for example, that young men's suicide rates have a strong negative correlation with female fertility, and that the same inverse relationship occurs between homicide and fertility. Easterlin argues that the three indexes all reflect a common causal factor, "variations in the ability of young persons to achieve their life-style aspirations." In a small generation one is more likely to marry and have children, and indexes of mental stress decline, with the converse taking place in crowded generations.

Once we look at matters in this way, it becomes evident that the rise in social pathology tells us little about the intrinsic nature of adolescence, except insofar as it is a period in which certain important life events do or do not take place. If the demographers are right, we shall soon be seeing a decline in these rates as generation size declines, and in fact we do seem to be seeing the first signs of that decline. And there is every reason to believe that we will be seeing a continuation of the pathology now affecting the young, though in older age groups, as that generation matures, and as the disadvantaged adolescents of today become the disadvantaged adults of tomorrow.

CONCLUSIONS

It may be useful to sum up, apodictically, some of the propositions emerging from this analysis.

1. The empirical evidence we have gives us no reason to believe that adolescence is marked by an unusual degree of psychopathology, or by excessive vulnerability to stress.

2. Psychological disturbances appearing in adolescence are unlikely to be transient, and are more likely the debut appearance of chronic or continuing disorders.

3. The more severe the symptom picture in adolescence, the more likely it is that later disturbances will be serious or disabling.

4. Strong feelings of resentment, anger, or alienation toward the parents will likely continue later in life, though they may be expressed differently. The same may be said for strong feelings of any kind—dependency, for example.

5. All of the propositions above reflect a conservative reading of the current evidence. It is quite possible to look at that evidence differently. The skeptical reader is advised to consult the brilliant analysis of longitudinal work on adolescence by Livson and Peskin (1980), who conclude that neither the "turmoil" nor the "stability" theories of the adolescent period are yet proven.

REFERENCES

Blos, P. *On adolescence*. New York: Free Press, 1962.

Blos, P. The second individuation process of adolescence. In R. S. Eissler *et al.* (Eds.), *Psychoanalytic study of the child* (Vol. 22). New York: International Universities Press, 1967.

Blos, P. The genealogy of the ego ideal. In R. S. Eissler *et al.* (Eds.), *Psychoanalytic study of the child* (Vol. 29). New Haven: Yale University Press, 1974.

Douvan, E., & Adelson, J. *The adolescent experience*. New York: Wiley, 1966.

Easterlin, A. *Birth and fortune*. New York: Basic Books, 1980.

Floyd, H. H., Jr., & South, D. R. Dilemma of youth: The choice of parents or peers as a frame of reference for behavior. *Journal of Marriage and the Family*, 1972, *34*, 627–734.

Freud, A. *The ego and the mechanisms of defense*. New York: International Universities Press, 1936.

Freud, A. Adolescence. In R. S. Eissler *et al.* (Eds.), *Psychoanalytic study of the child* (Vol. 13). New York: International Universities Press, 1958.

Gilligan, C. *In a different voice: Psychological theory and women's development*. Cambridge: Harvard University Press, 1982.

Greenberger, E., & Steinberg, L. D. The workplace as a context for the socialization of youth. *Journal of Youth and Adolescence*, 1981, *10* (3), 185–210.

Greenberger, E., Steinberg, L., & Vaux, A. Adolescents who work: Health and behavioral consequences of job stress. *Developmental Psychology*, 1981, *17*, 691–703.

Hall, G. S. *Adolescence*. New York: Appleton, 1904.

Hudgens, R. W. *Psychiatric disorders in adolescents*. Baltimore: Williams & Wilkins, 1974.

Jennings, M., & Niemi, R. G. *Generations and politics*. Princeton: Princeton University Press, 1981.

Kandel, D. B., & Lesser, G. S. Parental and peer influences on educational plans of adolescents. *American Sociological Review*, 1969, *34*, 213–223.

Kysar, J. E., Zaks, M. S., Schuchman, H. P., Schon, G. L., & Rogers, J. Range of psychological functioning in "normal" late adolescents. *Archives of General Psychiatry*, 1969, *21*, 515–528.

Lerner, R. M., & Knapp, J. R. Actual and perceived intrafamilial attitudes of late adolescents and their parents. *Journal of Youth and Adolescence*, 1975, *4*, 17–36.

Leslie, S. A. Psychiatric disorder in the young adolescents of an industrial town. *British Journal of Psychiatry*, 1974, *125*, 113–124.

Livson, N., & Peskin, H. Perspectives on adolescence from longitudinal research. In J. Adelson (Ed.), *Handbook of adolescent psychology*. New York: Wiley, 1980.

Masterson, J. F. *The psychiatric dilemma of adolescence*. Boston: Little, Brown, 1967.

Offer, D. *The psychological world of the teenager*. New York: Basic Books, 1969.

Offer, D., & Offer, J. B. *From teenage to young manhood*. New York: Basic Books, 1975.

Offer, D., Ostrov, E., & Howard, K. L. The mental health professional's concept of the normal adolescent. *Archives of General Psychiatry*, 1981, *38*, 149–152.

Petersen, A., & Taylor, B. The biological approach to adolescence. In J. Adelson (Ed.), *Handbook of adolescent psychology*. New York: Wiley, 1980.

Shapiro, J., & Wynne, E. O. Adolescent alienation: Evaluating the hypothesis. *Social Indicators Research*, 1982, *10*, 423–435.

Steinberg, L. D. Jumping off the work experience bandwagon. *Journal of Youth and Adolescence*, 1982, *11* (3), 183–205.

Steinberg, L. D., Greenberger, E., Garduque, L., Ruggiero, M., & Vaux, A. The effects of working on adolescent development. *Developmental Psychology*, 1982, *18*, 385–395.

Weiner, I. B. *Child and adolescent psychopathology*. New York: Wiley, 1982.

15

Midlife Change

JUDITH F. LASKY

The internal conflicts of adult men and women have provided the material for some of the world's greatest literature, such as *The Divine Comedy*, *Death in Venice*, and *Faust*. Most of the patients in a general psychotherapy practice are age 25 to 55, and thus much of the professional literature is about this age group. Yet it is only recently that a systematic inquiry into age-related change in adulthood has begun.

Numerous explanations of this fact have been offered. From a sociological perspective, Smelser (1980) suggests that it is when a phase of life becomes socially problematic that it becomes the object of scrutiny. Thus, it was precisely when the spread of compulsory education made it necessary for society to learn about how children develop that the study of this subject was stimulated (Stevens-Long, 1979, cited in Colarusso & Nemiroff, 1981). More recently, such factors as the growth of the middle-aged population, the concentration of childbearing in the early years of marriage, and the expansion in the number of professional and semiprofessional careers that require intellectual capacity as opposed to physical strength have combined to make middle age stand out as a distinct stage of life in a way that it had not before (Giele, 1980).

Another widely offered explanation is that Freud's understanding of the influence of childhood experience upon later life led to a concentration of research upon these early childhood experiences. This has been carried further by some authors (Abraham, Kocher, & Goda, 1980;

JUDITH F. LASKY • 257 Central Park West, New York, New York 10024.

Brim, 1976) who suggest that Freud's (1905/1953) statement that the chances of successful psychoanalytic treatment after 50 are slim may have contributed to the notion that later adulthood is a time without change. Yet as King (1974) points out, in his (1912/1958) paper on *Types of Onset of Neurosis* Freud alluded to the possibility that emergent conflicts of later life might stimulate the need for analysis.

The resistance of both society and individuals to recognizing developmental change in adulthood is cited by a number of authors as having contributed to the neglect of this area. Brim (1976) suggests that the fact that most leaders of our society are drawn from the middle-aged (particularly male) population encourages a desire to see them as stable, unchanging, and not subject to the same developmental vicissitudes as other age groups. Vaillant (1977) and Colarusso and Nemiroff (1981) suggest that most researchers are middle-aged and may find it easier to perceive in others rather than in themselves the process of change in response to biological, psychological, and social cues. As an example of this King (1974) cites the fact that Freud was 49 when he wrote (in *On Psychotherapy*) that change beyond the age of 50 seemed unlikely, as an indication of the difficulty we all face in accepting the reality that we, also, grow old.

On the other hand, we all do indeed age. And as Smelser (1980) suggests, the scholars who were trained in the golden era of American education (in the 1950s and 1960s) are now facing midlife themselves—which may be contributing to the emergence of this issue in many fields today, as an increasing number of books and articles dealing with it appear in today's popular and professional literature.

Both because it is a relatively new focus of research and because aging is a physiological, social, cultural, and psychological phenomenon and thus studied by a number of disciplines (Neugarten, 1968a), the problem of organizing the existing theory and research is a formidable one, as has been noted by all those who have attempted it (e.g., Chiriboga, 1981; Neugarten, 1968a). The particular organization selected ends up reflecting the particular interest of the reviewer. Likewise, it would be possible here immediately to narrow the focus to only those studies of midlife that are relevant for therapy. There are, however, two problems with this approach. Although one would not want to report on every article in the literature, one would not want to narrow the field down so as to exclude articles that may nonetheless be of interest. Second, although much has been written on the area of adult development, there are not very many research studies or even clinical reports that focus solely on the midlife period. This comes somewhat as a surprise when one considers the popular interest in this field. More of the studies that consider the midlife period do so in the context of the larger period between adolescence and old age. The approach to be taken here, then, will be to begin with some of the basic questions on adult development, and the ways in which the seminal theories have tried to answer them. Next, some of the large-scale research approaches to these questions will be outlined. Finally, the studies dealing with those aspects of the midlife period that are relevant for psychotherapy wil be presented. From there it will be possible both to present some clinical material and to discuss some clinical issues that can be clarified and deepened by an understanding of midlife developmental issues.

Perhaps the central issue in adult development is whether adults change over time in consistent ways. If so, do these changes constitute an orderly sequence of stages that are related to, even if not precisely determined by, age? If not age in itself, what is the impetus for change: biology, social status, psychological events, health? Are the stages the same for all cultures and historical periods, for men and women, for individuals of differing socioeconomic backgrounds? Where do individual differences fit in?

In general, the seminal theories of adult development have offered phenomenologically rich and useful descriptions of the differing phases of adulthood. They are generally based upon clinical work and biographical study, and do not include attempts to answer the more specific questions suggested above. It has been the task of research in adult development to try for more precision about the nature and timing of these changes.

SEMINAL CLINICAL THEORIES OF ADULT DEVELOPMENT

Jung is widely considered to be the first psychoanalyst to focus attention on development during the adult years. Based upon his work with adult patients, he described (1933) two stages of development with 40 as the midpoint between "the morning" and "the evening" of life. Before 40, one's attention is focused on the external world—one's family and work; after 40 one's inner world, including both religious and philosophical concerns as well as formerly suppressed aspects of the personality, come to the fore. A successful transition from the early to the later stage of life involves the acceptance of one's aging and eventual death and the surrender of one's youth, just as at an earlier stage the acceptance of the responsibility of adulthood involved the surrender of one's childhood.

Havighurst (1948) based his developmental model upon his research with children and adults. Although it is less familiar to clinicians than Erikson's and thus less often referred to in the clinical literature, it is important historically and also because of its emphasis on adulthood. He conceived of development as presenting a series of tasks to be solved by the individual in relation to the environment, motivated both by bodily changes that present the individual with new opportunities and needs and by the changing expectations of society as the individual grows. Tasks of middle age are: to achieve adult civic and social responsibility; to establish and maintain an economic standard of living; to help teen-agers become responsible and happy adults; to develop adult leisure-time activities; to relate to one's spouse as a person; to accept and adjust to the physiological changes of middle age; and to adjust to aging parents.

Erikson, in *Childhood and Society* (1950), presented his well-known description of the developmental stages. They arise from an interaction of psychological, physiological, and social events and require new, more mature resolutions. Although his work emphasizes the childhood and adolescent years, it does contain a description of the stages of adulthood which cover more than the last half of life. The sixth stage of life (twenties and thirties) centers around the conflict between intimacy and self-absorption. The individual must find a way to become truly involved with another person, or risk becoming isolated and self-involved. The seventh stage (forties and fifties) is characterized by the conflict between generativity and stagnation. The individual must find a way to move beyond concerns with the self and to become involved in establishing and guiding the next generation, or else face a sense of stagnation and personal impoverishment. In the eighth stage (sixties and beyond) the conflict is between ego integrity and despair, and can be resolved by an acceptance of one's life, place in the life cycle, and eventual death.

Buhler (1955) plotted curves of the dates of significant external events, inner experiences, works, and accomplishments for 300 subjects studied through their biographies. She found a generally regular sequence in the events, experiences, and attainments of life. She distinguished among: a period of expansion in opportunities, abilities, and accomplishments; a period in which a peak of functioning is reached; and a period in which positions are relinquished and restrictions emerge. She found that the more an individual is involved in spiritual and intellectual functioning the later in life the peak is reached, and the more in physical functioning, the earlier the peak is reached.

Jaques's (1965) work is important both because is appears to have been the first to use the term *midlife crisis* and because of its considerable influence. Based upon both his clinical work (he is a psychoanalyst of Kleinian orientation) and a study of 310 composers, poets, and writers who were no longer alive but about whom biographical data were available, he proposed that the middle and late thirties is a critical stage of development. For the creative individuals he studied, it was a time of great change: some became creative for the first time; others ended either their creative career or their life; those who had previously been creative and remained so often changed their mode of working from a "hot-from-the-fire creativity" to a "sculpted creativity." He suggested that for all individuals, arrival at the midpoint of life requires movement beyond early adult idealism, entailing the recognition of fate as a force in human life and the acceptance of death as more than a mere abstraction. "The paradox is that of entering the prime of life, the stage of fulfillment but at the same time the prime and fulfillment are dated" (p. 506).

Before proceeding to some of the more traditional kinds of research, it is important to define the midlife period. Although the term *midlife period* seems to refer to a generally agreed-upon age range, that is, 35 to 45 or 50, the precise boundaries of this range do sometimes differ. One reason for this difficulty is that there is no one physiological, psychological, or social event that signals the beginning of adulthood or its phases. Chronological age, which is most often used, is problematic because (1) it has different meanings for different people based on social class (Neugarten, 1968b); (2) the importance of physical and/or intellectual factors for self-concept varies between people (Buhler, 1955); and (3) age is also related to a host of other individual differences. Historical time, too, has a bearing: an individual who is 50 years old today might define himself as younger than did a 50-year-old 30 years ago. Nonetheless, when the term *midlife* is used today, it usually does not refer to a time before age 35 or after age 50, and it is the intervening range that will be referred to here. Based on the theories just discussed, some of the important issues of this time are: changes in one's perception of time and one's place in it; actual or anticipated changes in physical capacities; and changes in one's status in the world.

LONGITUDINAL RESEARCH

Since longitudinal studies are widely recommended as the way to study age-related differences, three of the more well-known studies that either directly address or include adult development will be presented first.

The Oakland Growth Study began in 1933 with a population of 212 seventh graders in five schools in the Oakland area. Of the subjects, 171 had been followed up through 1960, when they were 37, and it is these data, reported in 1972, that we shall look at (Haan, 1972). Based on interview and test data, a 100-item California Q sort (in differing adolescent and adult versions) was produced for each subject. A typologic analysis yielded five male and six female types, which were then examined for the stability and/or change in their descriptions over time. Differing patterns of continuity and discontinuity for the differing types were found, leading the author to conclude that there are no common processes of development from adolescence to early mid-adulthood that could be found by this method. An analysis of variable trends was then performed that involved singly analyzing 90 Q items for each time for stability and/or change with items selected according to their salience for each individual. When all subjects were considered, only intellectual brightness for men and concern for intellectual matters for women showed a correlation of .50 or more from junior high schol to adulthood. When only the half of the population showing continuity from adolescence to adulthood was considered, the following traits were found to persist: speed and resistiveness of reaction, pathological reaction, and miscellaneous strengths. More traits were found to substantially shift, with men in general moving to greater social control and certainty and women moving to less self-assuredness. A preliminary follow-up in 1972, when the subjects were 51, suggested the increased salience of philosophical concerns and psychological insightfulness, and a growth in productivity.

The Terman Gifted Group Study began in the California school system in 1921–1923 with 857 boys and 671 girls who had IQs of 135 and over. The original data were obtained through interviews, questionnaires, and parent and teacher ratings. Parent–teacher follow-ups were done in 1928 and 1940, and subject follow-ups were done approximately every 5 years. Sears took over the responsibility for the research after Terman's death, and it is his 1977 report on subjects whose average age was 62 upon which this description is based. Occupation and family life were considered to bear most directly on life satisfaction, and thus objective measures of life-cycle satisfaction with occupation, degree of work persistence in their sixties, satisfaction with family life, and unbroken marriage versus a history of divorce, were obtained and an attempt was made to find antecedents. Job satisfaction and work persistence were found to be most closely linked to enjoyment of work, ambition, and feelings of having lived up to one's potential, feelings that were found to be consistent over the three decades from 30 to 60. The antecedents to family satisfaction were more difficult to spell out, since many more factors contributed to the success of a marriage and family and many of these are beyond the direct control of the individual. Good mental health at ages 20 and 40, good early social adjustment, and a favorable attitude toward one's parents were found to have some bearing.

The Grant Study of Adult Development (Vaillant, 1977; Vaillant & McArthur, 1972) differs from the two aforementioned studies in that adulthood is its original focus. It was begun in 1938 when the subjects were 18, and published reports follow them through age 50. Its 268 subjects were selected from a larger group of Harvard College freshmen, eliminating those showing mediocre academic achievement, medical or psychological problems, or poor motivation. The attempt was to select a sample of "healthy" men. Subjects were initially interviewed on family background, career plans, and early sexual development; they were then tested and their families were also interviewed. From the time of their graduation until 1955 they received annual questionnaires, and biannual ones after that. In 1970, 95 men selected randomly were interviewed about the details of their adult life.

Vaillant (1977) both organizes his data in more clinical ways and makes larger speculative leaps than do the preceding authors. One approach he used was to present numerous case illustrations of forms of passage through the adult years, and then to present the general developmental model that he feels best describes this process. He suggests that evidence from the Grant Study confirms the adult life patterns outlined by Erikson, and offers more detail about the period from 20 to 40 (intimacy and generativity) that Erikson did not specify. For his subjects, the years from 25 to 35 were characterized by career consolidation (acquiring, using, and casting aside of career role models), family involvement, and continued (internal) separation from one's family of origin. The early forties were characterized by a reevaluation of one's earlier life, for some as a way of moving in new directions, for others as a way of more fully accepting emotional realities that could not previously be accepted.

A second approach Vaillant used was to devise a theoretical hierarchy of 15 ego defenses based upon a Freudian model, and then to assess their use by his subjects at various crisis points in their lives. In general, he found a decline in such immature defense mechanisms as fantasy and acting out, and an increase in such mature mechanisms as suppression, sublimation, and altruism over time. When his subjects were divided by blind raters into those who had achieved generativity (as defined by Erikson) and those who had failed to, 80% of those using the most mature defenses had, compared to only 16% of those using the least mature defenses. He also pointed out that even those who had achieved generativity had experienced considerable anxiety along the way, and concluded that adult developmental issues are no less a source of stress than are earlier issues.

Longitudinal studies present both theoretical and practical problems. When age-related differences are found, they cannot necessarily be traced to age *per se* but may be attributable to the aging of the particular generation; thus Schaie (1968) suggests a model combining longitudinal and cross-sectional features. As for the type of data collected, longitudinal studies that began as studies of children may look at variables more relevant to children than to adults (Neugarten, 1968b), or the intervening 20 or so years may have caused the data once deemed important to be considered no longer so (e.g., Vaillant's interesting example of somatotype information, considered useful when the Grant Study began in 1937, is no longer considered interesting or useful today). As a practical consideration, the number of years required for meaningful results to be obtained means that subjects may be lost and researchers change (Birren, 1968). Also, the amount of data obtained may be enormous, and thus it will often perforce be analyzed in broader ways than might be desirable if one's particular focus is narrow.

For our purposes, there are several points to be learned from the broadest of these studies. The finding that individuals vary in their consistency over time suggests that change with age means different things to different people. The prominence of different defenses at different ages suggests both differing stresses and differing means of coping over the years. The suggestion that those who adapt successfully at one point may still need to work at successful adaptation at a later point implies that adaptation is an ongoing process, necessitated in part by changing developmental demands. But, as Kohlberg (1973) suggests, it may turn out that in the attempt to establish stages, it is not large-scale longitudinal research but the careful analysis of a small number of longitudinal cases that is ultimately most useful.

Cross-Sectional Research

The cross-sectional studies to be reported here have the advantage of being able to focus specifically on the particular age group and traits that are the object of our interest.

The Kansas city studies (Neugarten, 1968a; Neugarten *et al.*, 1964) were specifically designed to study the aging process. As such, the midlife population is only a segment of their sample of 420 40- to 90-year-olds, and it is a segment at the young end of the scale, rather than the old, as was the case with the Oakland and Terman studies. The data obtained have been analyzed in numerous ways by a variety of researchers, presented in numerous publications, and used as the basis for later research. Two examples will be presented here.

Gutmann (1964) analyzed responses to seven TAT cards of subjects age 40 to 70, then grouped subjects according to styles of mastery: active, passive, or magical mastery. He found that, for men, the older groups tended to rely more on passive mastery and the younger on active mastery. Also, older men were more involved with control of themselves, younger men with control of the external world. For women, too, there was a trend for passive mastery to be more common in the old than the young. Another change was in the attitude toward the young, with younger women demonstrating a more nurturant, positive attitude and the older a more retentive, self-centered version of mastery.

As reported in Neugarten, Moore, and Lowe (1968), subjects from this group were questioned about a variety of age-related characteristics ("the best age for a man to marry," "prime of life for a man," "when a woman has most responsibilities"). Based on what appeared to be a consensus with regard to age-linked and age-appropriate behavior, a more precise questionnaire was constructed in which the respondent was asked on a series of items which of three ages he would regard as appropriate/inappropriate and which he would approve/disapprove. It was then administered to 400 subjects equally divided by sex (100 from 20 to 30, 200 from 30 to 55, 100 65 and over). It was found that older subjects placed more emphasis on the importance of age norms and constraints than did younger ones.

Lowenthal, Thurnher, and Chiriboga (1975) approached adult development from a somewhat different perspective. Instead of comparing subjects of different ages, they compared subjects who were facing differing normal transitions (high school seniors, young newlyweds, middle-aged parents, and a preretirement person). Each of their 216 middle- and lower-middle-class subjects was interviewed for 8 hr, with the structured part of the interview focused on their range of activities, leisure time, awareness of and involvement with social issues, and social role involvement. From their enormous amount of data, a few findings will be mentioned. Marital dissatisfaction and conflict with children were greater among middle-aged women than the other groups. A comparison of self-concept across the four time periods revealed a stronger and more positive self-image among men than women, with men moving toward mellowness and comfort with themselves as they aged. The middle-aged women were as conflicted and negative about themselves as the high school seniors. Several aspects of the individual's perspective on his life course were explored, including the "density" (or eventfulness) of time, past or future orientation, expectations of the future, and thoughts about death. It was found that although there were both young and old subjects who were oriented toward the past, the young subjects focused on specific events whereas the older tended to focus on larger patterns. Future orientation was found in some but not all young and old subjects; for the middle-aged and older subjects it seemed to be a critical factor in maintaining physical and mental health, whereas for the young it seemed to be linked to a wish to escape current problems. Although the number of thoughts of death did not vary with life stage, their context did, in that young people focused on accidental deaths and older people on specific and personal experiences (the actual or imagined death of someone close to them, the inevitability of death for everyone, etc.).

Gould (1972) observed UCLA outpatients who were assigned to therapeutic groups of homogeneously aged individuals. The 35- to 43-year-olds were characterized by a continued questioning of what they were doing with their lives and by a sense of time pressure, and the 43- to 50-year-olds by a greater sense of resignation to the passing of time. Gould then composed a questionnaire consisting of statements regarding attitudes toward time, toward parents, friends, children, and spouses, and toward self, job, and sex, which had to be rank ordered. It was administered to 524 nonpatient subjects (20 per year for ages 16 to 33, 20 for each 3-year period from 33 to 60). Curves for each statement for subjects of each age were obtained, and instances of stable and unstable curves presented. This method of analysis confirmed the view of the thirties as a period of transition and instability.

Cross-sectional research has been criticized for failing to distinguish age-related differences from differences due to cultural-historical changes (Schaie, 1968) or to illness, economic change, and so on (Neugarten, 1968a). Although this criticism is certainly valid, it may be less crucial to therapists, whose purpose is to understand more about the forces affecting particular individuals with whom they work. A more important criticism for our purposes concerns the fact that, as Brim and Kagan (1980) point out, research that concentrates on age periods makes it hard to connect a younger with an older age group. They point to the need to study transformations over the life span. A second criticism is the failure of some cross-sectional research to consider individual differences, or differences between groups of individuals when age (or age category) is the only basis for subdividing subjects. As Buhler (1955) suggested and some of the longitudinal studies seem to confirm, aging means many different things to different people depending upon a variety of factors.

It has already been suggested that longitudinal studies that began as studies of children are not always the best sources of data about adulthood. The same may hold for models of adult development based upon the close relationship between biological and psychological phenomena in childhood (Neugarten, 1968b). That is not to say that biological changes and/or their anticipation are not significant in adulthood, but that they may mean a greater variety of things to the adult than the child. It is also not to deny the psychoanalytic truism that for children as well as adults significant events will be interpreted both realistically and in ways that are colored by an unconscious, idiosyncratic set of meanings. But the fact that the reality situations adults find themselves in are more variable than those usually experienced by children makes it clear that the potential for realistically different meanings is greater. To give an obvious example, the approach of menopause will have very different meanings to a woman with two grown children, a woman who has chosen to remain single and childless, and a woman who has spent the greater part of her adulthood trying unsuccessfully to conceive a child. In addition, those dealing with adult development must perhaps take more account of historical, social, and cultural context. Thus Neugarten (1968b) states that

> if psychologists are to discover order in the events of adulthood, and if they are to discover order in the personality changes that occur in all individuals as they age, we should look to the social as well as to the biological clock. (p. 146)

Similarly, Clausen (1968), in his comments based in large part upon the Oakland Growth Study, states that

> once full physical maturity has been achieved, age is a basis for stage definitions to the degree that there are clear age norms for status transitions. . . . [In our society] these are differently patterned with the social structure, so that stage analyses of the adult years are most useful as we deal with sequences within roles. (p. 462)

RESEARCH ON THE INDIVIDUAL AT MIDLIFE

The studies presented thus far compared the midlife individual to individuals of other ages. The studies to be presented now focus specifically on midlife individuals, in terms of issues that have already been stressed. Clearly, such studies can be faulted on the same grounds as earlier-reported cross-sectional studies. Their findings may be specific to the particular cohort(s) studied, limiting their generalizability. A more important criticism is that whatever concerns are found in midlife individuals may also be found in younger and older groups. Two qualifying responses can be made to this. First, a psychoanalytic or indeed any developmental perspective would maintain that conflicts do indeed appear and reappear throughout the life span, although the form they take varies according to the developmental period. Second, our concern in this chapter is much narrower than adult, or even midlife, development. The specific focus is on characteristics of the midlife period that are relevant for psychotherapy either with midlife individuals or those approaching that stage. As such, it is less important to us that a particular concern is found only in the midlife individual, and more important that the concern actually characterizes the world of the individual in midlife. We shall look at subgroups, rather than an entire group, on many of the variables, because of what we have learned.

Perhaps the best-known study of the midlife period is Levinson, Darrow, Klein, Levinson, and McKee (1978). Its original focus was the midlife decade, during which "one made the shift from 'youth' to 'middle age' " (p.7), encompassing biological, psychological, and social changes. For reasons similar to those we saw for other studies, the study was limited to 40 men, age 35 to 45, coming from four occupational categories (hourly workers, executives, academic biologists, and novelists). Subjects were interviewed for 10 to 20 hr, and given a sample of five TAT cards; their wives were interviewed once; and the organizations (or, for the novelists, a large network of writers) within which they worked were studied. The task of the data collection was to construct the story of each man's life, out of which the authors' developmental theory emerged. Although their initial focus was on the midlife decade, they arrived at a more comprehensive theory of adult development. Since their data were obtained from midlife individuals, it is possible that formulations about earlier periods were colored by concerns of the current period. Vaillant (1977) gives some interesting examples of the problems of retrospective reconstruction. However, that need not affect what emerges about the midlife period.

Levinson *et al.* focused on life structure as the pattern of a person's life (incorporating relationship to one's self, family, peers, work, body) and as it evolves through an alternating sequence of structure-building and structure-changing periods. Their data are presented in the form of individual biographical sketches, and of more generalized descriptions of sequences within the various periods they define. They view the midlife transition—from about 40 to 45—as the transition from early to middle adulthood. During that time, the individual must evaluate the satisfactoriness of his current life structure as he contemplates the imminent future. For the individual to emerge from this transitional period in a satisfactory way, he needs to resolve the four polarities of young/old; destruction/creation; masculine/feminine; and attachment/separation. That is, he must be able to accept being older than young people, yet also at the young end of the middle-age spectrum. He must experience both his own mortality and his destructive potential more vividly than he has in the past, yet move beyond this awareness to both an enhanced creative potential and a greater concern for future generations. He must come to terms with feminine parts of the self, possibly entailing changes in his relations with his spouse or other important women. He must find a way to remain connected with others yet also accept both the need and the inevitability of the separateness of individuals. As for whether such reevaluation warrants the description *crisis*, Levinson *et al.* report that for about 80% of the men such reevaluation did evoke enormous conflict within the self and with the external world. The details varied with individual circumstance. So it differed for an already successful serious novelist who could no longer be content merely writing another novel, and for a worker who had reached a position in his company that did not allow for significant future change.

Given the nonrandom nature of the subject selection within each category, and the fact that there was no attempt to make the categories inclusive, the representativeness of the sample can clearly be questioned. Also, the fact that the data are not presented in quantified form but as patterns and descriptions limits the reader's ability to separate the actual data (details of a life) from their interpretations (i.e., their organization into the authors' developmental framework). On the other hand, the case study method has particular usefulness to the clinical practitioner who is concerned with a patient's passage through this period.

Thus far there are no similarly comprehensive studies of women, either through adulthood (as in the Grant study) or specifically at midlife. Existing studies have tended not only to be based upon small samples of subjects seen in clinical settings, but also to concentrate on only one or two issues. As will be noted, the studies vary in their usage of the term *midlife*, which they define anywhere from 35 to 55.

To approach the question of whether there is a specific biological trigger for middle age as there is for adolescence, Neugarten, Wood, Kraines, and Loomis (1968) studied women's attitudes toward menopause. They devised an Attitudes Toward Menopause Checklist, with 35 statements concerning the kinds of changes menopause leads to and subjects' feelings about it. It was administered to 100 women, age 45 to 55, and then to two groups of younger and one of older women, all of whom were married, had at least one child, and were not part of a clinical population. The middle-aged group had more accurate information about menopause, and saw it more positively and less as a source of discontinuity than did the younger groups.

Notman (1979) considered the implications of the menopause for midlife women according to life circumstances and personality characteristics. She views women's midlife concerns with mortality as related to reproductive potential and says:

> Although hormonal and physiological changes mark major periods of a woman's life, the characteristics and experiences of these periods may be less related to actual biological changes and more to social and psychological events than has been assumed. (p. 1270)

Based on her psychotherapeutic work at an outpatient clinic (the number of cases seen is not reported), she distinguishes several factors affecting response to menopause. Women with a heavy investment in their children (to the exclusion of other investments) may find it particularly traumatic. Women who have chosen not to have children may have to contemplate issues of finiteness and may need to find new ways of being generative at an earlier age, thus making midlife easier. Also, upper- and middle-class women with more ways of altering their lives seem to view menopause more favorably than lower-class women with more limited options. Similar to that of Neugarten, Wood, Kraines, and Loomis (1968) was Notman's finding that younger women anticipate more difficulties with menopause than older women report actually having had.

Barnett and Baruch (1978) are critical of research on women in midlife for focusing too much on childbearing without sufficient attention to the stage of the family life cycle in which it occurs, and for too little focus on work, which plays an important role for most women today. Bardwick (1971) suggests that a major difference in the lives of men and women is the consistency of their major role responsibilities. For men, major occupational role responsibility does not change; for women, it increases with the birth and decreases with the maturation of their children, leaving them more open to the reemergence of such motives as the wish for professional achievement at this time. She cites a study by Baruch (1967, in Bardwick, 1971) of Radcliffe women that explored the relationship between the number of years out of college and the motive to achieve professionally, and found it to be greatest for women who had been out of school for 15 or more years and planned no more children, and least for women with young children. This study was done in 1966, since which time changes in both economic and social conditions might have strengthened the achievement motive in mothers of young children; but one would still expect there to be a relationship.

Several authors have commented on therapeutic work with midlife women who confront these issues. Mogul (1979) sees midlife—the time when previous choices in important life areas are reviewed and reworked in the context of old wishes, current limitations, and the finiteness of opportunity and time itself—as occurring in the late thirties for women as opposed to the forties for men because of biological considerations. If a woman has had no children, it is her last chance to do so; if she has had them, their growth leading to a decline in the required involvement makes other possibilities more appealing and feasible. Mogul describes several possible combinations of family and/or career, each of which would entail different concerns at midlife and which would require different resolutions. Nadelson, Polonsky, and Mathews (1979) explore the impact on married couples of the midlife concerns and/or changes of a partner. Based on therapeutic work with 75 middle-aged couples, they report several sources of marital stress that are particularly problematic at midlife, including a wife's entry into a career after being a full-time mother and/or housewife, nonparallel growth in the marriage, and dual-career issues.

Applegarth (1977) reports on the number of women seeking treatment who, often though not always at midlife, attempt to establish a career only to find themselves facing work inhibitions. He discusses a number of sources of such inhibitions, including difficulty in learning associated with unrealistically high standards; the belief that one lacks a quality necessary for what is required; conflicts about being successful and showing up other women; guilt about aggressiveness and competitiveness; and the wish to continue the dependent gratifications one is entitled to because of one's "deprivation."

Finally, some studies have considered middle-aged persons' perceptions of the stage they are living in. The middle-aged population Neugarten (1968c) interviewed comprised 100 "well-placed" men and women, randomly selected from university alumni lists, business and professional directories. The form of the interview is not described, and results are given either anecdotally or in general terms. These individuals viewed middle age as a distinctive period in the life cycle, but not necessarily in chronological terms. Thus "middle-aged people look to their positions within different life contexts—body, career, family—rather than to their chronological age for their primary cues in clocking themselves" (p. 94). The only indicated difference between women and men that emerged was that women tended to define their age status in terms of timing of events within the family, whereas men used biological and work cues to signal the onset of middle age. The middle-aged individual differentiates himself from both the younger and the older generations. Time is perceived differently—in terms of "time-left-to-live" rather than "time-since-birth;" yet despite the new realization of mortality, middle adulthood is seen as a period of maximum functioning.

Lionells and Mann (1974) explored the conceptions of midlife, and their implications for their work and other aspects of their lives, of a small group of psychoanalysts. There were two groups of eight subjects each matched for sex and marital status; one group age 32 to 49 (here defined as early middle age) and the other age 44 to 64 (at or beyond midlife). The younger group defined midlife in terms of changes in their relation to their spouse, with marriage seen as a means of fulfillment at a time when aging is closing off certain other possibilities for fulfillment. They defined themselves as not yet middle-aged, and as preferring to work with younger people. The older group saw the aging process as furthering the growth of the individual toward greater inner freedom and less dependency on others. They described midlife as their favorite time of life, and prefered working with patients who were also at midlife.

This review of diverse research articles makes it clear that midlife is not defined consistently. Chronological age is not a sufficient indicator, nor is there one specific biological or social event to mark the onset of midlife. What advantage can be gained from defining this stage? Neugarten (1970, 1979) suggests that individuals in a particular society develop a concept of a normal, expectable life cycle in which certain events occur at predictable times. The feeling of being an early or late developer, of being off-time, is a difficult one, and it frequently impels people to seek psychotherapy. Although this feeling may exist at many ages it seems particularly distressing to the individual at midlife, who has (1) the perspective to see himself as off-time; (2) the awareness of the problems it can cause; and (3) the feeling that there is no longer unlimited time in which to catch up. Thus the attempt to define midlife involves giving precision to conceptions that strongly affect most people.

A number of midlife issues involve a changing perspective on time. Just as in adolescence people first become aware of the reality of death, so in midlife they become aware of the circumscription of their life span. In Neugarten's (1968c) conception time is viewed in terms of time-left-to-live rather than time-since-birth. With this changed awareness comes an increased involvement with old wishes and dreams that were laid aside in the struggle to become an adult. One is paradoxically at the prime of life, the peak of one's abilities and accomplishments, yet alert to the signs of impending change. There is an increased sense of actual or potential physical vulnerability, arising in part from seeing one's parents age and in part from seeing changes in one's own body. Depression must be mastered, and new ways found of moving beyond one's immediate personal concerns to a concern with the younger and future generations.

It is clear that there is a set of issues that are characteristic of midlife. What varies is the age at which they emerge within the individual, which is based upon the individual's intrapsychic, interpersonal, social, and biological situation. Thus midlife is perhaps best defined similarly, though not identically, to Levinson's stage of midlife transition: the stage when, after having established a stable adult life (based on family and/or profession), one reevaluates its satisfactoriness within the growing awareness that the time and potential for change are not unlimited. Whether such a reevaluation is experienced as a crisis, and whether it leads to dramatic changes in situation or to quieter, internal accommodations, will vary with the individual. What is unvarying is the need to respond to change.

CLINICAL IMPLICATIONS

The research literature on adulthood emphasizes that aging has biological, social, and psychological components, and that it has different meanings to different people. For the therapist this underscores the usual need to move from general principles to the specific, variable forms these take in the unique functioning of the individual. There is no doubt that knowledge about general principles, when properly applied, can only further one's understanding of the individual.

It is customary for the clinician to consider the presenting problem of an adult in terms of childhood developmental history. A similar attempt to locate the presenting problem within the framework of adult development is not always made. It seems clear that even if adult development is not important in precisely the same way as childhood development, there are sequences and stages that center around critical and commonly held concerns. Midlife seems to be one such stage. This is not to suggest that midlife concerns are unique to this period; however, the same concern or conflict coming at two different developmental periods will have differing meanings and differing means of resolution. This is a familiar notion as it applies to child development. For example, learning to recognize internal cues so as to obtain mastery over one's bodily functions is a very different process for the 2-year-old being toilet trained and the 14-year-old on a diet. Most clinicians implicitly take into account the developmental situation of the adult patient, and it might be useful to do so more explicitly.

Many of the concerns of midlife individuals can prompt them to seek psychotherapy: concern about actual or anticipated bodily changes; the need to accept responsibliity for aging parents; the feeling of being at or near the peak of one's career; the realization that one's children are growing and becoming a less central focus of one's life; the realization that the time available to make changes in one's life is not unlimited; and so on. These and similar concerns may bring an individual to psychotherapy although it may not necessarily represent the only solution. Numerous conflicts, with a wide variety of responses, are found throughout life and certainly are not restricted to childhood, adolescence, and midlife. Sometimes, however, such conflicts may be severe enough to disrupt psychic equilibrium, and to cause such severe anxiety that the individual feels incapable of resolving the difficulty without professional assistance. Two brief examples of such situations follow.

A 46-year-old woman who had struggled successfully for years to establish herself as a person who was unlike her mother reacted to the emergence of wrinkles on her face (causing her skin to look like her mother's) with intense anxiety. This stirred up unconscious regressive longings to be the same as her mother that she has successfully defended herself against. These unconscious wishes were associated with fears of being out of control and going crazy. Out of this complex picture, based upon the unique meaning that a particular sign of aging had for this woman, the desire for therapy arose.

A 39-year-old man had been able to accept his failure to establish long-term, satisfying relationships with relative equanimity, based on his belief that when he was ready he would find one. The approach of his fortieth birthday reminded him that the time available was not unlimited and that he might not be ready in time to find a relationship. This resulted in a new sense of urgency that led to, among other things, the wish for psychotherapy.

The decision to seek therapy may be based not only on a particular conflict, but on feelings about its timing. As has already been mentioned, the idea of being off-time in achieving a goal or resolving a conflict can motivate one to seek therapy. It may well, for instance, have been part of the motivation of the 39-year-old man. There may also be a more complicated motivational picture in which being off-time is part of the individual's self-image.

A young woman sought therapy because her entry into a graduate professional school was triggering conflict about the "regressive position" a student occupies. This, her third attempt at graduate study, was particularly poignant for her because she was in her mid-thirties and saw it as a last chance. On the one hand, obtaining a graduate degree was, in her mind, the sign of having finally obtained "adult" status (despite her success in a previous field). This woman was 11 months younger than her sister. As children, the two girls were sufficiently close in size and appearance that the same things were expected of each of them. Yet 11 months is a good deal of time in childhood, and often her older sister was capable of tasks that she could not yet accomplish. Later in life she found herself in a number of situations where she again was, at least as she perceived it, the "slow" one. She had always had the belief that one day the time would come to be "fast," and then she would be an adult. Each "failure" (which was for her a repetition of the childhood experience) made her feel more despairing and this was compounded by the sense of time passing as she entered her mid-thirties and, in reality, was no longer a child or even a young adult. Issues in therapy included separating herself from her older sister and overcoming her narcissistic difficulties with learning, as well as working to overcome her view of herself as defective by virtue of being off-time.

I have suggested that the same conflict coming at different times in an individual's life may have different meanings, and may motivate recourse to therapy at one time but not another. Certain phases of life may also be more or less optimal for working through particular lifelong problems in treatment.

Depression appears to be a major occurrence for midlife individuals. It is reported as a fundamental symptom of midlife distress that may motivate attempts at resolution (Brim, 1976; Davidson, 1979), and as the maladaptive outcome of a failure to resolve midlife issues in a constructive, generative way (Benedek, 1975; Erikson, 1950). Jaques's (1965) view of the midlife crisis as a depressive crisis incorporates each of these aspects. As a symptom it is a response to the emergence of previously suppressed thoughts about aging and dying, separation and loss that make one question the path one has taken and whether one has the strength to face what is to come. Midlife can be seen as a time when it is both necessary and possible to work through earlier depressive struggles in a more profound way than was previously possible. A successful resolution may lead to an increase in one's capacity to love oneself and others without being blind to their shortcomings, and an ability to be hopeful about life's possibilities without shutting out awareness of death. An unsuccessful resolution may lead to chronic depression.

Kernberg (1975) has raised a similar point regarding narcissistic disorders and the midlife period, stating that

> throughout an ordinary life span most narcissistic gratifications occur in adolescence and early adulthood, and . . . even though narcissistic triumphs and gratifications are achieved throughout adulthood, the individual must eventually face the basic conflicts around aging, chronic illness, physical and mental limitations, and above all, separations, loss and loneliness [which the narcissistic individual has thus far managed to avoid]. (p. 311)

It may require such a confrontation with limitations to genuinely convince narcissistic individuals that another success is not the answer to their problems, and that intensive psychotherapeutic treatment may be required. It is for this reason that Kernberg (1982) sees middle age as the best time to treat certain narcissistic patients. King (1974) makes a similar point.

The discussion thus far has suggested how a developmental perspective can contribute to a fuller understanding of the patient at midlife. Next I shall consider the impact of the therapist's age and developmental stage upon the treatment process. Two aspects can be differentiated: the meaning for the patient of the therapist's age and the effect of the therapist's age on his or her work with the patient.

The age of the therapist is not often referred to in clinical articles. The more one adheres to a psychoanalytic perspective, the more attention one pays to the role of transference (rather than reality factors) in the patient's reactions to and perceptions of the therapist. Transference feelings often use a detail of current reality that will provide the "reality justification" for the full flowering of the transference itself. Or, current reality may aid in the emergence of one set of transference feelings rather than another. The age of the therapist may be such a detail. When it does have an impact, it may be because of the patient's expressed choice to work with a therapist of a particular age, or because of issues that emerged naturally in response to the age of the therapist.

Patients may express the wish to work with a therapist who is near their own age because they feel they will speak the same language. On the other hand, they may not want to work with a same-age therapist because of fear that they will compare themselves unfavorably to the therapist and that their envy and competitiveness will complicate the treatment. Patients may seek an older male therapist out of a wish to find a respected authority to whom they can submit and who they imagine will take charge of their lives. Or, they may seek an older female therapist because their grandmother was the one positive figure in their life. In all of these instances, the wishes that the patient brings to treatment regarding a therapist of a particular age may or may not be fulfilled, and will probably be succeeded by other, deeper wishes. Nonetheless, the patient's desires concerning the age of the therapist are an important source of information about the partient's dynamics.

Even when the patient has not requested a therapist of a particular age, the therapist's age may evoke certain responses or encourage certain transference responses to emerge sooner rather than later. For example, Abraham *et al.* (1980) describe the tendency of certain elderly patients to transfer onto their therapists feelings they have had toward their children, before a maternal or paternal transference becomes crystallized. Similarly, a number of elderly therapists report the regular appearance of grandmother or grandfather transference material that may only occur in treatment with a younger therapist if it is of particular importance in the mental life of the patient.

The impact of the therapist's age on the therapy is worth examining. Presumably, a well-trained therapist can work with patients varying in age as well as in other personal characteristics. Yet examples in the research and clinical literature suggest some ways in which the researcher's or therapist's own age and developmental status may affect his or her understanding or treatment of a patient of a particular age. Vaillant (1977) began studying the Grant study men when he was 33 and they were 46. He describes his initial impression that all of the men were deeply depressed, and the way in which this confirmed his own feelings about aging. Later, he realized that in fact these men had attained a kind of maturity that enabled them to acknowledge the real pain in their lives, and that his seeing this as depressing was the equivalent of the elementary school child seeing sex as disgusting. Another example is the tendency to be particularly attentive to patients' developmental struggles that resemble one's own, which Lionells and Mann's (1974) analyst-subjects report.

There are also examples of the therapist's countertransference or more general neurotic reactions to patients of a particular age. Abraham *et al.* (1980) suggest that in working with considerably older patients the therapist may feel like a child toward a weakened parent in need of help, and thus be unable to be helpful. A parallel problem may involve envious feelings in older therapists working with younger patients who are anticipating new accomplishments and experiences.

The point is not that patients of a certain age are best seen by therapists of that or any particular age. Rather, the awareness of potential age-related countertransference, neurotic difficulties, and limitations in understanding can help diminish their impact just as awareness of one's other neurotic conflicts can keep such conflicts from having damaging effects.

SUGGESTIONS FOR FUTURE RESEARCH

I began by noting that the area of adult development (and its subdivisions into phases) has only relatively recently become a field of systematic research. As such, the questions that remain unanswered are numerous and cannot all be addressed here. Rather, I shall concentrate on those research questions that arise from the practice of therapy in general, and whose answers will have particular relevance to the practice of therapy.

Many of the questions a clinician would pose for the researcher in adult development have to do with discovering, and finding some of the factors governing, individual differences. For example, midlife has different meanings for different individuals, and may cause a psychological crisis in some but not in all. It would be useful to be able to spell out some of the variables that make an individual more or less vulnerable to such crisis. From a social psychological point of view, Brim (1980) and Neugarten (1970) comment on the occurrence of expected events at an unexpected time. A clinician's interest would lie in the intrapsychic factors that cause midlife to be experienced as particularly stressful or particularly growth producing. Kernberg's (1975) remarks about the difficulty of this time period for the narcissistic individual have already been mentioned, and are a good example of the kind of information the clinician would appreciate. Several of the case examples presented individuals for whom issues of timing and being off-time were central and directly related to their self-concept, and who were thus particularly vulnerable to feelings of being off-time at midlife. Such material sugests the usefulness of research comparing individuals who differ in diagnosis, adaptive styles, life circumstances, sense of being on- or off-time, and so on in order to distinguish between those who are likely, and those who are unlikely, to undergo midlife stress. A further step would be a comparison of those for whom such stress is a precipitant of growth and those for whom it leads to maladaptive behaviors. Such research would be of use to the clinician working with the individual at midlife, and could also have broader implications for preventive mental health.

A second area of importance for future research is the effect of the researcher's or therapist's age and developmental status upon the work he or she does. Several comments by researchers about their possible age biases have been noted, and some of their implications for clinical work were considered. Lionells and Mann (1974) began a consideration of this issue with a small sample of therapists; a larger-scale inquiry would be useful.

The midlife change has two different but related meanings. The first involves intrapsychic, biological, and social changes that, occurring outside the times most commonly noted for radical change (such as childhood and adolescence), require new modes of adaptation for the midlife individual. The second meaning is that voluntary change is possible at midlife. Such change can be effected by a variety of means, including psychotherapeutic intervention. In fact, midlife may indeed be an ideal time for such changes to be attempted. With the current popular interest in the developmental tasks of adulthood and aging, we can look forward to a continuing proliferation of research on the midlife period.

REFERENCES

Abraham, G., Kocher, P., & Goda, G. Psychoanalysis and aging. *International Review of Psycho-analysis*, 1980, *7*, 147–155.

Applegarth, A. Some observations on work inhibitions in women. In H. Blum (Ed.), *Female psychology*. New York: International Universities Press, 1977.

Bardwick, J. *Psychology of women*. New York: Harper & Row, 1971.

Barnett, R., & Baruch, G. Women in the middle years: A critique of research and theory. *Psychology of Women Quarterly*, 1978, *3*, 187–197.

Benedek, T. Depression during the life cycle. In E. Anthony & T. Benedek (Eds.), *Depression and human existence*. Boston: Little Brown, 1975.

Birren, J. Principles of research on aging. In B. Neugarten (Ed.), *Middle age and aging*. Chicago: University of Chicago Press, 1968.

Brim, O. Theories of the male mid-life crisis. *Counseling Psychologist*, 1976, *6*, 2–9.

Brim, O. Types of life events. *Journal of Social Issues*, 1980, *36*, 148–157.

Brim, O., & Kagan, J. Constancy and change: A view of the issues. In O. Brim & J. Kagan (Eds.), *Constancy and change in human development*. Boston: Harvard University Press, 1980.

Buhler, C. The curve of life as studied in biographies. *Journal of Applied Psychology*, 1955, *19*, 405–409.

Chiriboga, D. Developmental psychology of middle age. In J. Howells (Ed.), *Modern perspectives in the psychiatry of middle age*. New York: Brunner/Mazel, 1981.

Clausen, J. The life course of individuals. In M. Riley, M. Johnson, & J. Foner (Eds.), *Aging and society* (Vol. 3). New York: Russell Sage, 1968.

Colarusso, C., & Nemiroff, R. *Adult development: A new dimension in psychodynamic theory and practice*. New York: Plenum Press, 1981.

Davidson, L. Preventive attitudes toward midlife crisis. *American Journal of Psychoanalysis*, 1979, *39*, 165–173.

Erikson, E. *Childhood and society*. New York: W. W. Norton, 1950.

Freud, S. On psychotherapy. In J. Strachey (Ed. and Trans.), *Standard edition of the complete psychological works of Sigmund Freud* (Vol. 7). London: Hogarth Press, 1953. (Originally published, 1905)

Freud, S. Types of onset of neurosis. The standard edition (Vol. 12). London: Hogarth Press, 1958. (Originally published, 1912)

Giele, J. Adulthood as transcendence of age and sex. In N. Smelser & E. Erikson (Eds.), *Themes of work and love in adulthood*. Cambridge: Harvard University Press, 1980.

Gould, R. The phases of adult life: A study in developmental psychology. *American Journal of Psychiatry*, 1972, *129*, 521–531.

Gutmann, D. An exploration of ego configurations in middle and later life. In B. Neugarten with H. Berkowitz, W. Crotty, W. Gruen, D. Gutmann, M. Lubin, D. Miller, P. Peck, J. Rosen, J. Shukin, & S. Tobin, (Eds.), *Personality in middle and later life*. New York: Atherton Press, 1964.

Haan, N. Personality development from adolescence to adulthood in the Oakland Growth and Guidance Studies. *Seminars in Psychiatry*, 1972, *4*, 399–414.

Havighurst, R. *Developmental tasks and education*. Chicago: University of Chicago Press, 1948.

Jaques, E. Death and the mid-life crisis. *International Journal of Psychoanalysis*, 1965, *46*, 502–514.

Jung, C. *Modern man in search of a soul*. New York: Harcourt, Brace, 1933.

Kernberg, O. *Borderline conditions and pathological narcissism*. New York: Jason Aronson, 1975.

Kernberg, O. In M. Pines, New focus on narcissism offers analysts insight into grandiosity and emptiness. *New York Times,* March 16, 1982, p. C1.

King, P. Notes on the psychoanalysis of older patients. *Journal of Analytical Psychology,* 1974, *19,* 22–37.

Kohlberg, L. Continuities in childhood and adult moral development revisited. In P. Boltes & K. Schaie (Eds.), *Life-span development psychology.* New York: Academic Press, 1973.

Levinson, D., Darrow, C., Klein, E., Levinson, M., & McKee, B. *The seasons of a man's life.* New York: Alfred A. Knopf, 1978.

Lionells, M., & Mann, C. *Patterns of mid-life in transition.* Unpublished manuscript, 1974.

Lowenthal, M., Thurnher, M., & Chiriboga, D. *Four stages of life.* San Francisco: Jossey-Bass, 1975.

Mogul, K. Women in midlife: Decisions, rewards, and conflicts related to work and careers. *Psychiatry,* 1979, *136,* 1139–1143.

Nadelson, C., Polonsky, D., & Mathews, A. Marriage and midlife: The impact of social change. *Journal of Clinical Psychiatry,* 1979, *40,* 292–298.

Neugarten, B. (Ed.). *Middle age and aging.* Chicago: University of Chicago Press, 1968. (a)

Neugarten, B. Adult personality: Toward a psychology of the life cycle. In B. Neugarten (Ed.), *Middle age and aging.* Chicago: University of Chicago Press, 1968. (b)

Neugarten, B. The awareness of middle age. In B. Neugarten (Ed.), *Middle age and aging.* Chicago: University of Chicago Press, 1968. (c)

Neugarten, B. Dynamics of the transition from middle to old age. *Journal of Geriatric Psychiatry,* 1970, *4,* 71–81.

Neugarten, B. Time, age and the life cycle. *American Journal of Psychiatry,* 1979, *136,* 887–894.

Neugarten, B., with Berkowitz, H., Crotty, W., Gruen, W., Gutmann, D., Lubin, M., Miller, D., Peck, P., Rosen, J., Shukin, J., & Tobin, S. *Personality in middle and later life.* New York: Atherton Press, 1964.

Neugarten, B., Moore, J., & Lowe, J. Age norms, age constraints and adult socialization. In B. Neugarten (Ed.), *Middle age and aging.* Chicago: University of Chicago Press, 1968.

Neugarten, B., Wood, V., Kraines, R., & Loomis, B. Women's attitudes toward menopause. In B. Neugarten (Ed.), *Middle age and aging.* Chicago: University of Chicago Press, 1968.

Notman, M. Midlife concerns of women: Implications of the menopause. *American Journal of Psychiatry,* 1979, *136,* 1270–1274.

Schaie, K. Age changes and age differences. In B. Neugarten (Ed.), *Middle age and aging.* Chicago: University of Chicago Press, 1968.

Sears, R. Sources of life satisfaction of the Terman Gifted Men. *American Psychologist,* 1977, *32,* 119–128.

Smelser, N. Issues in the study of work and love in adulthood. In N. Smelser & E. Erikson (Eds.), *Themes of work and love in adulthood.* Cambridge: Harvard University Press, 1980.

Vaillant, G. *Adaptation of life.* Boston: Little, Brown, 1977.

Vaillant, G., & McArthur, C. Natural history of male psychological health. I: The adult life cycle from 18 to 50. *Seminars in Psychiatry,* 1972, *4,* 415–427.

IV

SPECIAL TOPICS

The selections in this part are varied, representing different approaches to research that have in common a link to psychotherapy. In Chapter 16, Freedman and Grand describe a programmatic research effort. They are concerned with shielding, a concept used by Freud, referring to a boundary between the internal and external that serves to protect the inner person from being overwhelmed by external stimulation. They describe the gradual evolution and refinement of the concept through a series of research projects, and then provide a link between the concept and the clinical situation.

The topic of fantasy is discussed by Pope in Chapter 17. He defines fantasy, approximately, as mental activity that is not directed at any immediate goal. An interesting historical survey of the concept is provided, followed by a brief summary of a wide variety of research findings. Links between these findings and clinical practice are then spelled out, with the practice being eclectic in nature, in contrast to the psychodynamic approach represented in the majority of the chapters. Most interestingly, Pope concludes with a model of psychotherapy based on the research findings, a generalization that is rarely achieved by either clinicians or researchers.

In Chapter 18 Nachmani uses the theory derived from a basic area of inquiry, structural psycholinguistics, to expand upon the understanding in an area of clinical psychopathology, namely, the development of a form of schizoid behavior. By analyzing the original mother–child matrix, which gives rise to both language development and the foundation of personality structure, it is possible to note correlations in the two paths of development. In doing this, it is shown how the knowledge of one area can add greater depth to our understanding of a seemingly unrelated clinical area.

Puhakka, in Chapter 19, approaches the problem of investigating religion, a world view that is of enormous consequence to our appreciation of the human condition. Psychoanalysis has a historical interest in religion; it was the object of some of Freud's early works. She contrasts the objectification of the religious experience, which is common to psychoanalytic and social psychology, with the more phenomenological approach of William James. She then draws parallels between the approach of James and of psychoanalytic psychotherapy, in that both value the inner experience rather than the objectification of the central concern. This metainquiry shows how an understanding of the appropriate approach to religion can help us in the clinical area.

In the final chapter, Keisner shows how three related disciplines—social psychology, sociology, and psychoanalysis—have explored the self-fulfilling prophecy process. These perspectives on the same phenomenon are respectively illustrated by research on behavioral confirmation strategies, observations of the effect of labeling on role behaviors, and the conceptualization of projective identification in the psychoanalytic relationship. Keisner concludes by arguing for a human similarity hypothesis and then offers two paradigms to help explain how and why patients use self-fulfilling prophecy strategies in the therapy encounter.

16

Shielding
An Associative Organizer

NORBERT FREEDMAN AND STANLEY GRAND

INTRODUCTION

It is one of the hallmarks of Freud's genius that he takes a concept, a single word, and endows it with myriad meanings and expanding significance. For example, the words *shield, sheath,* and *crust* can be found through five decades of Freud's writings. In tracing the use of these words, one witnesses both the evolution of psychoanalytic thought and its consistent adherence to principles.

THE CONCEPT OF SHIELDING[1]

The division of psychic experiences into the realms of outer and inner worlds is an abiding conviction in Freud's writings. The notion of a shield, that is, of a very rudimentary psychic structure that maintains this separation between inner and outer, can be found as early as 1895 in the *Project for a Scientific Psychology* (Freud, 1895/1966). There Freud

[1]The psychoanalytic interpretation of shielding, both its metapsychology and its clinical interpretation, is discussed in *The Shielding Phenomenon: Metapsychological and Interpretive Aspects* (N. Freedman and S. Grand, book in preparation, 1984).

NORBERT FREEDMAN AND STANLEY GRAND • Research Unit for the Study of Communicative Process, Department of Psychiatry, Downstate Medical Center, State University of New York, Brooklyn, New York 11203.

had already established the notion of a protective screen that shields the system against overwhelming masses of stimulation, and had established two mechanisms for accomplishing this task: the contact barriers between neurons, and a class of impermeable neurons that inhibited the flow of stimuli through the psychic systems. Not only protection against painful stimuli but the constancy of stimulation was at issue here and in the subsequent translation of this early neurophysiological model into its psychological counterpart in the *Interpretation of Dreams* (1900/1953). This conception of a protective screen for maintaining the constancy of stimulation and the avoidance of pain was further elaborated in his 1915 paper on the *Instincts*, where Freud highlighted the role of muscular action (flight) as the hallmark of the first distinction between that which is inside the organism and therefore cannot be avoided, and that which is outside and can be avoided.

A major step in the development of Freud's concept of the shield came in 1920 with *Beyond the Pleasure Principle*. Freud turned his attention to psychic trauma and to the role of consciousness in the formation of a "protective shield against the enormous energies at work in the external world" (1920/1955, p. 27). Consciousness is here seen as having a "crust" lying between the external stimulus energies and the more vulnerable "living substance . . . which would be killed by the stimulation if it was not provided with a protective shield. . . . Protection against stimuli is an almost more important function for the living organism than reception of stimuli" (p. 27). Whereas the shield protects consciousness from the energies in the external world, there is no barrier between consciousness and internal stimuli. That the shield can be turned toward internal stimuli as well is evident from the fact that when internal stimuli become too unpleasurable,

> there is a tendency to treat them as though they were acting not from the inside, but from the outside, so that it may be possible to bring the shield against stimuli into operation as a means of defense against them. This is the origin of projection. (p. 29)

Thus, Freud expanded his conception of the shield to include the notion of defense from both real and projected external sources of stimulation.

In 1925, Freud used the mystic writing pad as a model for the protection of the perceptual apparatus (Freud, 1925/1961). Here, the outer celluloid layer of the mystic pad serves as a "protective shield" for the thinner, more vulnerable paper behind it, just as the

> perceptual apparatus of our minds consists of two layers, of an external protective shield against stimuli whose task it is to diminish the strength of excitations coming in, and of a surface behind it which receives stimuli, namely the system Pcpt-Cs. (p. 230)

And, in 1926, Freud related the protective shield to the concept of primal repression and the earliest outbreak of intense anxiety, which results from the excessive degree of excitation and "the breaking through of the protective shield against stimuli" (Freud, 1926/1959). Finally, in his *Outline of Psychoanalysis* (1940/1964), Freud related the protective shield against stimuli to the development of the ego with its characteristics of avoiding excessive stimuli by flight, dealing with moderate stimuli by adaptation, and also "by learning to bring about expedient changes in the external world to its own advantage" (p. 145).

Thus, we see in Freud's conception of a shield or protective barrier to external stimulation an abiding concern with the enormous power of reality and the absolute necessity to deal with this power through biological and psychological adaptation and, through expedient action, to bring about change in the external world. Although the conceptual forces shifted from neurophysiological mechanisms involved in the regulation of quantities of energy, to the constancy of stimulation and defense against unpleasure, to expedient action and mastery, the problem remained the same throughout Freud's writings—the task of establishing psychic integrity, a boundary for shielding the mental apparatus from the trauma and power of reality.

It is the purpose of this chapter to reintroduce the notion of the shield as one of the earliest forms of structuring palpably visible in an interchange. We take as our cue Freud's observation that "it is in the efficacy of the muscular system that the division between inner and outer is sustained" (Freud, 1915/1957). First, however, we shall consider the context in which shielding occurs, the associative process, and shall introduce our concept of shielding as an associative organizer.

THE CONCEPT OF ASSOCIATIVE ORGANIZERS

The hypothesis we wish to advance here is that in the associative act of every patient, there exists a sensorimotor matrix—a matrix of bodily sensations and motor experiences, rooted in the coenesthetic and vestibular experience of earliest childhood. This sensorimotor matrix is present in the most mature contemporary dialogue and exerts an organizing, regulatory, and structuralizing impact on the communicative process. One is the process of verbalization, important in the transformation of image into symbolic form and the elaboration of the experience of consciousness. The other is a process that is not only presymbolic but preimagistic as well. This process is linked to the most rudimentary psychic organizations and operates outside of consciousness. It is the recognition of these different levels of sensorimotor organization, transposed into adult discourse, that forms the basis for what we have termed *associative organizers* (Freedman, 1980, in press), that is, structures that either inhibit or facilitate the shaping of psychic experiences, and hence the capacity to put thought into words.

The term *associative organizer* has been adapted by us from Spitz's (1965) work on psychic organizers in early development. Spitz postulates specific forms of activity at specific periods of development that occur in spatial arrangements. These function as catalysts for the further development of psychic structures. According to Spitz, psychic organizers are phenomena of psychic integration. These organizers coalesce out of the matrix of maturational and developmental forces occurring within the context of the infant–mother dialogue. Moreover, Spitz has described each of these centers of integration as precursors or stepping-stones preparatory to the establishment of higher levels of psychic organization.

By analogy to Spitz, functionally similar forms of sensorimotor organization can be observed in the behavior of a patient during the clinical interview or during the analytic hour. From our research, we have identified two organizers that represent two anchoring points in the developmental sequence. One is an organizer that is essentially concerned with defining the boundaries between inner and outer experiences. It may be presumed to have its origins long before the onset of language acquisition or symbol formation. This activity we call shielding. The second organizer is concerned with reflectiveness and those ego functions preparing for the capacity to generate language. It refers to a meaning-generating process and specifically to the transformation of image into words. This we have termed *transforming activity.*[2]

Shielding and transforming activities are present in any associative narrative. Shielding is concerned with context setting. Any time there is a major assault on the integrity of the psychic apparatus, shielding is likely to become dominant in the communicative situation. If there is only shielding, that is, a preoccupation with boundary definition, then transformational activity is likely to be curtailed, and with it comes a paucity of new symbolic representations. Optimally, shielding and transforming activities oscillate in rhythmic fashion throughout the associative process. This chapter is concerned only with the shielding process.

[2]In our previous work (Freedman & Bucci, 1981) we have called this the "contrasting response." The term *transforming* adheres more closely to the presumed process, that is, transforming image into words.

The notion of associative organizers emerged from the cumulative data of our studies of recorded behavior over the past 15 years. Throughout these studies, largely paradigmatic, we focused both on the verbal channel, developing codes for the analysis of language behavior (Steingart & Freedman, 1975), as well as on the nonverbal channel, developing codifications of kinetic behavior (Freedman, 1971). The concepts that guided our understanding of this work went through successive stages of gradually progressing understanding. It is through such a process that the concept of shielding has emerged. Observations in the clinical domain led to certain abstract ideas and inferences that could then be tested or validated in the same, as well as more controlled domains. Confirmation of these ideas and inferences by repeated observation led to the elaboration of the notion of associative organizers and the sharpening of the concept of shielding upon which this notion is based.

In developing the concept of shielding as an associative organizer, we shall describe our initial uncovering of the clinical phenomena that occurred in the context of relatively naïve clinical observations. We shall then describe a program of systematic empirical research linking shielding to various conditions of psychological functioning. Finally, we shall turn to issues of clinical application showing how the appearance of shielding in the treatment setting may serve as a guide to therapeutic intervention.

THE PHASE OF "NAÏVE" OBSERVATION

Our first observations of certain regularities in patients' tactile self-stimulation during clinical interviews led us to postulate that such behavior has relevance as a regulator of tensional and excitement states generated during the clinical interaction. We tested this assumption by introducing some quasi-experimental conditions into our clinical observations. Essentially, we organized the clinical interviews into two distinct phases defined by the kind of restrictions we imposed upon the patients' thought processes. In one phase, labeled "secondary process," patients were directed to talk about their daily work routines, that is, the real events of daily life. In the second phase of the interview, labeled "primary process," the patient was encouraged to say whatever came to mind, that is, the moment-by-moment flow of thoughts regardless of the logic or meaningfulness of these thoughts. These earliest observations were made on acutely psychotic patients and so the material often included delusional or hallucinatory experiences, as well as many other evidences of disordered thought. Our intent was to create conditions that would promote either unrestricted, primitive, "drive dominated" thought, or tightly restricted "reality oriented" thought. We shall illustrate our findings by describing the behavior of one patient as he underwent these two conditions of clinical communication.

This man was a psychiatric patient suffering from a persecutory delusion that he had been poisoned by a mysterious gas while sitting on a park bench with his girl friend. He felt harassed by an unknown persecutor. He was eager to tell his story, and did so in the expressive primary-process phase of the interview. He was then asked to talk about reality, his daily activities on the ward.

The videotaped record emerging from these two conditions presented dramatically different pictures (Freedman & Hoffman, 1967). The differences could be discerned even with the sound off. In the primary-process phase, the man looked alive. His hands moved in rhythmic motions with the beat quality of his speech, his head moving in unison, his gaze fixed on the interviewer and his voice volume up. In the secondary process phase, his gaze was downcast, speech was a monotone, his head was motionless, and both hands engaged in continuous, repetitive self-stimulation, sometimes caressing his thighs and sometimes rubbing each other while resting in his lap. It is this particular form of continuous self-stimulation that is of primary interest here.

The self-stimulation observed during the secondary-process phase appeared to be far from a random motor discharge phenomenon. Indeed, it had a very specific structural organization linked to the formal requirements of a very specific type of thought processing. The form of body-focused activity this patient exhibited was a special type of tactile self-stimulation. It was bilateral and continuous over long durations of the communicative monologue. It cut across sentence structure and articulatory boundaries and had no relation to the rhythm or content of the verbal flow. In appearance, we seem to be observing two simultaneous but discrete channels of behavior, one having to do with the verbal articulation of thought, the other with continuous and repetitive innervation of the skin surface. The relationship between such continuous bilateral self-stimulation and the restricted and limiting conditions of reality-oriented secondary-process thought suggested that a specific function was served by such self-stimulation: Namely, through tactile or kinesthetic innervation, the patient was able to create a shield to ward off interference from both inner and external experience.

Shielding activity, as a continuous, somatosensory innervation independent of speech,is a reliable and quantifiable observation that has been noted by other observers. Krout (1954) labels such motor phenomena accompanying speech as autistic. Sainsbury (1958) labels it noncommunicative gesturing and manipulating behavior. Ekman and Friesen (1969) refer to self-adaptors, and Mahl (1968), like Krout, conceives of such behaviors as autistic.

These observers of self-stimulation during communication all view such behavior as nonfunctional with respect to communicated thought. We, on the other hand, view continuous self-stimulation during speech as a reflection of an inner process that the speaker engages in simultaneous with, but relatively independent of, his efforts to encode his thoughts into meaningful symbols. Thus, what is at issue in the various accounts of this form of self-stimulation during communicative activity is not the phenomenon itself, but its conceptualization. Observations of this patient had the quality of discovery for us and led to the hypothesis that continuous self-stimulation reflects the patient's effort to create a shield against internal and external interference. It is toward the validation of this hypothesis that our empirical work was directed.

THE PHASE OF CLINICAL GENERALIZATION

Moving from the level of naïve observation to that of systematic clinical studies, our investigative focus shifted to establishing a relationship between shielding and a variety of clinical states and patient variables. We began with studies of psychopathological conditions in order to document our initial observations of the aforementioned patient. We then extended our observations to the more normal range of communicative behavior both within and out of the clinical setting as a way of testing the limits of the shielding phenomena. Next, we turned to a direct experimental manipulation in the laboratory to specify the function of shielding activity. And, finally, we returned to clinical issues both in psychopathology and in the treatment situation to reinterpret the function of shielding as an associative organizer.

SHIELDING IN PSYCHOPATHOLOGY

In a general sense, shielding behavior has been observed to be significantly related to clinical states or conditions in which the integrity of the psychic apparatus is at stake (Grand, 1982). Shielding could be inferred from the "autoerotic" rhythmic bodyrocking and pelvic thrusting noted in the more pathologically disturbed nursery school children observed by Spitz and Wolf (1950). Head banging and other bizarre forms of self-stimulation noted by Mittleman (1960) in schizophrenic children may also serve a similar function. Furthermore, the revival of early autonomous kinesthetic, thermal, acoustic, vestibular, and tactile sensitivities in regressive states has been noted by Bak (1939), Greenacre (1953), Jacobson (1954), Rado (1956), and Peto (1959). According to Peto (1959) as well as Grand (1982), these experiences serve as sensory precursors to the establishment of the body ego, and are important in maintaining the integrity of the psychic apparatus. Finally, shielding could be noted in the defensive physical distancing seen in children whose mothering fails to provide sufficient somatosensory stimulation, and thereby interferes with psychic development. According to Shevrin and Toussing (1965), regressive states often "defensively raise thresholds for all stimuli emanating from the environment or from inside the body." All these forms of behavior have the common property of continuously altering the sensory conditions of experience by means of self-regulatory maneuvers on the body surface. All appear to serve the function of maintaining the integrity of the mental apparatus by screening out intrusive and, in more primitive and regressive states, overwhelming innervation, and all play an important role in the integrity of the cohesive body self.

These ideas were corroborated in successive studies on chronic schizophrenic patients. In the first study by Grand, Freedman, Steingart, and Buchwald (1975a), chronic schizophrenic outpatients were observed in audiovideo-recorded clinical interviews, and the prevalence of shielding activity during verbalization was evaluated. The patients were classified according to whether or not they exhibited characteristics of "proneness to rehospitalization" (Freedman, Rosen, Engelhardt, & Margolis, 1967). That is, we had previously, on the basis of interviews with significant relatives, determined whether the patient's mode of social living was typically isolated, with the maintenance of inhibitory controls over expressions of aggression (prone), or whether they were nonisolated and often aggressive in their mode of social participation (nonprone). The two groups revealed very different forms of communicative organization, both kinetic and verbal, in their recorded clinical interviews. It was the chronic isolated and aggressively inhibited patient rather than those overtly aggressive and nonprone to hospitalization who exhibited extremely high levels of self-stimulation in the form of continuous finger/hand motions, very much as in our initial observations of the psychiatric patient. Not only was shielding activity high in such patients, but when we considered the organization of language (syntactic forms) we noted that these patients exhibited significantly higher rates of simple narrative language construction. This category of language construction is a form of speech reflecting a narrow breadth of experiencing the world (Steingart & Freedman, 1975). In contrast, the more overtly aggressive chronic patients exhibited more complex language structure, which is consistent with a broader breadth of experiencing the world. Thus, those patients who shield in the motor realm also restrict the range of ideas to which they attend, whereas patients who do not shield motorically seem to allow a more varied range of ideation into consciousness.

In a subsequent study by Grand, Marcos, Freedman, and Barroso (1977), our goal was to establish shielding behavior as an attribute of chronic schizophrenia and to reject the alternative hypothesis that shielding might be a function of cultural or linguistic factors in the patient's background. A study of bilingual patients was initiated to clarify the issue. Audiovideo-recorded interviews were conducted in patients' native "dominant" language (Spanish) and second language (English). Patients' psychopathology was evaluated by psychiatrists using the Brief Psychiatric Rating Scale. The recorded behavior was scored for an entire range of kinetic activity using our previously developed scoring system.[3]

In the interview condition where nondominant language prevailed, the patient under a linguistic handicap showed an upsurge of communicative, object-focused gestures, that is, those movements reflecting a representational and articulatory process. However, when the patients communicated in their dominant, native language, shielding activity clearly emerged.

There appears to be convincing evidence from various sources that chronic schizophrenic patients, even under nonstressful and supportive communicative conditions, will exhibit a high incidence of shielding activity when they are asked to engage in a communicative interchange. Thus, our studies are consistent with data relating similar kinds of body

[3]As part of our general research objective, the more comprehensive analysis of nonverbal behavior entailed two broad categories of kinetic activity—object-focused movements and body-focused movements. Object-focused movements are speech-related communicative gestures. There are various levels of object-focused movements, those in which movements are subordinated to speech (speech primacy), and those in which content is represented in gesture (representational movements). There are also various levels of body-focused activity involving the touching of the hands on the body surface. These range from discrete, instrumental activity to continuous forms of self-stimulation. Shielding refers to a particular subcategory of body-focused activity (Freedman, 1971).

touching to psychopathological stress (Ekman & Friesen, 1969), as well as data published by Condon and Ogston (1966) showing asynchronies, such as occur during shielding, in the speech–body movement relation in schizophrenic patients.

SHIELDING AND FIELD DEPENDENCY

Is the shielding response peculiar to chronic schizophrenia, or is it an adaptation to be found in nonpsychopathological groups struggling with boundary problems? The possibility that shielding may be a transitory adaptation in any person when the integrity of the self–nonself barrier is threatened was evaluated. Here, we move closer to the idea of shielding as an associative organizer, that is, a momentary adaptation that serves to structure the associative process at points of stress.

College students participated in an audiovideo-recorded associative monologue, a situation in which they were asked to talk for 5 min about whatever came to mind (Freedman, O'Hanlon, Oltman, & Witkin, 1972). Subsequently, language organization and shielding behavior were evaluated. Subjects were classified as high or low on the dimension of field dependency as measured by the Embedded Figures Test (EFT) and the Rod and Frame Test (Witkin & Goodenough, 1976). The measure of field dependency served as a way of defining the individual's capacity to delineate not only the me from the not-me, but, according to Witkin, the capacity for defining an individual's separate identity. Both field independent and field dependent subjects participated in two kinds of communicative contexts, one a "warm" empathic condition, the other a "cold" and detached condition.

The important finding that emerged in this study was that shielding occurred specifically in the field dependent individuals and only under the cold, stressful interview condition. Considering that the field dependent person is particularly vulnerable to the stress of field forces, that is, extremely sensitive to the coercion of environmental pressures, the appearance of shielding in field dependent subjects under interview stress conditions affirms our notion that shielding functions to ward off intrusive and interfering stimulation. There was an additional aspect to these observations when the relationship between shielding and the verbalization of threatening content was appraised (Freedman, Blass, Rifkin, & Quitkin, 1973). Although the field dependent subjects did adapt to the environmental demands, that is, produced an associative monologue, they had greater difficulty in articulating aggressive content in their verbalization. Thus, one might say that they screened out not only unpleasant stimulation from without (the rejecting interviewer), but stressful experiences from within. In this sense, shielding was probably a transitory adaptation to a nonempathic interview situation.

The appearance of shielding among less differentiated subjects exposed to environmental stress bears some similarity to shielding in chronic schizophrenic patients. There, however, the shielding activity was a characteristic of the patient regardless of the communicative situation generally, whether it was stressful or not. The field dependent subjects, on the other hand, did not exhibit shielding when interviewed by a nonthreatening warm interviewer, but only under the cold rejecting interview condition. In light of Witkin's view that the field dependency dimension sensitively taps the subjects' capacity to differentiate inner versus outer, or self versus nonself experience, the data presented here, taken together with the findings among chronic schizophrenic patients, suggest that shielding may be operating as a boundary-defining behavior. When boundaries are severely impaired, as in chronic schizophrenia, shielding is pervasive even under the slightest threat of merging. When boundaries are already established, but perhaps less firm than might be optimal for managing environmental stress, shielding again occurs, but now as a momentary way of maintaining the integrity of the boundary between self and other under the transitory stress of the cold interview condition.

SHIELDING AND DEVELOPMENT

Next, we directed our attention to developmental considerations. Theoretically, we could postulate antecedents of shielding phenomena in early infancy where the meeting of the hands at midline may be viewed as a prototype of shielding. Indeed, this phenomenon has been suggested to be a marker in which the infant delineates the me from the not-me. Although we did not study infants, we wanted to determine the point in development at which a child is likely to take recourse to continuous self-stimulation during verbalization in order to see whether there is an age saliency for shielding activity.

In a developmental study (Freedman, 1977), children 4 to 16 were observed under videotaped conditions while they were performing simple cognitive tasks. One task was vocabulary definition. We were interested in the typical organization of bodily activity at each age level as the child attempted to define a word.

An orderly progression of patterns of self-regulation was observed across the age spectrum. However, shielding, as we have defined it in our studies of adult subjects, peaked at ages 8 and 10—the latency years, and at later age levels, other forms of bodily activity were more prevalent. What is the meaning of shielding during latency?

If we grant that latency-age children are most concerned with the developmentally appropriate and phase-specific task of repression and transformation of direct drive expression into age-relevant skill, the appearance of shielding at this age level is consistent with the need to maintain the boundaries between internal and external demands. Although shielding may have its onset in the earliest phase of development before boundaries have been established, it is called upon at the very point when these boundaries are being threatened in these children, or, for that matter, throughout the life cycle.

SHIELDING AS AN EFFORT AT REGULATION AND COMPENSATION

So far, we have shown that shielding occurs among individuals with known vulnerabilities in ego boundaries, sometimes more intense, sometimes in transitory fashion. It has been suggested that the individual's self-created experience of sensory innervation during discourse *functions* as a shield to screen out unwanted information. Hence, it performs a regulating function. However, an alternative interpretation would be feasible. One might say that shielding is simply a discharge or an expressive response that arises under conditions of stress.

The Stroop Color-Word Interference Test provides a way to ascertain the extent to which a person is able to screen out unwanted or interfering thoughts while sustaining attention on task requirements. Grand, Freedman, Steingart, and Buchwald (1975b) studied the group of schizophrenic patients referred to earlier, but now evaluating the relation between language and Stroop performance. There was a negative correlation between Stroop performance and language construction: the greater the patients' difficulty in warding off interfering cues on the Stroop, the more restricted their language. Since a restrictive use of language had previously been associated with high levels of shielding, the possibility is suggested that shielding functions as a compensatory mechanism warding off interference.

This functional interpretation has since been tested in a series of direct laboratory studies. In these studies, subjects were given the Stroop task but this time their hand movements during Stroop performance were videotaped. In the Stroop task, names of colors are printed in noncorresponding colored inks, and the subject is asked to disregard (suppress) the impact of the word and name the color of the ink. In this study it was possible to observe whether or how the subject shielded while coping with the stress of color–word interference.

Twelve-year-olds were observed under these stressful conditions (Barroso, Freedman, Grand, & van Meel, 1978). The level of shielding was exceedingly high throughout performance and for all subjects. It was higher than that observed for vocabulary definition. This observation, again, demonstrates that the demand to stay on the beam and exclude extraneous cues creates sensorimotor tension and hence, shielding. Yet the observation is consistent with the discharge interpretation. If shielding is to be an organizer and regulator as well, then it should bear a relationship to performance. In a further study (Barroso, Freedman, & Grand, 1980), the relationship between the upsurge of bilateral self-stimulations (i.e., shielding) and the number of errors in color naming was ascertained, and it was found that the higher the level of shielding activity the fewer the number of insertion errors (i.e., complete or partial color-naming responses interposed between items on the task). Here, then, is a demonstration of shielding as a process that regulates attention and functions to improve performance.

A systematic testing of this regulatory interpretation was carried out by Grand, Breslow, and Freedman (1980), who studied the relationship between self-stimulation and Stroop color–word errors under normal and hearing-occluded conditions. It was hypothesized that a reduction in the auditory feedback from the vocal-motor performance on the Stroop task would significantly disturb the perceptual conditions for effective cognitive processing. Effective performance would require compensatory stimulation in order to maintain the attentional focus on the task requirements (i.e., to name colors and disregard the incongruous words). The findings indicated that not only did self-stimulation increase when voice feedback was reduced, but this increase was associated with a reduction in the specific type of color–word performance errors determined earlier (i.e., insertion errors). Furthermore, a test of individual differences in continuous self-stimulation revealed that high kinesic shielders made significantly fewer errors in performance than did low shielders. Our understanding of these results is that not only does continuous self-stimulation serve to shield cognitive performance from the intensive stress of the interfering words, but that shielding has adaptive and compensatory significance in regard to aiding the self-editing process when the normal cues for this process given by auditory feedback are reduced.

In this sense, the experimental studies provide a conceptual base for an understanding of how shielding activity may sustain the division of psychological experiences emanating from within or without.

PSYCHOPATHOLOGY REVISITED

The clinical observations derived from our empirical studies have led to several conclusions: shielding occurs primarily in individuals whose body boundaries are vulnerable or undergoing some current threat; it occurs as a response to a severe strain on attention deployment; it is most prevalent in individuals with immature or less differeniated ego organization; and it serves as a compensatory regulator of attentional processes, thereby augmenting boundary integrity. Grand (1977) had already suggested such a view with respect to high rates of self-stimulation observed during communicative activity in congenitally blind and chronic schizophrenic patients, two populations that are each, in their own ways, deprived of adequate stimulus nutriment for maintaining the integrity of ego boundaries. As a compensatory regulator, shielding could provide the stimulus nutriment necessary for the optimal titration of a faltering sense of body boundedness.

More recently, Grand (1982) reviewed a wide range of clinical, theoretical, and empirical literature on schizophrenic pathology, all supporting the cogency of this view of shielding as a compensatory regulator of disturbances in the articulation of the body ego. From this conclusion, Grand developed a model for schizophrenic psychopathology that places the core disturbance in this illness in the failure to articulate a cohesive body ego, and posits a key compensatory mechanism to reside in the shielding response. As a result of a very early failure of the vestibular system to adequately integrate the multiple sensory inputs from coenesthetic and gravitational experiences that form the basis for orientation and focus in the experience of reality, body cohesiveness or boundedness does not develop. Hence, clear distinction between inner and outer experience, as well as self and object representations do not emerge. However, compensatory mechanisms that shield the mental apparatus as well as support impaired body ego integration do apparently develop in the areas of somatosensory, proprioceptive, and thermal experience. It is such shielding mechanisms that may be useful for sustaining the integrity of the schizophrenic patient's mental ego and prevent its complete fragmentation during regressive episodes in this disorder.

EXTENSION TO THE TREATMENT PROCESS

Shielding, as we have held throughout, is not just a signpost of psychopathology, but an adaptive boundary-defining maneuver in all phases of the life cycle. Nowhere is this demonstrated more convincingly than in the associative process of any patient in the treatment setting or in the listening process of the therapist or analyst. One of the primary demands made upon the listening therapist is the regulation of attention. Under conditions of reduced motility, the listening therapist often comes under stress. In therapists' efforts to receive, and at the same time ward off interferences, they are likely to invoke some of the earliest sensorimotor adaptations, one of these being shielding.

In an earlier study, we observed that shielding appeared to be associated with a state of receptive attention among a group of college students engaged in a conflict resolution discussion (Freedman, Barroso, Bucci, & Grand, 1978). Following these observations, we applied our analysis of nonverbal behavior to a group of medical students conducting an interview. In a group termed "rhythmically attuned listeners," we identified an optimal organization of clinical performance where there was transitory shielding followed by rhythmic gestures at the transition from listening to intervention. There was also a second group termed "discursive listeners" who used less than optimal forms of interviewing, where the usual rhythmicity was absent and shielding was used throughout the interview.[4]

These findings have relevance to our understanding of psychotherapeutic and psychoanalytic listening. Therefore we set out to evaluate the impact of different transference states in the patient on the therapist's listening stance as observed on the nonverbal level. The design was that of a single therapist's caseload. The impact of two patients on "Dr. S's" listening behavior was traced. Although the patients were similar in regard to age, sex, and diagnosis, they differed in transference manifestations reflected on Gill and Hoffman's (Gill, 1982) scoring system. One patient showed a high level of transference manifestations, the other a great difficulty in invoking the transference relationship. It was this latter patient who induced continuous shielding in the therapist. With the former patient, the rhythmicity that was posited for optimal listening was maintained.[5] Thus, the most attuned listener can become a discursive listener under appropriate transference stress.

[4]Ongoing work with Dr. Felix Barroso.
[5]Ongoing work with Dr. Michael Berzofsky.

Listening is clearly a condition essential for treatment. The studies have given rise to a conceptualization of optimal listening in psychoanalytic therapy (Freedman, 1983). In listening, there is a phase of receiving and a phase of restructuring. Optimal listening involves a rhythmic alternation between the two, and reveals an orderly progression observable on the sensorimotor level. Continuous shielding across communication transitions would then suggest a disruption of the rhythmicity of listening. Although such loss in rhythmicity may be viewed as a visible manifestation of countertransference, it may also be seen as an adaptive response in the therapist serving to maintain the integrity of his or her psychological boundaries.

Having identified a phenomenon in the realm of psychopathological communication, we examined its generality by exploring a wide range of normal communicative behavior in both children and adults. We were then able to adduce a specific functional and adaptive point of view with respect to its use in both normal and psychopathological communication. Finally, we extended this understanding to an examination of the treatment process itself. We shall now consider the shielding process in the clinical setting.

Shielding in Clinical Practice

From the foregoing review of our clinical-experimental research, we can draw two major inferences about the appearance of shielding during the associative process that have relevance for clinical work. First, when a clinician observes shielding behavior in his patient, he is witness to a psychological crisis in intrapsychic function. That is, he is confronted with a breakdown in the smooth transformation of thought into its verbal representation, a breakdown in a process essential to the work of psychoanalytic psychotherapy. Theoretically, then, shielding can be understood as the de-differentiation of psychic structure, or as a genetically regressive response, or as an indicator of the insufficiency of the ego to master conflict, or as the upsurge of overwhelming tension. Thus, from the various metapsychological perspectives shielding represents a crisis in the associative process.

A second inference that can be drawn from our research is that when the clinician observes shielding in his patient, he is also witnessing an effort in the direction of repair, that is, an automatic self-regulatory effort whose function it is to restore the boundary-defining process. By implicating touch and the experience of bodily stimulation, shielding generates a sense of bodily integrity, a heightening of the boundaries between self and object, as well as between inner and outer experience. As such, it sets in motion an adaptive process leading to self-repair. The body, the body ego, and the ego—these are concepts that reflect an important developmental line that is recapitulated in the microgenetic sequence from self-touching to verbalization. In this sense, shielding is structure building.

Shielding, then, presents the clinician with a paradoxical phenomenon. On the one hand, it is a signpost of de-differentiation; on the other hand, of self-repair. What, then, is the clinical task when observing shielding during the associative flow? When a clinician observes shielding in his patient, he must make a judgment of emphasis—does shielding reflect a crisis, or an emerging capacity for self-repair? We state this in relative terms since both aspects are present in all occurrences of shielding, and it is the balance between the two that determines the clinician's response. In our view, this judgment of the relative importance of either crisis or self-repair is based upon the intensity of the shielding process, that is, the degree to which it is pervasive and invariant.

Much of the data reviewed here have been obtained from patients with chronic and pervasive psychopathology (schizophrenia). The predominant form of self-stimulation in such states is a continuous and pervasive one, whether tactile, kinesthetic, or vestibular. We understand this to signify not only the presence of severe boundary disturbance, but of an early developmental failure as well. Shielding in such states is a reflection of a deficiency in the earliest infant–mother dyadic relationship, an indicator of a failure in those conditions that make possible boundary formation and more adequate internalization, as this has been described by Kohut (1971), Gedo (1979), and others. Khan's (1974) position on the role of cumulative trauma, which formulates the concept of a shield as central, is at the heart of our argument. Cumulative trauma, according to Khan, ensues from a failure by the mother to provide an adequate shield for her child that would enable structure building. Thus, in regressive states, it is the task of the therapist to create a compensatory shield. Patients in clinical states of extreme and pervasive pathology, who exhibit shielding, demonstrate little capacity to process stimuli from within (i.e., imagery and symbol formation is barren). They are preoccupied with warding off projected intrusions from the external world. Under such circumstances, the therapeutic stance should be directed toward providing the patient with a firm sense of the therapist's presence. Whether through resonating, mirroring, or structuring of reality, the patient must be helped to reconstitute a boundary between self and other, inner and outer experience. In this way, the therapist provides a compensatory and supplementary shield against the experience of external threat. Shielding, here, is a palpable expression of the underlying deficiency in psychic structure, and requires interventions appropriate to such deficit conditions.

Turning now to shielding under "low intensity" conditions, we have noted an upsurge of self-stimulation in a variety of stressful situations. Of greatest interest to the clinician is the transitory appearance of shielding either in the patient or the therapist. In patients, it may occur during prolonged silences as a manifestation of a particular transference crisis. In the listening therapist, it is likely to be a signpost of a countertransference reaction. In both instances, it signifies an inability to sustain unacceptable imagery. Shielding in such instances reflects a disruption of the patient's adherence to the basic rule or of the therapist's evenly suspended attention. Here, the stance is different from that of shielding under pervasive pathology. We suggest that transitory shielding is structure building, and therefore should be allowed free play so that spontaneous regeneration of the verbalization process can take place. Freud (1923/1961) noted that touch is adaptive insofar as it stimulates perception of inner states, that is, makes unconscious experience move toward consciousness. From our own perspective, this suggests that therapeutic silence in the face of transitory self-touching would allow a process of image formation and verbalization to take hold.

There is one final clinical implication for an understanding of shielding. So far, our discussion has been very data linked. Shielding under high-intensity "deficit" conditions calls for an active establishment of a facilitating environment; whereas shielding under transitory conditions of internal stress calls for the spontaneous regeneration of the symbolizing process. Yet shielding may also be viewed as a metaphor, and here we would like to go beyond the empirical observations that have been presented.

Shielding is not always visibly present in self-stimulating behavior. Nevertheless, we often see examples of shielding in people who surround themselves with somatosensory experiences throughout the day. These forms of somatosensory experience have a structure identical to the visible shielding response (although the sensory modality may not always involve touch). The person who walks around with a radio and earphones, surrounding himself with continuous sound, the patient who comes to the treatment room wanting the air-conditioning on because of the hum, the person who engages in overeating, or who maintains a pace of restless activity and vigilance, creating conditions of psychic numbness—these and many others are metaphorical examples of shielding. They have the structure of continuous sensory innervation. In a previous discussion of shielding as a metaphorical phenomenon, we liken shielding to a metaphor for a beleaguered state, a respite under conditions of impending fragmentation creating an at least momentary sense of cohesiveness (Freedman, 1980). As a very minimum, these maneuvers constitute a defense against the impingement of the external world and also a numbing of the internal world. When a person is asked, "What do you experience?" the answer is usually, "I don't know. Nothing at all." Imagery and symbolization tend to be temporarily lost.

The clinical stance when there is evidence of metaphorical shielding in our patients would be identical to that which we have noted from our empirical observations. In a deficit state, we are faced with a crisis condition and the therapist will feel impelled to provide the appropriate environmental supplies that may serve to reestablish the cohesiveness of boundaries. And, as we discussed above, in less intense pathological states we are called upon to maintain the usual condition of psychoanalytic neutrality so that the regeneration process may take its course. In light of our psychoanalytic goal of facilitating meaning construction, the question we ask ourselves is whether meaning has been paralyzed or only temporarily suspended.

In concluding, the clinician is a constant witness to the efficacy of the motor system to define that which is inner and that which is outer, to protect that which is vital, to regulate that which is ominous, and to move toward the attainment of new meanings. Shielding assumes a significant place as one of the earliest somatosensory defense mechanisms essential to the building of psychic structure. In this sense, the shielding process that was recognized in Freud's writings in the 1890s, and persisted in his writings into the 1930s, has a very practical and clinical relevance in the 1980s.

REFERENCES

Bak, R. C. Regression of ego orientation and libido in schizophrenia. *International Journal of Psychoanalysis*, 1939, *20*, 64–71.

Barroso, F., Freedman, N., Grand, S., & van Meel, J. The evocation of two types of hand movements in information processing. *Journal of Experimental Psychology*, 1978, *4*, 321–329.

Barroso, F., Freedman, N., & Grand, S. Self-touching, performance, and attentional processes. *Perceptual and Motor Skills*, 1980, *50*, 1083–1089.

Condon, W. S., & Ogston, M. B. Sound film analysis of normal and pathological behavior patterns. *Journal of Nervous and Mental Disease*, 1966, *143*, 338–347.

Ekman, P., & Friesen, W. The repertoire of nonverbal behavior: Categories, origins, usage, and coding. *Semiotica*, 1969, *1*, 49–98.

Freedman, N. The analysis of movement behavior during the clinical interview. In A. Siegman & B. Pope (Eds.), *Studies in dyadic communication*. New York: Pergamon Press, 1971.

Freedman, N. Hands, words, and mind: On the structuralization of body movements during discourse. In N. Freedman & S. Grand (Eds.), *Communicative structures and psychic structures*. New York: Plenum Press, 1977.

Freedman, N. On splitting and its resolution. *Psychoanalysis and Contemporary Thought*, 1980, *3*, 237–266.

Freedman, N. On psychoanalytic listening: The construction, paralysis, and reconstruction of meaning. *Psychoanalysis and Contemporary Thought*, 1983.

Freedman, N. On depression: The paralysis, annihilation, and reconstruction of meaning. In J. Masling (Ed.), *Empirical studies of psychoanalytic theories* (Vol. 2). Hillsdale, N.J.: Analytic Press, in press.

Freedman, N., & Bucci, W. On kinetic filtering in associative monologue. *Semiotua*, 1981, *34*, 225–249.

Freedman, N., & Hoffman, S. P. Kinetic behavior in altered clinical states: An approach to the objective analysis of motor behavior during clinical interviews. *Perceptual and Motor Skills*, 1967, *24*, 525–539.

Freedman, N., Rosen, B., Engelhardt, D. M., & Margolis, R. Prediction of psychiatric hospitalization: 1. The measurement of hospitalization proneness. *Journal of Abnormal Psychology*, 1967, *72*, 468–477.

Freedman, N., O'Hanlon, J., Oltman, P., & Witkin, H. The imprint of psychological differentiation on kinetic behavior in varying communicative contexts. *Journal of Abnormal Psychology*, 1972, *79*, 239–258.

Freedman, N., Blass, T., Rifkin, A., & Quitkin, F. Body movements and the verbal encoding of aggressive affect. *Journal of Personality and Social Psychology*, 1973, *26*, 72–85.

Freedman, N., Barroso, F., Bucci, W., & Grand, S. The bodily manifestations of listening. *Psychoanalysis and Contemporary Thought*, 1978, *1*, 157–194.

Freud, S. The interpretation of dreams. In J. Strachey, (Ed. and Trans.), *Standard Edition of the complete psychological works of Sigmund Freud* (Vols. 4 and 5). London: Hogarth Press, 1953. (Originally published, 1900)

Freud, S. Beyond the pleasure principle. *The standard edition* (Vol. 18). London: Hogarth Press, 1955. (Originally published, 1920)

Freud, S. Instincts and their vicissitudes. *The standard edition* (Vol. 14). London: Hogarth Press, 1957. (Originally published, 1915)

Freud, S. Inhibitions, symptoms and anxiety. *The standard edition* (Vol. 20). London: Hogarth Press, 1959. (Originally published, 1926)

Freud, S. The ego and the id. *The standard edition* (Vol. 19). London: Hogarth Press, 1961. (Originally published, 1923)

Freud, S. A note upon the "mystic writing pad." *The standard edition* (Vol. 19). London: Hogarth Press, 1961. (Originally published, 1925)

Freud, S. An outline of psycho-analysis. *The standard edition* (Vol. 23). London: Hogarth Press, 1964. (Originally published, 1940)

Freud, S. Project for a scientific psychology. *The standard edition* (Vol. 1). London: Hogarth Press, 1966. (Originally published, 1895)

Gedo, J. E. *Beyond interpretation: Toward a revised theory for psychoanalysis*. New York: International Universities Press, 1979.

Gill, M. M. *Analysis of transference, Vol. 1: Theory and technique*. New York: International Universities Press, 1982. (Psychological Issues Monograph 53)

Grand, S. On hand movements during speech: Studies on the role of self-stimulation in communication under conditions of psychopathology, sensory deficit, and bilingualism. In N. Freedman & S. Grand (Eds.), *Communicative structures and psychic structures*. New York: Plenum Press, 1977.

Grand, S. The body and its boundaries: A psychoanalytic view of cognitive process disturbances in schizophrenia. *International Review of Psychoanalysis*, 1982, *9*, 327–342.

Grand, S., Freedman, N., Steingart, I., & Buchwald, C. Communicative behavior in schizophrenia: The relation of adaptive styles to kinetic and linguistic aspects of interview behavior. *Journal of Nervous and Mental Disease*, 1975, *161*, 293–306. (a)

Grand, S., Freedman, N., Steingart, I., & Buchwald, C. Organization of language behavior in chronic schizophrenia. *Journal of Abnormal Psychology*, 1975, *84*, 621–628. (b)

Grand, S., Marcos, L. R., Freedman, N., & Barroso, F. Relation of psychopathology and bilingualism to kinesic aspects of interview behavior in schizophrenia. *Journal of Abnormal Psychology*, 1977, *86*, 492–500.

Grand, S., Breslow, R., & Freedman, N. On the role of reduced auditory feedback and kinesic self-stimulation during Stroop Color-Word performance. *Journal of Personality*, 1980, *48*, 173–189.

Greenacre, P. Certain relationships between fetishism and the faulty development of the body image. *Psychoanalytic Study of the Child*, 1953, *8*, 79–98.

Jacobson, E. Transference problems in the psychoanalytic treatment of severely depressive patients. *Journal of the American Psychoanalytic Association*, 1954, *2*, 595–606.

Khan, M. *The privacy of the self*. New York: International Universities Press, 1974.

Kohut, H. *The analysis of the self: A systematic approach to the psychoanalytic treatment of narcissistic personality disorders*. New York: International Universities Press, 1971.

Krout, M. H. An experimental attempt to produce unconscious manual symbolic movements. *Journal of General Psychology*, 1954, *51*, 94–120.

Mahl, G. F. Gestures and body movements in interviews. In J. Shlien (Ed.), *Research in psychotherapy* (Vol. 3). Washington, D.C.: American Psychological Association, 1968.

Mittleman, B. Intrauterine and early infantile motility. *Psychoanalytic Study of the Child*, 1960, *15*, 104–127.

Peto, A. Body image and archaic thinking. *International Journal of Psychoanalysis*, 1959, *40*, 223–231.

Rado, S. Hedonic control, action-self, and the depressive spell. *Psychoanalysis of behavior: Collected papers of Sandor Rado* (Vol. 1). New York: Grune & Stratton, 1956.

Sainsbury, P. Gestural movement during psychiatric interview. *Psychosomatic Medicine*, 1958, *17*, 458–469.

Shevrin, H., & Toussing, P. W. Vicissitudes of the need for tactile stimulation in instinctual development. *Psychoanalytic Study of the Child*, 1965, *20*, 310–339.

Spitz, R. A. *The first year of life: A psychoanalytic study of normal and deviant development of object relations*. New York: International Universities Press, 1965.

Spitz, R. A., & Wolf, K. M. Autoerotism: Some empirical findings and hypotheses on three of its manifestaitons in the first year of life. *Psychoanalytic Study of the Child*, 1950, *11*, 85–120.

Steingart, I., & Freedman, N. The organization of body-focused kinesic behavior and language construction in schizophrenic and depressed states. *Psychoanalysis and Contemporary Science*, 1975, *4*, 423–450.

Witkin, H. A., & Goodenough, D. R. *Field dependency revisited*. Princeton, N.J.: Educational Testing Service, 1976. (Research bulletin)

17

Fantasy

KENNETH S. POPE

DEFINITION AND OVERVIEW

So great is the power of fantasy that few readers, if any, will be able to complete this chapter with their minds entirely concentrated upon the text. Rare are the stretches of time when our minds remain unwaveringly upon the task at hand in the here and now. The mind wanders. Even as we work away at a given assignment, we mentally dart ahead to tonight's dinner, to tomorrow's deadline, to next week's visit to the doctor. We become aware of how our body feels in the chair, and that reminds us of how we feel with our new jogging program, of our plan to buy some new furniture, of the need to turn up the furnace. Fragmentary images, distant recollections, the jingle from an advertisement, and habitual daydreams vie for our attention.

Our moment-to-moment flow of awareness is often filled with such fantasy material. It is puzzling that only recently have careful, systematic, scientific studies of fantasy been undertaken. Perhaps part of the problem has been the difficulty of definition (see Pope, 1983b). Some fantasy material is best characterized as daydreams, those stories we spin for ourselves. At other times we focus on imagery, which may strike us as attempts at symbolic description (Pylyshyn, 1973), as a way of perceiving in the absence of the object (Richardson, 1969), or as a planned construction (Neisser, 1972). At still other times, we emphasize

KENNETH S. POPE • 11747 Sunset Boulevard, Suite 226, Los Angeles, California 90049.

a process called imagination: "imagining a complex object or event in concrete symbols as images, whether or not the object or event exists; or the symbols or images themselves: for example, a daydream" (English & English, 1958). And this brings us back to daydreams again.

Two definitions have been most influential. One, by Singer (1966), describes mental activity that is not preoccupied with the current, immediate environment, with the here and now:

> Generally the word is used to mean a shift of attention away from an ongoing physical or mental task or from a perceptual response to some external stimulation towards a response to some internal stimulus. (p. 3)

The second, by Klinger (1971), describes mental activity (or rather, the reports of mental activity) that is not preoccupied with accomplishing a task:

> Fantasy is defined as verbal reports of all mentation whose ideational products are not evaluated by the subject in terms of their usefulness in advancing some immediate goal extrinsic to the mentation itself; that is, fantasy is defined as report of mentation other than orienting responses to, or scanning of, external stimuli, or operant activity such as problem-solving in a task situation. (pp. 9–10)

Although this chapter will review research based upon widely varying, and sometimes contradictory, definitions of fantasy, I shall propose a definition that may be useful in guiding future research, and that rests on the groundwork laid by Singer and Klinger: *Fantasy is mental activity, other than orienting responses and non-directed perceptual scanning, which is not preoccupied with accomplishing a task (whose goal is extrinsic to the mentation itself) in the current environment*. This definition rules out orienting responses. If, while trying to solve a mathematical puzzle, a car backfires, one may jump, becoming aware of the car as one processes the *bang*!; but this is clearly not fantasy. It also rules out nondirected perceptual scanning. Walking into a new room, one becomes aware of how the room looks, of what is in it, and later, while carrying on a conversation, one may happen to "notice" a particular lamp or the way the sunlight hits the floor at a certain angle. One is not directing one's attention to these matters to accomplish a given task, or for any particular, definable reason. These nondirected perceptions are not fantasy. Finally, mental activity associated with a task, whose goal is extrinsic to the mentation itself, in the present environment is ruled out. Thus, keeping one's mind on one's work as one tries to fix one's car, cook food, or analyze the statistics on one's latest research program is not fantasy.

Ruled in according to this new definition would be mental activity that would qualify as fantasy under both Singer's and Klinger's definitions: an idle memory of what our first day at school was like, wondering what it will be like when we grow very old, a daydream about winning the Olympics or living on Mars, an unbidden image of the scream of tires and the shattering of glass from an automobile wreck we were recently in, a remembered melody of a song our parents sang for us as they put us to bed, and so on. But unlike Singer's definition, this proposed definition clearly specifies as fantasy some mental activity preoccupied with the current array of external stimuli. For example, if our attention is riveted to a showing of *Casablanca* or a performance of *Rhinoceros*, we are engaged in fantasy, even though we are fully tuned in to our immediate environment. And unlike Klinger's definition, the proposed definition makes fantasy some operant, problem-solving, or goal-directed mental activity. For example, while driving around to do some errands, we may begin to wonder how we can get the boss to give us a raise. We may mentally rehearse several possible approaches, evaluating each in terms of its usefulness in achieving our goal. Although this qualifies as operant activity according to Klinger's criteria, it still appears to be engaging in fantasy.

Despite the difficulties of definition, fantasy and imagination have fascinated humankind since the early Greeks and Chinese of approximately 500 B.C., though it is only relatively recently that systematic, scientific studies have been accomplished. In the subsequent sections of this chapter, I shall be looking at some historical trends, major areas of research, and finally some implications for clinical practice.

HISTORICAL TRENDS

As early as Longinus, we find the idea that fantasy and imagination are somehow closely connected with the fundamental perceptual systems.

> In this sense some call them mental representations. In a general way the name of *image* or *imagination* is applied to every idea of the mind, in whatever form it presents itself, which gives birth to speech. But at the present day the word is predominantly used in cases where, carried away by enthusiasm or passion, you think you see what you describe. (Longinus, *On the Sublime*, quoted in Bate, 1970, p. 69)

Current cognitive theory (see Pope & Singer, 1978b, 1980) describes perception as an active, constructive process of matching incoming stimuli to neural models or schemata, and describes fantasy and imagination as using parts of this system. "Perception is a cyclic activity that includes an anticipatory phase; imagery is anticipation occurring alone" (Neisser, 1976, p. 147).

And as early as Plato, we find the idea that fantasy and the imagination are part of a destructive process. "All poetical imitations are ruinous to the understanding of the hearers" (Plato, *The Republic,* quoted in Bate, 1970, p. 43). The idea that there is something wrong with fantasy, that it is somehow associated with psychopathology, culminated with popular misconceptions about Freud's work that regarded fantasies as "primary process" (i.e., much less sophisticated or useful than "secondary process"), and briefer recourse to the imaginal processes as the "psychopathology of everyday life."

The first systematic, comprehensive exploration of fantasy and imagination was undertaken in the seventeenth century by Hobbes (1651/1968, 1642/1972). Beginning with the notion that our imagination results from the cumulative effects of experience, Hobbes analyzed the ways in which our fantasies can serve to motivate us, and also the ways in which we can use the imaginative arena to rehearse and play out possible behaviors in order to evaluate them for their effects.

In the eighteenth century, Addison (1712/1970) began focusing on the inherent pleasures of fantasy, and along with his Swiss followers, Bodmerd and Breightinger, led British and Continental critics alike to use British empirical psychology as a frame of reference for their writings. But it was Samuel Johnson who, throughout the breadth of his writings, developed a specific and detailed theory of the imagination and its function in human life, which today seems well borne out by research data. He writes of a philosopher who, trying to comprehend why the great pyramids were erected, concludes: "It seems to have been erected only in compliance with that hunger of the imagination which preys incessantly upon life. . . . Those who have already all that they can enjoy, must enlarge their desires" (Johnson, *The History of Rasselas,* quoted in Bronson, 1958, p. 671). According to Johnson, the imagination is one of the central functions of the human personality, operating almost constantly, whether we wish it to or not, pulling us into the future or back into the past, never allowing us to be content for long with the present. We can always imagine more of what we have (or something different from what we have), and thus soon come to want what we can imagine. The imagination, therefore, accounts for almost all human motivation, excepting a very few physical needs. Our hope is not to shut down the workings of the imagination (which would in any case be impossible), but rather to regulate it with reality. It is the source not only of our discontents but also of what is best in human nature, our ability to sympathetically identify with other people and with sound values. But regardless of what we attempt to do with our tendency to imagine, the process continues. "The truth is that no mind is much employed upon the present: recollection and anticipation fill up almost all our moments'(Johnson, *The History of Rasselas,* quoted in Bronson, 1958, p. 668).

Nineteenth-century writers such as Hazlitt, Wordsworth, and Keats elaborated the ideas of Hobbes and Johnson. Like Johnson, Coleridge held imagination to be perhaps the central human process, or, as he puts it, the "prime agent of all human perception" (*Biographia Literania,* quoted in Bate, 1970, p. 387). Hazlitt, like both Hobbes and Johnson, wrote of the interplay among imagination, feelings, and motivation. "The imagination is that faculty which represents objects, not as they are in themselves, but as they are moulded by other thoughts and feelings, into an infinite variety of shapes and combinations of power" ("On Poetry in General," quoted in Bate, 1970, p. 304). Writers stressed the power of the imagination to allow us to sympathetically identify with others. Writing of Shakespeare, Hazlitt asserted, "He had only to think of anything in order to become that thing." ("On Shakespeare and Milton," quoted in Bate, 1970, p. 307). This concept had enormous impact, finding its way into the notion of *Einfuhlung* developed by Theodor Lipps as well as the school of method acting initiated by Konstantin Stanislavsky.

We come to the close of the nineteenth century and the beginning of the twentieth full of hope and promise for a systematic, scientific exploration of fantasy and imagination. William James calls our attention to the "stream of thought" (James, 1890), and Wundt's structuralist approach to consciousness is taken up in America by E. B. Titchener. "All human knowledge is derived from human experience . . . and there can be no essential difference between the raw materials of physics and the raw materials of psychology" (Titchener, 1909, p. 6). But this new initiative is soon dropped by American psychology in favor of behaviorism. Although artists such as Dorothy Richardson, James Joyce, Virginia Woolf, and Marcel Proust continued to explore the stream of consciousness, psychologists lost interest or nerve. As Brown (1958) described it: "In 1913 John Watson mercifully closed the bloodshot inner eye of American psychology. With great relief the profession trained its exteroceptors on the laboratory animal" (p. 93). It was not until 1966 that any book was published in the United States that was devoted completely to a comprehensive, systematic study of fantasy, daydreaming, or the imaginal processes (Singer, 1966). Only in the last several decades has there been a scientific study of fantasy undertaken with vigor by various branches of psychology. The next section will describe research findings in 12 major areas.

MAJOR RESEARCH FINDINGS

NORMATIVE ASPECTS OF THE STREAM OF CONSCIOUSNESS

We are aware that our stream of consciousness is characterized by shifts of attention from one topic to another, from the business at hand to various fantasy material and then back again. Recent research has gauged the frequency of such shifts, and the amount of time spent in fantasy.

The pioneering work of Singer (1966, 1980) gathered evidence, based on questionaires administered to a wide variety of people, that almost everyone reports engaging in daydreaming at least once a day.

To supplement Singer's retrospective data, Pope (1978) asked 54 undergraduates to "think aloud" into a portable tape recorder for a total of 15 min each (5 min while standing or walking, 5 min while sitting, and 5 min while lying down). Each person was asked to say "shift" when he or she noticed a shift or substantial change in the focus, content, tone, organization, or direction of conscious experience. The resultant transcripts were then examined by independent raters who, not knowing where the subjects had indicated shifts, identified shifts. There was close agreement between the two raters, and also between the raters and the subjects, concerning where the shifts in thought occurred. The raters then evaluated each thought segment according to whether it was focused on the immediate current environment ("present thoughts") or on the past, future, or an otherwise imaginary topic ("fantasy thoughts").

During a 15-min period, subjects averaged about 31 shifts of thought. About 11 of the thought segments were focused upon the immediate, current environment. About 5½ min of the 15-min period were cumulatively spent in thoughts focused upon the current environment. This experiment suggests that subjects in a "free" environment (they were given no specific task to accomplish during the reporting periods), while reporting their thoughts aloud, have thought segments lasting about 30 sec (between shifts) and spend about two-thirds of the time in fantasy.

Believing that "thinking aloud" might considerably slow (or otherwise distort) the flow of consciousness, Pope (1978) recruited 36 subjects and asked them to report their stream of thought using a simple key press. Participants were asked to press one key while their thoughts were focused on the here and now, and another key while thinking of the past, the future, or any imaginary topic. To indicate shifts, subjects were asked to briefly release the key they were currently pressing and then quickly to either repress the same key or press the other key, depending upon the nature of their thoughts. During a 15-min period, subjects averaged about 167 shifts, about 89 "present thought" segments, and spent about 6 min with their minds focused upon the current environment. While not reporting their thoughts aloud, then, people seem to have thought segments of about 5 sec between shifts and to spend about 40% of their time focusing on the current environment.

The latter (key press) data are more likely to be approximations of normal thought segments. Klinger (1978) obtained similar data from subjects trained to monitor their thought processes, who estimated the length of their thought segments at means of 8.8 sec in a laboratory situation (listening with moderate attention to a tape recording) and 14.5 sec in a sample of everyday situations.

THE RELATIONSHIP OF FANTASY TO SCHIZOPHRENIA AND VIOLENCE

There is a popular notion that frequent fantasy is characteristic of the abnormal mind, that people can become more and more drawn into a fantasy world until they become either schizophrenic or violent. The data, however, do not support such assumptions.

Rabinowitz (1975) made a careful study of people's hostile fantasies, examining the likelihood that they would be acted upon. Similarly, Beit-Hallahmi (1971) explored the sexual and aggressive fantasies of violent and nonviolent prison inmates. Although both Rabinowitz and Beit-Hallahmi found that these fantasies were often related to general attitudes about violence, neither found a systematic relationship between the fantasies and overt aggressive or violent behavior. Klinger (1971) reviews the available evidence on the subject and concludes that if there is a strong, monotypical aggressive fantasy that appears continually, then there is increased probability of overt aggression. But if the aggressive fantasy is part of a wider range of fantasy, or is not dominant, or if there are indications of qualifications of the fantasy (such as awareness of the risks of acting it out), then the translation of the fantasy into action is contraindicated. He raises a theme that repeats itself in much of the research literature: that violent (and psychotic) individuals, rather than having a greater fantasy life than "normals," in fact seem to have a constricted, barren, or inhibited fantasy life. For example, Klinger (1971) reports that under some circumstances "overtly nonaggressive subjects tell more aggressive TAT stories than do overtly aggressive subjects" (p. 327).

Similarly, research indicates that although there may be specific patterns of fantasy characteristic of various disorders, there appears to be no greater frequency of fantasizing among schizophrenic and other psychotic populations, and that the general fantasy process is not at the heart of psychosis (Singer & Brown, 1977; Starker, 1978, 1979; Streissguth, Wagner, & Eschler, 1969). In a particularly careful series of six interrelated studies, Starker (1979) concludes:

> Results of the studies failed to support the idea that psychotic patients have particularly frequent or vivid daydream activity, and indicate instead that psychotic patients tend to inhibit aspects of normal fantasy. Hallucinatory schizophrenics, in particular, were blocked in their emotional-interpersonal imagery. (p. 25)

THE BIOLOGICAL RHYTHM OF FANTASY

Given our initial normative data about the stream of consciousness, and our failure to find evidence that increased fantasy is predictive of violence or psychosis, is greater precision possible in specifying when fantasies are likely to occur? Kripke and Sonnenschein (1978) asked 11 volunteers to report individually after breakfast to a well-lighted room containing a bed and a chair. Each participant was fitted with electrodes for measuring EEG and horizontal eye movements. The volunteers stayed alone in the room without communicating with anyone for 10 hr. Every 5 min, the subject wrote down what he or she had been thinking about during the previous 5-min period. Each of these 5-min reports was rated both by the subject and also by an independent rater according to the following scale:

1. Vivid dreamlike fantasies with bizarre or symbolic content and strong emotions such as wishes or fears
2. Abstract and fanciful thinking, especially past or future
3. Concrete, realistic, and present-time thinking or planning
4. Awareness of present surroundings and thinking about what is happening
5. Present-time perceptual scanning and perceptual-motor activities

What emerged from this study was a reliable pattern of fantasy for each subject: Fantasy tended to peak during a cycle that repeated every 90 min. The most intense fantasy was associated with the most continuous alpha and with the least rapid eye movement. It appears that our waking fantasy follows a cycle similar to that of our night dreams (or sleeping fantasies), and that the cycle lasts about 90 min (one cycle every 72–120 min).

To determine whether the findings held outside the laboratory, Kripke and Sonnenschein (1978) recruited eight new volunteers who agreed to report their mental activities into a dictaphone every 10 min throughout a 12-hr day. These subjects did as they pleased, engaging in a variety of activities (e.g., working as a computer programmer, working as a hotel clerk, going to the beach, taking a shower, eating, watching television). These activities precluded the physiological measures of the first experiment. Since the second experiment also revealed fantasy cycles of approximately 90 min duration (though there were fewer well-formed daydreams reported than in the laboratory setting), Kripke and Sonnenschein believe it is likely "that these cycles were of endogenous biologic origin, and that they were expressions of an oscillation in brain functional state" (p. 328).

THE EFFECTS OF POSTURE ON FANTASY

A variety of studies have provided evidence that posture affects the thought processes. Kroth (1970) discovered that people who were reclining rather than sitting could free associate more freely and spontaneously. Morgan and Bakan (1965) found that people in a sensory deprivation situation experienced much more vivid imagery while reclining than while sitting. Berdach and Bakan (1967) report that, in a memory study, earlier and more memories occurred when people were in a reclining rather than a sitting position. Segal and Glickman (1967) found that people were much less likely to recognize an external signal (in an experiment on the Perky effect, i.e., the tendency to confuse internal images with external stimulation) while lying than while sitting.

In the experiment described earlier, Pope (1978) obtained reports of mental activity from subjects using either a "thinking aloud" or a key press method. Participants reported during three 5-min periods, while standing or walking, while sitting, and while lying down. The effects of posture were significant. There were more shifts, more thought segments concerned with the present environment, and more cumulative time spent with thoughts focused on the present environment when participants were walking or standing than when they were reclining. Sitting showed an intermediate effect. It is likely that standing or walking increases the amount or variety of stimulation from the environment, thus serving to decrease fantasy activity. The effects of information load will be discussed in a later section.

INDIVIDUAL STYLES OF FANTASY

The Imaginal Processes Inventory, developed by Singer and his associates (see Singer, 1978, 1980, 1981–1982), provides evidence that each of us has a personal, characteristic fantasy style. There are three stable factors indicating dimensions of thought along which most people can be measured. One factor is called "poor attentional control," which reflects a lack of ability to sustain extended sequences of fantasy or even extended sequences of attention to environmental stimuli. People who score high on this factor often find themselves bored and distractible. A second dimension is labeled "guilty and dysphoric fantasy." Those who score high on this factor have an uncomfortable inner life, and frequently engage in guilty or depressing fantasies. The third dimension is termed "positive-constructive" and is characterized by future-oriented, planful, vivid, and generally enjoyable fantasies.

THE EFFECTS OF AGE AND GENDER ON FANTASY

A number of important studies have provided evidence that women report more frequent and more vivid imagery than men (Durndell & Wetherick, 1975; Michael, 1966; Schmeidler, 1965; Sheehan, 1967). But it is the amazingly comprehensive and detailed work of Giambra (1979–1980), involving 1200 people age 17 to 92 from Ohio, Maryland, and Pennsylvania, that suggests with some clarity the effects of age and gender upon fantasy. As Giambra summarizes his findings:

> Females reported higher levels of daydreaming and nightdreaming frequency and emotional reactions to daydreaming as well as more daydreams of a problem-solving nature. Females also reported lower levels of daydreams of a sexual, bizarre-improbable, heroic and achievement-oriented nature. Most sex differences persisted over the life span with the difference for sexual daydreams increasing with increasing age level. Except for problem-solving daydreams, all daydreaming contents investigated decreased with increases in age. Across the lifespan problem-solving daydreams were the most likely for both sexes except for seventeen to twenty-nine year old males where such daydreams were second most likely; from age seventeen to twenty-nine sexual daydreams were most likely for males. After age forty, achievement-oriented daydreams were relatively more important for females than for males. Females across the lifespan also reported more interpersonal curiosity than males while males exhibited more curiosity about things. Unexpectedly,

male curiosity about things was equally strong as their curiosity about people. (p. 1)

FANTASY AND INFORMATION LOAD

In experiments described above, subjects produced less fantasy when out and moving about than they did while in the laboratory (Kripke & Sonnenschein, 1978), and likewise produced more fantasy when lying down than while standing or moving around (Pope, 1978). It was suggested that the effects could be because of the greater variety of environmental stimulation available in the former conditions than in the latter. The idea that fantasy production is influenced by the "information load" available to or being processed by the subject was systematically explored by Antrobus (1968; Antrobus & Singer, 1964) and his colleagues. Using signal detection tasks, he found that as the complexity of the task is increased (thereby increasing the information load), the reports of stimulus-independent thought decrease. In fact, he was able to generate a mathematical equation relating the increase of information load to the decrease in frequency of fantasy activity.

FANTASY AND EYE MOVEMENT

The Kripke and Sonnenschein (1978) studies revealed that the greatest fantasy activity was associated with the least eye movement, a finding also reported by Marks (1972) and Klinger, Gregoire, and Barta (1973). This makes intuitive sense if we consider that, while fantasizing, the less the eyes are in motion, the less new visual information will be coming in for the mind to process.

A complex set of interrelated experiments was conducted to specify more precisely the activity of the eyes during imaginal activity, and to be in a better position to make inferences about some of the underlying processes of this activity (Meskin & Singer, 1974; Rodin & Singer, 1977; Rosenberg, 1977). A variety of subjects were videotaped while they were asked questions. Some questions emphasized verbal clichés. In light of the split-brain experiments, it was hypothesized that questions involving logical-verbal-sequential processing might make subjects more likely to shift their eyes to the right. Other questions involved visual imagery that would, the experimenters hypothesized, make left-shifting more likely. For some subjects, the questions were asked face-to-face, the hypothesis being that during imaginal activity, the eyes would be shifted away from this face and toward a blank area of the wall, thus clearing the visual field of much of the "stimulus load." For other subjects, questions were asked over a loudspeaker, supposedly eliminating the need for the subject to shift toward a blank area. The findings of these experiments, which analyzed a wide variety of variables, are as follows. When subjects are attending to privately generated material, their eyes tend to show reduction in movement and, in effect, some degree of blotting out external stimulation. Questions involving visual-spatial imagery led more likely to left-shifting of the eyes, whereas more verbal material led more likely to right-shifting. While figuring out the answers to the questions, individuals in general tended to shift away from confronting the face of someone. In addition, daydreaming style appears related to habitual patterns of eye shifting. People who report frequent daydreaming on the Imaginal Processes Inventory are more likely to be left-shifters. In summary, during fantasy, eye movement is held to a minimum. Whether eyes are initially shifted at the beginning of a fantasy segment as well as the direction in which they tend to shift seem to be a function of: (1) the degree to which a person reports frequent fantasizing, (2) the nature of the material to be processed in fantasy, and (3) the stimulus array of the environment.

CURRENT CONCERNS, OPERANT THOUGHT, AND FANTASY

As mentioned earlier, Klinger defines fantasy as mental activity that is not directed to accomplishing a task or achieving a goal, and says that while fantasizing we are engaging in "respondent thought." The mental activity associated with goal- or task-directed activity is termed "operant thought."

> In very general terms, operant thinking appears to differ from respondent thinking in that it is accompanied by a sense of volition, is checked against feedback concerning its effects, is evaluated according to its effectiveness in advancing particular goals, and is protected from drift and distraction by the thinker's deliberately controlling his or her attention. (Klinger, 1978, p. 235)

Operant thought has been reliably distinguished from respondent thought in an experiment (Klinger, 1974) in which people were asked to think aloud under four conditions: (1) solving a manual puzzle, (2) solving a logic problem, (3) letting their minds wander, and (4) reclining with their eyes closed. The content analysis of the resultant transcripts revealed significantly more utterances indicating that the participant was evaluating the effectiveness of the previous thought segments or controlling the direction of attention in the first two conditions (task conditions) than in the last two. Attention-controlling utterances tended to occur at points where the participant was stuck during problem solving (46%), after a mind-wandering episode during a task condition (32%), and after successfully completing a problem-solving step while attempting to figure out what to try next (16%).

The flow of operant and respondent thoughts is greatly influenced by the *current concerns* of the individual, as they interact with stimuli or cues from the environment.

> At any moment, concerns and cues jointly induce relevant thoughts that best bridge them. In other words, as people travel through their world of stimulation they ignore most of it and process those features of it that fit into one or more of their current concerns. Where one's mind goes from one moment to the next depends on which cues one encounters that relate to one's various current concerns. (Klinger, 1979, p. 8)

Klinger has found, through systematic studies, that the influence exerted by various concerns varies as a function of commitment, probability of attainment, and imminence.

THE CAPACITY TO BECOME ABSORBED IN FANTASY

When biofeedback was initially used to help people relax, a curious phenomenon occurred. Some people ("high absorption"), without feedback, were easily able to mentally withdraw from the environment, engage in peaceful fantasies, and become mentally and physiologically relaxed. Other people ("low absorption"), without feedback, had difficulty becoming absorbed with their imagery, and had subsequent difficulty with relaxation. Yet in a biofeedback condition, the "high absorption" people had difficulty relaxing, whereas the "low absorption" people tended to be able to utilize the feedback to increase their relaxation. In a series of careful studies, Qualls and Sheehan (1979, 1981a, 1981b) discovered the nature of the problem and its solution. During normal biofeedback trials, "high absorption" people became preoccupied with the feedback signal itself, and this attention to the environment inhibited their normal fantasy capabilities. The solution was a simple change in the instructions: "High absorption" people were encouraged to use their imagery capabilities and to feel free of a constant preoccupation with the signal. As a result, the "high absorption" participants were able to make the expected gains, many of them weaving their fantasies around the periodic sound of the feedback signal.

SELF-AWARENESS

While engaging in voluntary, intrinsically motivating activities—such as fantasy, playing basketball, or composing a symphony—we may have an optimal or "peak" experience that Czikszentmihalyi characterizes as the sense of "flow" (Csikszentmihalyi, 1978; Csikszentmihalyi & Figurski, 1982). During these times we center our attention upon a limited internal or external stimulus field, and are not concerned about our "control" over the situation. Recent research (Csikszentmihalyi & Figurski, 1982; Klinger, Barta, & Glas, 1981) lends support to a notion likely suggested by our own personal experience: that self-awareness interferes with the process of flow and its associated positive affect. When we suddenly become self-aware, the "trance" is broken—as, for example, when our engagement in watching a play is interrupted by someone leaning over to ask what time it is and we are suddenly aware that we are seated in a crowded theater. The research of Csikszentmihalyi and Figurski (1982), in fact, suggests that self-awareness not only interrupts the experience of flow but, furthermore, tends to be associated with aversive experience.

SEXUAL FANTASIES

The topic of sexual fantasy has already been touched on briefly, but the research in this area has done so much to change popular ideas about the norms that the findings of several additional studies are worth mentioning.

Davidson (1974), in a study of 202 undergraduate women, found that 80% reported moderate- to high-frequency sexual daydreaming, and that 100% reported at least some sexual fantasy. Hariton (1972), in a study of 141 suburban housewives, found that 65% reported sexual fantasies during intercourse with their husbands, and that 37% reported very frequent sexual fantasies during sexual relations with their husbands. The most common fantasies involved a man other than the husband (often an imaginary lover), submission fantasies (many involving force), and making love in a different place (e.g., at the beach, in a car).

Guido (1981) studied the masturbation fantasies of 104 heterosexual and homosexual cohabiting and single males who were matched according to age, gender, sexual orientation, marital status, race, education, income, religion, and pathology. There were virtually no differences in the fantasy activities of homosexual and heterosexual men, aside from the sex objects being consistent with sexual orientation. The most frequent themes involved "unknown people or situation" and "past sexual experience." High frequencies of sexual fantasies were positively correlated with high frequencies of sexual behaviors. Giambra and Martin (1977) studied the sexual fantasies of 277 men ranging from 24 to 91 years old.

> The frequency and intensity of sexual daydreams declined with increasing age, and after age 65 virtually disappeared. The occurrence of sexual daydreams varied directly with each of three behavioral indicators of sexual vigor for all age groups through age 64. (p. 497)

Hoyt (1977) studied an area of sexual fantasy otherwise entirely neglected by systematic research: the so-called primal scene fantasy (the experience of observing one's parents engaging in intercourse). His data indicate a much higher frequency of such fantasies among young adults than might have been suspected from the clinical literature.

In none of these studies was there evidence that an active sexual fantasy life (in general, during masturbation or intercourse) was associated with psychopathology or self-defeating behaviors. Sexual fantasy frequency seems to be positively related to frequency of sexual behavior.

IMPLICATIONS FOR CLINICAL PRACTICE

Although the findings described in many of the major research areas, taken singly or in combination, suggest self-evidently their implications for clinicians, to what extent can we specify more exactly their contribution to clinical practice?

THE CLIENT'S ACCEPTANCE OF FANTASY

Many clients are fearful or ashamed of their fantasy life. The client who experiences an aggressive daydream or a sexual fantasy about someone who is not his or her spouse may suppose that something is amiss, perhaps even that mental illness is immanent. The research data indicate that fantasy is a normal part of life, an important human function with great potential for enriching experience.

THE CLINICIAN'S ACCEPTANCE OF FANTASY

Many clinicians are likewise uncomfortable or not knowledgeable when dealing with fantasies. Behaviorally speaking, at least, most psychologists and other researchers avoided or neglected the topic for the better part of this century, as the section on "Historical Trends" described. The degree to which clinicians can be alert to, knowledgeable about, and comfortable with fantasy material (their own as well as the client's) may have a profound impact on the therapeutic process.

As a specific example, Schover (1981) studied 12 men and 12 women from each of the following categories: psychiatrists, psychologists, and social workers. She found that male therapists were much less comfortable with sexual material, particularly with female clients.

> Male therapists with liberal sexual attitudes were sexually aroused by, and verbally encouraged, the seductive female client. Conservative men were aroused by female dysfunction . . . but reacted with anxiety and verbal avoidance of the material. (p. 477)

Such issues have obvious implications for training programs as well as for clinical practice.

INFLUENCING AFFECT

Singer's research regarding daydreaming styles (e.g., "guilty and dysphoric," "postive-constructive"), Klinger and Csikszentmihalyi's research regarding self-awareness, and similar studies reveal the interaction between fantasy and affect. Clients may be taught to interrupt and discontinue those fantasies associated with negative feelings and to learn how to engage in more positive daydreams. Schultz (1978) has demonstrated the feasibility and effectiveness of this in a careful study involving 60 people hospitalized for depression. The work of Beck (1970) and Lazarus (1968) shows the variety of ways in which interventions into the fantasy system can therapeutically influence the affect system.

IMPROVING PHYSICAL AND ATHLETIC PERFORMANCE

Fantasy can be used to improve our performance of physical activities. The benefits result both from our mental practice (Beasley, 1978; Rawlings & Rawlings, 1974) and from altering our mental approach to the task (Suinn, 1976).

RISK-FREE EXPERIMENTING WITH BEHAVIOR

As discussed earlier, fantasy provides an arena in which possible avenues of future behavior may be tried out and examined in terms of likely consequences. Clients can be taught to use this method of risk-free experimentation when they confront major decisions. Janis and

Mann (1977) have explored various methods for using this approach when arriving at a decision that will be irreversible.

OVERCOMING MENTAL BLOCKS

Clients will occasionally report being "blocked" in a particular undertaking. For some reason (usually psychological) the person is unable to make progress. The most frequently mentioned form of this phenomenon is, of course, writer's block.

The use of fantasy (envisioning positive results, confronting fears and doubts, etc.) can be effective in overcoming this condition. Barrios and Singer (1981–82) assigned 48 volunteers who were experiencing difficulty (i.e., were "blocked") in relation to artistic, literary, scientific, or professional projects to one of four conditions: (1) waking imagery, (2) hypnotic dreams, (3) rational discussion, and (4) control (these subjects were offered encouragement). Those in the first two conditions, unlike the others, were most effective in resolving their creative blocks. Research by Aranosian (1981–82) has shown the value of fantasy in the initial learning of such artistic, literary, scientific, or professional skills, and the tendency of such fantasy-enriched learning to lead to greater creativity.

BLOCKING HABITUAL BEHAVIORS

Rather than being "blocked," some clients may find themselves unable to cease engaging in a particular behavior (e.g., excessive drinking, smoking, overeating). As Cautella and others (Cautella & McCullough, 1978) have demonstrated, fantasies can be used to increase the individual's perceived and actual control over such compulsive or habitual behaviors. Such techniques appear to be effective whether the behavior draws the person toward (as in addictions) or away from (as in phobias) a particular object.

ENHANCING INTERPERSONAL SKILLS

Clients may increase their understanding of other people and their ability to deal effectively with those in their environment through fantasy. In an elaborate study, Frank (1978) demonstrated that participants' ability to empathize and to identify sympathetically with others was improved through exercises of the imagination. The work of Kazdin (1978) has shown the ways in which "covert modeling" can increase a person's ability to deal assertively with others.

COMMUNICATION

Effective communication is a necessary aspect of the therapeutic process. Singer and Pope (1978) have described in detail the ways in which clients communicate with their therapists through the modality of fantasy. Crucially important is the therapist's willingness to use his or her own imagination to reconstruct the client's environment and the people described by the client as they play a role in the client's life.

ENRICHING LIFE

Finally, clinicians can help their clients to become aware of and to utilize the enormous capacity of fantasy to enrich life. Fantasy is available to us during the vast majority of our waking hours, and can serve as an alternative environment to our current surroundings (Pope & Singer, 1978a, 1978b, 1978c). At the most basic level we can use fantasy to entertain ourselves or to maintain alertness at times of boredom. It can be used to reduce stress and to increase relaxation. It can give us a "time-out" period or a "minivacation" from particularly arduous or tedious tasks.

In an ingenious study, Emery and Csikszentmihalyi (1981) examined the lives of 30 men. One-half of them had grown up to become blue-collar workers; the other half, carefully matched to the first group along a wide range of variables, to become university professors. Results showed a significant difference on the dimension of cultural models. Books, and the fantasy models they provided, can serve as a projective indicator.

> Books provide examples of previously successful responses to existential problems. . . . Persons who face particularly intense problems in their lives, such as poverty, oppression, marginality, homelessness, and so on, may learn to model their response to these problems on successful examples provided by the cultural heritage transmitted in books. (p. 17)

A RESEARCH-BASED MODEL FOR THERAPY

The preceding section presented discrete principles, derived from nonclinical research, with implications or applicability for a wide variety of therapeutic approaches. Yet nonclinical research in the area of fantasy can also serve as the basis for specific models of therapy. The following model focuses on sexual fantasies and is used most frequently in the treatment of sexual dysfunctions or the more general enhancement of sexual enjoyment and fulfillment. Presented in more detail elsewhere (Pope, 1983a), this model—in which the client is lead through progressive stages of work with sexual fantasies—is here described only in terms of major stages, rationale, and principles derived from research.

DESCRIPTIONS

The client creates a sexual fantasy of the kind he or she characteristically experiences. Then the client attempts to address the following questions with the help, support, and nonintrusive guidance of the therapist.

(a) What are the elements that make up your fantasies? Is a complex plot important? Does the setting play a major role?

(b) How would you describe the people in your fantasies?

(c) Do you play a major role in your fantasies? Are you different in your fantasies from the way you are in your "real" life?

(d) How frequently do you experience such fantasies in your day-to-day life? Are there particular conditions that make sexual fantasies more likely?

(e) Does the occurrence of sexual fantasy tend to influence you to engage in particular behaviors (e.g., discontinuing other activities to enjoy the fantasy; attempts to terminate the fantasy; masturbation; initiating sex with someone else)?

(f) Is there a typical duration to your fantasies?

(g) How do your typical fantasies relate to other aspects of your life and experience?

(h) Are you active in planning and directing your fantasies, or do they seem to unfold spontaneously?

(i) Do you ever share your fantasies with anyone else?

(j) To what extent does each of your senses play a role in your fantasies?

Session 1 serves a variety of purposes. Clients become more aware of their fantasy processes, of their own personal norms, rhythms, and patterns. For some, this is a monumental revelation. They become more adept not only at creating fantasies but also at reflecting upon and discussing them. By providing information, developed from the research reviewed earlier in this chapter, the therapist can help the client to accept his or her own fantasy processes.

WISHES

Session 2, for many clients, is a move into newer territory: creating "ideal" fantasies, imagining something truly enjoyable and fulfilling. Clients' habitual fantasies may be unduly constrained, barren, or in some cases painful. An image of a past sexual disaster may appear as an "unbidden image"; a stilted, rigid, repetitive image may be replayed over and over to the exclusion of any new or spontaneous sexual fantasy; attempts to envision a delightful sexual episode may be interrupted by the pressures of work, a feeling of inadequacy, or an overwhelming sense of guilt and anxiety.

Some clients will use this session to give themselves permission to form a clear goal that, once perceived, will be easy to implement. Others will use this sort of fantasy as "covert rehearsal," "covert modeling," or "covert conditioning," that is, as a mediating variable to help enable themselves to alter their sexual experiences. Just as research-based strategies may be used in other areas to enable clients to discontinue habitual behaviors, move beyond the feeling of being "blocked," or acquire or improve new behaviors, such strategies may be applied to human sexuality.

VIVIDNESS AND DIRECTION

I described earlier how behaviorism led to a denial of the prevalence, significance, and influence of fantasy in the ways people lead and experience their lives. Until relatively recently, psychology had little understanding of fantasy, and consequently did not know how to use this resource therapeutically.

In a similar manner, individuals who have denied or neglected their fantasy capacities may find themselves having difficulty using this resource. Imagined scenes may lack vividness. The "action" in a fantasy may seem beyond the control of the client.

In Session 3, the client practices imagining scenes and creating vividness, immediacy, realism, and a control over the action or "plot." The capacity to become absorbed in fantasy improves. As with other skills, imagining scenes clearly and directing their movement improves with repeated, serious practice.

FEAR AND TREMBLING

The three major styles of fantasy (see "Individual Styles of Fantasy") are associated with specific affects, and fantasies themselves may be used to influence affect. Sexual fantasies in particular tend to arouse at least some negative feelings—anxiety, fear, shame, guilt—that, in this culture, are often associated with sexuality.

Paradoxically, this culture also tends to put forth such a positive image of sexuality that people are sometimes reluctant to acknowledge any negative feelings, particularly about something so seemingly ideal as sexual fantasies.

In Session 4, the client re-creates the fantasies from the previous sessions while paying particular attention to any negative feelings such as apprehension, sadness, panic, shame, boredom, uncomfortable tension, or guilt. The client then works with the therapist to find out what aspects of the fantasies are evoking these negative reactions. Sometimes simply recognizing, identifying, and openly discussing such uncomfortable feelings is therapeutic. Sometimes the client benefits by education, acceptance, or support provided by the therapist. Sometimes the client is able to learn from these negative reactions, and can move toward more enjoyable forms of sexual fantasy and experience. And, of course, sometimes an openness to these negative feelings leads to a discovery of deeper conflicts or sources of distress, which benefit from careful exploration.

DISCOVERY

In this session, the client is encouraged to explore new fantasy possibilities, something different from what is habitual (the emphasis of Session 1) or what is obviously ideal (Session 2). New settings, new behaviors, new styles and patterns are experimented with. In this way, the client not only enlarges the scope of what is possible (at least in terms of fantasy experiences) but also can practice what was referred to above as "risk-free experimenting with behavior."

FEELINGS

Returning again to the theme of feelings, stressed (although with the emphasis on negative feelings) in Session 4, this session invites the client to re-create the previous fantasies as well as to generate new ones while, at each point, asking the question, What feeling is this fantasy evoking?

Here, a standard technique of sex therapy, "sensate focusing," is adapted and becomes "emotional focusing." The premise is that emotional responses (even—or especially—in fantasies) constitute powerful resources for enriching sexual experiences, and that a person's ability to integrate emotions with physical sensations, sexual arousal, and imaginative behavior can contribute to sexual enjoyment and fulfillment.

INTIMACY AND VALUES

Some of the most important implications of nonclinical research for clinical practice involve the enhancement of interpersonal skills, communication, and the enrichment of life. Session 7 focuses on aspects of sexual fantasy that enhance a sense of closeness to others and to those deep and personal values that most enrich one's life.

The client is encouraged to re-create the previous fantasies and to create new ones, while asking the following questions.

(a) Does this scene make me feel actually or potentially intimate with someone else?

(b) If so, does the intimacy feel desirable, nonintrusive, self-affirming, and fulfilling?

(c) If not, does the distancing feel positive, neutral, or negative?

(d) Does this fantasy—actually or potentially—express, support, fulfill, supplement, or at least peacefully coexist with my deepest values?

In summary, the knowledgeable and skillful use of fantasy provides clinician and client alike with valuable resources for dealing effectively with a vast array of problems in living, from unwanted habits and creative blocks to major crises and decisions. Few techniques are useful for such a range of difficulties. Unlike medications and biofeedback equipment, fantasies are free. There are few reports of destructive side effects associated with their use. Furthermore, using fantasies to cope with life's problems is something clients can be taught to do on their own. Therapists have available few approaches that are more useful than fantasy in making life more rich and enjoyable, or at least less miserable.

REFERENCES

Addison, J. The pleasures of the imagination. In W. J. Bate (Ed.), *Criticism: The major texts.* New York: Harcourt Brace Jovanovich, 1970. (Originally published, 1712)

Antrobus, J. S. Information theory and stimulus independent thought. *British Journal of Psychology*, 1968, *59*, 423–430.

Antrobus, J. S., & Singer, J. L. Visual signal detection as a function of sequential task variability of simultaneous speech. *Journal of Experimental Psychology*, 1964, *68*, 603–610.

Aranosian, C. M. Musical creativity: The stream of consciousness in composition, improvisation, and education. *Imagination, Cognition, and Personality: The Scientific Study of Consciousness*, 1981–82, *1*, 67–88.

Barrios, M. V., & Singer, J. L. The treatment of creative blocks: A comparison of waking imagery, hypnotic dream, and rational discussion techniques. *Imagination, Cognition, and Personality: The Scientific Study of Consciousness*, 1981–82, *1*, 89–110.

Bate, W. J. (Ed.). *Criticism: The major texts.* New York: Harcourt Brace Jovanovich, 1970.

Beasley, W. F. *The effect of physical and mental practice of psychomotor skills on chemistry student laboratory performance.* Unpublished doctoral dissertation, University of Maryland, 1978.

Beck, A. T. Role of fantasies in psychotherapy and psychopathology. *Journal of Nervous and Mental Disease*, 1970, *150*, 3–17.

Beit-Hallahmi, B. Sexual and aggressive fantasies in violent and non-violent prison inmates. *Journal of Personality Assessment*, 1971, *35*, 326–330.

Berdach, E., & Bakan, P. Body position and free recall of early memories. *Psychotherapy: Theory, Research, and Practice*, 1967, *4*, 101–102.

Bronson, B. H. (Ed.). *Samuel Johnson.* New York: Holt, Rinehart and Winston, 1958.

Brown, R. W. *Words and things.* New York: Free Press, 1958.

Cautella, J. R., & McCullough, L. Covert conditioning: A learning-theory perspective on imagery. In J. L. Singer & K. S. Pope (Eds.), *The power of human imagination.* New York: Plenum Press, 1978.

Csikszentmihalyi, M. Attention and the holistic approach to behavior. In K. S.Pope & J. L. Singer (Eds.), *The stream of consciousness.* New York: Plenum Press, 1978.

Csikszentmihalyi, M., & Figurski, T. J. Self-awareness and aversive experience in everyday life. *Journal of Personality*, 1982, *50*, 14–26.

Davidson, A. D. *The relationship of reported sexual daydreaming to sexual attitude, sexual knowledge, and reported sexual experience in college women.* Unpublished doctoral dissertation, University of Cincinnati, 1974.

Durndell, A. T., & Wetherick, N. E. Reported imagery and social desirability. *Perceptual and Motor Skills*, 1975, *41*, 987–992.

Emery, O. B., & Csikszentmihalyi, M. The socialization effects of cultural role models in ontogenetic development and upward mobility. *Child Psychiatry and Human Development*, 1981, *12*, 3–18.

English, H. B., & English, A. C. *A comprehensive dictionary of psychological and psychoanalytical terms.* New York: McKay, 1958.

Frank, S. J. Just imagine how I feel: How to improve empathy through training in imagination. In J. L. Singer & K. S. Pope (Eds.), *The power of human imagination.* New York: Plenum Press, 1978.

Giambra, L.M. Sex differences in daydreaming and related mental activity from the late teens to the early nineties. *International Journal of Aging and Human Development*, 1979–1980, *10*, 1–34.

Giambra, L. M., & Martin, C. E. Sexual daydreams and quantitative aspects of sexual activity: Some relations for males across adulthood. *Archives of Sexual Behavior*, 1977, *6*, 497–505.

Guido, P. A. *A comparison of heterosexual and homosexual single and cohabiting male masturbation fantasies*. Unpublished doctoral dissertation, California School of Professional Psychology, 1981.

Hariton, E. B. *Women's fantasies during sexual intercourse with their husbands*. Unpublished doctoral dissertation, City University of New York, 1972.

Hobbes, T. *Leviathan*. Harmondsworth, U.K.: Penguin Books, 1968. (Originally published 1651)

Hobbes, T. *Elementa philosophica de cive*. Garden City, N.Y.: Anchor Books, 1972. (Originally published, 1642)

Hoyt, M. *The primal scene: A study of fantasy and perception regarding parental sexuality*. Unpublished doctoral dissertation, Yale University, 1977.

James, W. *The principles of psychology*. New York: Henry Holt, 1890.

Janis, I., & Mann, L. *Decision-making*. New York: Free Press, 1977.

Kazdin, A. E. Covert modeling: The therapeutic application of imagined rehearsal. In J. L. Singer & K. S. Pope (Eds.), *The power of human imagination*. New York: Plenum Press, 1978.

Klinger, E. *Structure and functions of fantasy*. New York: Wiley-Interscience, 1971.

Klinger, E. Utterances to evaluate steps and control attention distinguish operant from respondent thought while thinking out loud. *Bulletin of the Psychonomic Society*, 1974, *4*, 44–45.

Klinger, E. Modes of normal conscious flow. In K. S. Pope & J. L. Singer (Eds.), *The stream of consciousness*. New York: Plenum Press, 1978.

Klinger, E. *Therapy and the flow of thought*. Paper presented at the annual meeting of the American Association for the Study of Mental Imagery, Los Angeles, 1979.

Klinger, E., Gregoire, K. S., & Barta, S. G. Physiological correlates of mental activity: Eye movements, alpha, and heart rate during imagining, suppression, concentration, search, and choice. *Psychophysiology*, 1973, *10*, 471–477.

Klinger, E., Barta, S. G., & Glas, R. A. Thought content and gap time in basketball. *Cognitive Therapy and Research*, 1981, *5*, 109–114.

Kripke, D. F., & Sonnenschein, D. A biologic rhythm in waking fantasy. In K. S. Pope & J. L. Singer (Eds.), *The stream of consciousness*. New York: Plenum Press, 1978.

Kroth, J. A. The analytic couch and response to free association. *Psychotherapy: Theory, Research, and Practice*, 1970, *7*, 206–208.

Lazarus, A. A. Learning theory and the treatment of depression. *Behavior Research and Therapy*, 1968, *6*, 83–89.

Marks, D. F. Individual differences in the vividness of visual imagery and their effect on function. In P. W. Sheehan (Ed.), *The function and nature of imagery*. New York: Academic Press, 1972.

Meskin, B. B., & Singer, J. L. Daydreaming, reflective thought and lateral eye movements. *Journal of Personality and Social Psychology*, 1974, *30*, 64–71.

Michael, L. I. *A factor analysis of mental imagery*. Unpublished doctoral dissertation, Auburn University, Auburn, Ala., 1966.

Morgan, R., & Bakan, P. Sensory deprivation hallucinations and other sleep behavior as a function of position, method of report, and anxiety. *Perceptual and Motor Skills*, 1965, *20*, 19–25.

Neisser, U. Changing conceptions of imagery. In P. W. Sheehan (Ed.), *The function and nature of imagery*. New York: Academic Press, 1972.

Neisser, U. *Cognition and reality*. San Francisco: W. H. Freeman, 1976.
Pope, K. S. How gender, solitude, and posture influence the stream of consciousness. In K. S. Pope & J. L. Singer (Eds.), *The stream of consciousness*. New York: Plenum Press, 1978.
Pope, K. S. Enhancing sexual enjoyment, intimacy and fulfillment. In A. Sheikh (Ed.), *Uses of imagery in the healing arts*. Milwaukee: Marquette University Press, 1983. (a)
Pope, K. S. *Implications of fantasy and imagination for mental health: Theory, research, and interventions*. Report commissioned by the National Institutes of Mental Health, 1983. (b)
Pope, K. S., & Singer, J. L. The flow of human experience. In K. S. Pope & J. L. Singer (Eds.), *The stream of consciousness*. New York: Plenum Press, 1978. (a)
Pope, K. S., & Singer, J. L. Regulation of the stream of consciousness: Toward a theory of ongoing thought. In G. E. Schwartz & D. Shapiro (Eds.), *Consciousness and self-regulation: Advances in research and theory* (Vol. 2). New York: Plenum Press, 1978. (b)
Pope, K. S., & Singer, J. L. (Eds.). *The stream of consciousness*. New York: Plenum Press, 1978. (c)
Pope, K. S., & Singer, J. L. The waking stream of consciousness. In J. M. Davidson & R. J. Davidson (Eds.), *The psychobiology of consciousness*. New York: Plenum Press, 1980.
Pylyshyn, Z. W. What the mind's eye tells the mind's brain: A critique of mental imagery. *Psychological Bulletin*, 1973, *80*, 1–24.
Qualls, P. J., & Sheehan, P. W. Capacity for absorption and relaxation during electromyograph biofeedback and no-feedback conditions. *Journal of Abnormal Psychology*, 1979, *88*, 652–662.
Qualls, P. J., & Sheehan, P. W. Imagery encouragement, absorption capacity, and relaxation during electromyograph biofeedback. *Journal of Personality and Social Psychology*, 1981, *41*, 370–379. (a)
Qualls, P. J., & Sheehan, P. W. Trait-treatment interactions: Reply to Tellegen. *Journal of Experimental Psychology: General*, 1981, *110*, 227–231. (b)
Rabinowitz, A. Hostility measurement and its relationship to fantasy capacity. *Journal of Personality Assessment*, 1975, *39*, 50–54.
Rawlings, E. I., & Rawlings, I. L. Rotary pursuit tracking following mental rehearsal as a function of voluntary control of visual imagery. *Perceptual and Motor Skills*, 1974, *38*, 302.
Richardson, A. *Mental imagery*. London: Routledge & Kegan Paul, 1969.
Rodin, J., & Singer, J. L. Eyeshift, thought, and obesity. *Journal of Personality*, 1977, *44*, 594–610.
Rosenberg, B. *Mental activity and environmental responsiveness: Optokinetic nystagmus during mental tasks associated with left and right cerebral hemispheres*. Unpublished doctoral dissertation, Yale University, 1977.
Schmeidler, G. R. Visual imagery correlated to a measure of creativity. *Journal of Consulting and Clinical Psychology*, 1965, *29*, 78–80.
Schover, L. R. Male and female therapists' response to male and female client sexual material: An analogue study. *Archives of Sexual Behavior*, 1981, *10*, 477–492.
Schultz, D. Imagery and the control of depression. In J. L. Singer & K. S. Pope (Eds.), *The power of human imagination*. New York: Plenum Press, 1978.
Segal, S., & Glickman, M. Relaxation and the Perky effect: The influence of body position and judgements of imagery. *American Journal of Psychology*, 1967, *60*, 257–262.
Sheehan, P. W. A shortened form of Bett's questionnaire upon mental imagery. *Journal of Clinical Psychology*, 1967, *23*, 386–389.

Shepard, R. The mental image. *American Psychologist*, 1978, *33*, 125–137.
Singer, J. L. *Daydreaming*. New York: Random House, 1966.
Singer, J. L. Experimental studies of daydreaming and the stream of thought. In K. S. Pope & J. L. Singer (Eds.), *The stream of consciousness*. New York: Plenum Press, 1978.
Singer, J. L. Romantic fantasy. In K. S. Pope (Ed.), *On love and loving*. San Francisco: Jossey-Bass, 1980.
Singer, J. L. Towards the scientific study of the imagination. *Imagination, Cognition, and Personality: The Scientific Study of Consciousness*, 1981–1982, *1*, 5–28.
Singer, J. L., & Brown, S. L. The experience-type: Some behavioral correlates and theoretical implications. In M. A. Rickers-Orsiankina (Ed.), *Rorschach psychology*. Huntington, N.Y.: Krieger, 1977.
Singer, J. L., & Pope, K. S. Imagery and fantasy techniques in psychotherapy. In J. L. Singer & K. S. Pope (Eds.), *The power of human imagination*. New York: Plenum Press, 1978.
Starker, S. Dreams and waking fantasy. In K. S. Pope & J. L. Singer (Eds.), *The stream of consciousness*. New York: Plenum Press, 1978.
Starker, S. Fantasy in psychiatric patients: Exploring a myth. *Hospital and Community Psychiatry*, 1979, *30*, 25–30.
Streissguth, A. P., Wagner, N., & Eschler, J. D. Effects of sex, illness, and hospitalization on daydreaming. *Journal of Consulting and Clinical Psychology*, 1969, *33*, 218–225.
Suinn, R. M. Body thinking: Psychology for Olympic champs. *Psychology Today*, 1976, *10*, 38–43.
Titchener, E. B. *A textbook of psychology*. New York: Macmillan, 1909.

18

Psycholinguistics and Psychopathology

GILEAD NACHMANI

Psycholinguists and psychoanalysts both note areas of similarity in their two fields of study (Chomsky, 1978; Edelson, 1972; Freud, 1900/1953; Smith, 1978; Spence, 1979). In this chapter knowledge from structural linguistics will be brought to bear on specific psychodynamic phenomena, namely, a schizoid syndrome, in the hope of understanding it better. In the psychodynamic literature much more attention has been paid to semantics than to grammatics. This is evident from the vast literature on such subjects as symbol formation and dream interpretation. Manifest content signifies what is meant (latent content). The mechanisms that generate and organize meaning, such as condensation, displacement, and metaphor, likewise are more often studied insofar as they signify meaning, as opposed to their properties as fundamental organizing principles, rules, or as psychodynamic grammars. Furthermore, as Edelson (1972) points out, writers in the field of psychoanalysis often fail to draw a distinction between language as a capacity and speech as its realization in performance; psychoanalysts are generally concerned with semantics and not syntax.

By syntax what is meant is structure, and grammar is the set of rules that organize the structure. The grammar of a language is the way in which knowledge of a language is represented. It determines or generates how language sounds (phonetics) and what it means (semantics) as well as the form it takes (syntax). Fundamental ideas from linguistics

GILEAD NACHMANI • 115 East 87th Street, Suite 32F, New York, New York 10128.

are an important part of the fundamentals of psychoanalysis. As Ricoeur (1978) shows, for Freud symptoms were mnemonic symbols of traumata. The semiotic nature of mnemonic symptoms is realized through the psychoanalytic process itself where symptoms are eventually replaced (substituted) by discourse. What is of particular importance here is not the meaning of the symptom, but that through the grammatic principle of substitution a symptom ceases to be: It is transformed from symptom via substitution into discourse. Implicit here is the idea that the unconscious has a capacity to be realized as spoken language that is comprehensible through transformations. This point is addressed by Chomsky (1978) in his explication of grammar as a kind of unconscious knowledge—not unlike the system Ucs (Freud, 1925/1961). For psychoanalysts the process of transformation that renders the unconscious conscious is called psychoanalysis. From a structural psycholinguistic point of view, psychoanalysis could be defined as a basic capacity, a fundamental organizing principle, which is realized in a type of performance that demonstrates the competence of the underlying grammar of rules by which these transformations occur. Abstractly, language is a capacity realized as speech. Conflict is a capacity realized as resolution and compromise formation.

As mentioned above, a distinction can be made between language as a capacity, and speech as an indication of competence in an already existing capacity. This distinction poses an interesting challenge, particularly for psychodynamic psychotherapy. In this field, a psychotherapist may be confronted with several questions of linguistic competence (honesty, security operations/censorship, and verbalisms and "as if" phenomena). First, there is the question of conscious honesty. Does a patient say what he means? Is he saying something to achieve a particular purpose (a deception) and not a therapeutic goal mutually agreed upon by both therapist and patient? If there is a question of honesty, then the challenge of linguistic competence is a thorny issue. Are the liar's sentences grammatically unsound (not likely if it is a credible lie: Implicit in the lie is that the listener/therapist believes it because there is not yet a source of contradiction, doubt, or mistrust)? Most likely a good liar is grammatically competent and this is revealed by his initial credibility. In deceitful speakers, then, both language and performance competence may be quite intact, although questions might be raised about the speaker's motivation and integrity.

With regard to censorship and security operations (Sullivan, 1953, 1956), several areas from psycholinguistics must be considered. Censorship may have a grammatic quality not simply insofar as it represents the hypothetical boundary between conscious and unconscious expressibility, but also as a reflection of several grammatic processes such as condensation (the capacity to represent several thoughts simultaneously). With regard to the former, that is, the boundary between conscious and unconscious, Chomsky (1978, p. 28) presents an intriguing linguistic exegesis of the systems Cs, Pcs, and Ucs wherein admissibility to consciousness is considered in terms of a grammatic accessibility to consciousness, or system of transformations. In interpreting and amplifying Freud, Chomsky states that the system Ucs has no access to consciousness except through preconsciousness, and what is preconscious becomes conscious. These, then, are two distinct areas of grammatic transformations: The first concerns that which is preconscious; and these matters or contents become conscious simply because it is the grammatical "fate" of preconscious stuff to become conscious. The preconscious has easy accessibility to consciousness. The transformations involved in rendering the unconscious conscious—the second area of grammatic transformations—vary with the magnitude of the resistances: These transformations—that is, the psychoanalysis of the resistances—demonstrate that the content of the unconscious is accessible to consciousness both in principle and in fact. Yet the processes by which these contents emerge differ considerably, most notably by the intervention of a psychoanalyst (an "extragrammarian"?) when through the psychoanalysis of the resistances, unconscious material may become conscious material: Symptoms may become discourse. This can be considered a structural grammatical definition of a psychoanalytical process: a description of outcome. What, though, of a psychopathology where there is a seeming order to the grammar, yet not unlike the symptom, the meaning is obscure if not singularly absent?

This is a common state of affairs in the speech of obsessive-compulsive personalities. In such personalities spoken language plays a prominent role in their symptomatic behavior. They are often regarded as articulate speakers to the point of being narcissistic perfectionists, yet when it comes to the expression of affectively charged matters they are often incomprehensible. Their speech is grammatically intact insofar as their sentences are properly organized, but a listener can be hard pressed to understand them. They speak not to convey meaning, but to move away from meaning (Sullivan, 1953), to become obscure, and they use a certain kind of organization to create disorganization and distraction. In areas of conflict, obsessive defenses create relief through undoing, taking apart, disassembling, and in such a fashion that disorder is created in an orderly way. Obsessive phenomena involving splitting of grammatics from semantics where disorder is created for the purpose of relief are characteristic of schizoid and narcissistic conditions, not simply the obsessive neuroses. When it occurs, however, this confusion-making speech must be considered symptomatic, and as a type of incompetent performance. This performance reflects a singular structural pathology where the very principle of organization, which has the expected purpose of enhancing the expression of meaning, is seemingly split away from it, and, as if by a psychological dissection, one can see that grammatical organization and semantic organization can exist separately, in parallel, and perhaps even at cross purposes. Furthermore, by examining these phenomena one can see how grammatic structure, the rules of organizing communication, can in act come to represent shields or screens behind which a private (autistic) organization exists and in front of which a mindlessness exists for a purpose other than that of communication, or interpersonal relatedness and attachment, or intimacy for which spoken language is so often the means. Organization in this latter instance represents facade—just enough of a social and psychological ingredient to permit a person to be noticed enough to be not noticed.

This grammar, then, conveys a very different kind of meaning. It is structure for the purpose of being "as if" one were something else. This kind of "as if" phenomenon has been addressed elsewhere by such authors as Sullivan (1953), Deutsch (1965), and Meltzer (1975), who make it quite clear clinically and developmentally that language can have two distinct purposes: to inform and to deceive, and both purposes can be formidably organized. In deception as purpose, conscious cunning and conscious deceit need not be implied or even present at all. Deceit can be an unconscious but organized response to an intolerance of anxiety. It can also represent a particular type of developmental arrest where parents have unrealistically great expectations of their children. Most often these expectations constitute an omnipotent fantasy in which the child possesses special qualities, such as being not a child but a parent to its own parents. This can be a case of unresolved and unevolved symbiotic psychosis in the mother, who has a child not to become a parent but to be reparented. This is not an uncommon unconscious fantasy among many parents-to-be, but it is psychotic when practiced in reality. Such parents often possess an intolerance for dependent behavior, of course a *sine qua non* of infancy and childhood. Children born into such families are often high risks for early and profound emotional disorders. Among the pathological yet adaptive modes of experience that children of such parents develop is a particular precocious use of spoken language—they sound like adults so that the mother will not feel overwhelmed by the child's normal needs, and therefore her own incapacity to meet them. How can a pathologically dependent child whose dependency needs were not met, meet the needs of a dependent child?

My hypothesis here is that there is a linguistic response in children to very high expectations and very anxious mothering from parents.

This response takes the form of grammatically intact spoken language which is typically semantically void. This spoken language is often meaningless to listeners other than "the" mother, and its meaning for "the" mother is most likely that her child does not need her and does meet her expectations. By the time we see these children as patients they are already involved in profound object relations with essentially one object—their mother. They often come into therapy because their "as if" language did an effective job of reassuring their mother, and very few others. As a result, they are often at a major turning point in their lives where they have encountered several genuine failures in both working and loving. They have tried to live through the expectations of others and have foresaken their own selves and become insubstantial. Diagnostically they are most often schizoid narcissistic personalities with distinct capacities to become yet more disintegrated: They can be "as if" schizophrenics with distinct pathological issues of hostile dependency. Initially their capacity to disintegrate or get worse is no simple matter. One does not know if it is a reliving through reenactment of the psychosis of the parent, or a representation of the dread that they characteristically feel when they cannot meet or know the need one expects them to meet (they expect one to expect this), or generally the background level of psychosis that can develop in a person who from early childhood has lived with basically one object relationship—meeting the expectations, through a false facade, of a psychotic parent whose psychosis appeared diminished, at least less threatening to the child, if the child acted or said certain things in certain ways. The clinical picture so far presented appears grave and does need to be modified. Among the qualities that just as readily characterize these patients are several talents, skills, or areas of conflict-free competence or knowledge that they have begrudgingly "used" to survive in life; yet they consider *this* talent to be a tragic abuse of their self, which they perceive as exploitive and fraudulent. This particular split in self-perception appears to accompany the grammatic-semantic structure split, and they may both represent facets of some very fundamental pathological split.

On the one hand, these people are chronically motivated to behave as one wishes them to. In fact, anticipating one's expectations of them is something they consider elementary and failing to do so is associated with a precipitous loss of self-esteem. They expect to read one's mind. This grandiosity pervades virtually all of their interpersonal expectations: They expect minds to be read and likewise expect one to read theirs. Spoken language for the purpose of informing another about who they are is regarded with contempt and annoyance. Their expectations and behavior suggest that they are committed to relating in a primitive, empathic mode and that their object relatedness is well within the domain of the paranoid-schizoid position. Their condemnation of spoken, consensually valid language is in one respect a bold statement of their psychopathology, whereas in another respect it is an insistence on empathic preverbal relatedness indicative of the failure in their mothering and the need for "repair." That their spoken language is grammatically intact, well-organized, and socially acceptable may indicate a precociousness in the area of appearing normal through speech with one or two people, such as mother. But they seem to have a private lexicon that upon psychoanalytic reconstruction, forces one to conclude that form and not meaning was critical to the others upon whom they depended: Appearance was important, substance irrelevant. This theme emerges early in their psychotherapies, for the therapist and not necessarily for them. For them, the theme of structure organizing substance, that is, what mommies can do for babies, when it emerges does so with an explosiveness and a genuine air of crisis or of imminent catastrophe.

Such an event is not so much an insight into what was repressed out of awareness, but an introduction into awareness of something that until recently could not even be conceptualized. Specifically, they come to realize in the present tense the terror that they probably experienced as children in the face of psychotic parents. As children they could not reflect upon it and call it experience; what they could do was develop a precocity for the use of form. The capacity to feel and to reflect upon what one feels involves a good-enough other to help one postpone, until one is capable of, the more cognitive transformations of affect into language. There is a kind of precocity for adapting to forms and expectations of the other that indicates that auxiliary ego functioning from mother to child was rarely extended. This extension, particularly during periods of total dependency in childhood, is the process that sets off the capacity to transform feelings into thoughts and words. Mothers repeatedly identify, empathically via projective and introjective identification, with their children. When they can do so, some feelings become familiar and they eventually name them. Early on mothers can differentiate types of crying (e.g., hunger, anger, frustration) and respond accordingly. When mothers cannot respond, perhaps because they become panicked, the crying of the child rapidly loses its communicative potential and the child may precociously learn not to cry, simply because not crying means less anxious mommy—or, in the symbiotic minds, less bad breast, less dread, less bad-me and less not-me!

When a mother can "hold onto" her child's feelings, she can eventually come to name them. When she cannot then she will not know what the child feels and will not be able to inform the child, not simply by naming, but, most critically, by acting accordingly and then naming when the child can appreciate the name. If this does not happen, then the child is exposed early on to chronic helplessness in the face of certain feelings. Among the most important of these feelings is that of its own mother's madness—the empathic introjective identification of this madness: an experience that in its own right might be called "nameless dread" (Bion, 1967). The psychotherapeutic achievement of sensing what it was like to be with mother when she was mad brings attention to yet another psycholinguistic phenomenon, specifically, a singular and definitive confusion or misidentification of tense. This reexperience, or to be more accurate, this experience of the thing-in-itself for the first whole time, is generally felt to be a "fear of breakdown" (Winnicott, 1974).

By "fear of breakdown," Winnicott has a specific type of experience in mind. Breakdown means a failure of organized defenses during a time, infancy, of absolute dependency on the mother. Thus, failure of defense means failure of auxiliary ego-functioning. This fear of breakdown, this expectation of some dreadful event in the future, had already occurred. The paradoxical quality by which even the past is confused with a future expectation is a striking experience for analyst and patient. A patient is convinced that something dreadful—a breakdown—will occur. It has a phobic organization, and avoidance behavior of it is common even though the "it" is history rather than future. Now the confusion of tense is something specific to this particular experience and does not characterize such patients' thinking in general. They clearly know the distinction between tomorrows and yesterdays. In the realm of experience characterized by total dependency, a certain dismantling occurs through which sense is taken apart (Meltzer *et al.*, 1975). This loss of the capacity to tell past from future is no small effect: It is stunning. It is an insult to the very organizing potentials that grammars possess. From what experiences can grammar be disrupted?

If, as Winnicott (1974) hypothesizes, "fear of breakdown" occurs in a state of total dependency, then the infant has not yet differentiated the "me" from the "not-me": that is, has not developmentally advanced beyond total dependency. The infant's mind and ego are not able to gather phenomena into their own omnipotence. The infant cannot yet even in fantasy localize the sources of its pleasures and pains. The original experience, then, of the mother's agony and the infant's resulting suffering simply cannot be placed into the past tense until it can first be placed into the present tense, and then into omnipotent control, and then finally into a past tense. This can only occur when there is adequate auxiliary ego-support. An analyst or therapist may provide just that support, and only then can the patient look for his or her own past, which has not yet been experienced, in the future. Epistemologically and interpersonally that makes some sense; but what of the original grammar-disruptive process? Clearly, a totally dependent child is quite vulnerable; yet these patients have in some respects grown up with some successes in their lives. In a way, they have developed asynchronously.

At their most vulnerable they have lost the capacity to relate to objects—a secondary autism, a withdrawal from reality. This is especially evident if one considers that the appreciation of temporal relatedness (i.e., tense) is a primal object relationship. Where are the psychological mechanisms that can explain the disruption of a fundamental organizing principle? In the research by Meltzer *et al.* (1975) on autistic children, the process of dismantling is strong enough, effective in outcome, developmentally timely, as well as passive—a criterion that should not be minimized in dealing with a totally dependent child. Dismantling is a passive process, perhaps the precursor of undoing, which allows the various senses to attach themselves to the momentarily most stimulating object, as opposed to conjointly focusing on the same object. By conjointly focusing on the same object, common sense evolves—all senses focus upon and confirm the same object (one hears what one sees). Attention holds the senses together. Dismantling undoes attention. If attention is suspended, the senses are not consensually focused upon an object and an object relation cannot exist. This, by definition, is a condition of autism. Since there is no object relation in a dismantled state of mind, there are neither persecutory nor depressive anxieties: A remarkably economical solution to an agonizing object relationship is to unrelate it. This may be why in other aspects of their lives, these patients have some capacity to get on with living—they do not agonize such as in persecutory or depressive anxieties. Likewise, this may also account for the sense of imminent catastrophe associated with the fear of breakdown: Once dismantling has begun to be relinquished as a mode of relating and replaced by attention to an object through an object relationship, the novel experience of anxiety (persecutory or depressive) with regard to object loss is startling.

I have attempted to outline a possible relationship between structural grammatics and psychodynamics, in order to gain a better grasp of an essentially psychotic syndrome with autistic and obsessional characteristics. Intrinsic to this syndrome are "as if" phenomena and a confusion of future with past tense. Both of these factors reflect early developmental disruptions of the capacities to generate meaning. In "as if" phenomena, there is a confounding of structure with function, of appearance with substance, and difficulty in creating interpersonal transitions from the superficial to the intimate. In future–past confusion, one sees again a fundamental grammatic disruption where phobic and obsessional preoccupation are symptomatic results of exposure to maternal psychopathology. In both cases, one can psychoanalytically reconstruct the symptomatic asset to a highly anxious mother–child relationship—in particular, where the child is very dependent, if not totally dependent, upon an anxious mother. The disruption that results can be seen in incapacities to organize thought and express it through language. In fact, there is a singular incapacity to use grammatic organizing principles in order to become more articulate and consensually valid.

REFERENCES

Bion, W. R. *Second thoughts.* New York: Jason Aronson, 1967.

Chomsky, N. *Psychoanalysis and language.* New Haven: Yale University Press, 1978.

Deutsch, J. Some forms of emotional disturbances and their relationship to schizophrenia. In *Neurosis and character types.* New York: International Universities Press, 1965.

Edelson, M. Language and dreams: The interpretation of dreams revisited. *Psychoanalytic Study of the Child,* 1972, *27,* 203–282.

Freud, S. The interpretation of dreams. In J. Strachey (Ed. and Trans.), *Standard edition of the complete psychological works of Sigmund Freud* (Vols. 4 and 5). London: Hogarth Press, 1953. (Originally published, 1900)

Freud, S. Some additional notes on dream-interpretation. *The standard edition* (Vol. 19). London: Hogarth Press, 1961. (Originally published, 1925)

Lacan, J. *Ecrits I.* Paris: Seuil, 1966.

Meltzer, D., Bremner, J., Hoxter, S., Weddell, D., & Wittenberg, I. *Explorations in autism.* Perthshire, Scotland: Clunie Press, 1975.

Ricoeur, P. Image and language in psychoanalysis. In J. Smith (Ed.), *Psychoanalysis and language.* New Haven: Yale University Press, 1978.

Smith, J. *Psychoanalysis and language.* New Haven: Yale University Press, 1978.

Spence, D. Language and psychotherapy. In D. Arronson & R. W. Rieber (Eds.), *Psycholinguistic research.* Hillsdale, N.J.: Lawrence Erlbaum, 1979.

Sullivan, H. S. *The interpersonal theory of psychiatry.* New York: W. W. Norton, 1953.

Sullivan, H. S. *Clinical studies in psychiatry.* New York: W. W. Norton, 1956.

Winnicott, D. W. Fear of breakdown. *International Review of Psychoanalysis,* 1974, *1,* 103–108.

19

Religious Experience and Psychoanalytic Psychotherapy

KAISA PUHAKKA

Religion is a world view that imbues thought, feeling, and action with significance and value. Such a characterization is, of course, too broad to be considered a definition of religion. Its purpose is merely to call attention to some basic features that religion shares with other forms of human thought and action—in particular, with science. Scientific activity, like religious activity, is embedded in a world view that furnishes its rationale and purpose. The religion of the average person is seldom articulated as a coherent system of value and belief. But the same holds true for the average scientist. As Kuhn (1971) has pointed out, the practitioners of science are usually unaware of the presuppositions that guide their scientific practice. Long before Kuhn called attention to the similarity between this aspect of scientific and religious practice, William James (1902/1979) had noted another obvious similarity, namely, that not only religious ideas and sentiments but also scientific ones are rooted in the psychological and physiological conditions of the human mind.

Religion thus poses a unique challenge to its scientific investigation, in that the investigating science and the object of inquiry are both instances of human thought and action. The two world views enter into a peculiar reflexive encounter in which human thought and action looks at itself. Such an encounter is not unlike the psychotherapeutic situation in which

KAISA PUHAKKA • Department of Psychiatry, Medical College of Ohio, Toledo, Ohio 43699.

the thought and action (i.e., both overt and covert communication) of the therapist reflect upon those of the patient. In both cases, the reflexive interplay of thought and action generates a creative, self-modifying process.

A major task of this chapter is to unravel the encounter between psychology and religion from the presuppositions of research that have obscured it. Thus, the epistemological assumptions and methodological prescriptions based upon them are at the focus of concern in the following review of the literature. Such an approach is justified by the growing dissatisfaction, if not disillusionment, with the progress of research that has become evident in the field during the past decade (Becker, 1971; Friedrichs, 1982; Tisdale, 1980; Wuthnow, 1981). Unwieldy presuppositions of theory and research are to blame for the current impasse. It is part of the purpose of this chapter to argue that an alternative approach is available in the psychological tradition. The roots of this approach go back to James's *Varieties of Religious Experience.* James's "radical empiricism" is free of the presuppositions that have burdened the investigation of religion from Freud and psychoanalysis down to contemporary social psychological approaches. As such, radical empiricism offers a model approach at a time when the investigation of religion is in search of a viable scientific paradigm. More importantly, this approach, unlike the others, does justice to the irreducible subjectivity of religious experience. Thus it becomes possible to address the unique issues that arise when the object of investigation is human thought, feeling, and action. This in turn enables one to examine the parallels between psychoanalytic psychotherapy and the investigation of religion, which will be taken up in the last part of this chapter. Such parallels provide the context in which the contributions of one field to the other can be fruitfully explored.

OBJECTIFICATION OF RELIGION: PSYCHOANALYTIC THEORY AND SOCIAL PSYCHOLOGY

The two approaches that have dominated the psychological investigation of religion are psychoanalytic theory and social psychological research. These approaches rarely see eye to eye on basic philosophic and methodological issues. Even such fundamental notions as "truth" and "objectivity" mean different things to them. Nevertheless, psychoanalytic theory and social psychological research share certain epistemological presuppositions the uncovering of which is crucial to understanding the problems that face contemporary psychology of religion. These presuppositions are traceable to a peculiar version of Cartesian dualism that has dominated social sciences since Auguste Comte. It combines Descartes's notion that reality is divided into two categories, the subjective (mental) and the objective (physical), with the positivistic notion that only the objective is knowable (Wuthnow, 1981). The task of science, according to such Cartesian dualism, is to eliminate the dichotomy between subject and object by somehow "objectifying" the subjective. In the Cartesian schema, the ideas in the mind of the scientist belong to the category of the subjective, and the transformation of these ideas into repeatable procedures is what enables science to capture the object of investigation, presumably as it is, untainted by subjectivity. But when the object of investigation is itself a subjective phenomenon, such as a world view or belief system, the quest for objectification faces a dilemma: Either the phenomenon under study is preserved as it is, in which case it remains subjective and hence unknowable, or it is somehow rendered objective and thus knowable, in which case it is no longer the original phenomenon. The psychological investigation of religion has been caught up in the second horn of the dilemma.

The objectification of religion in psychoanalytic theory has been accomplished in two steps. The first step is to focus exclusively on the cognitive content of religious belief. In phenomenological terms, this is to reify the object pole in the subjective experience associated with religion. Thus Freud equated "religion" with "religious ideas," which he defined as

> teachings and assertions about facts and conditions of external (or internal) reality which tell one something one has not discovered for oneself and which lay claim to one's belief. (1927/1964, p. 37)

This definition was challenged by a friend of Freud's who held that religion could be based solely upon an "oceanic feeling" that is wholly devoid of ideational content. Freud's disclaimer to any personal experiences with such an "oceanic feeling" is well known. More important in the present context is that Freud backed away from the "oceanic" as well as other kinds of religious feeling on scientific grounds: "It is not easy to deal scientifically with feelings," he wrote and concluded, "nothing remains but to fall back on the ideational content which is most readily associated with the feeling" (1930/1962, p. 12).

The second step in the objectification of religion consists of tracing the ideational content of belief to the unconscious dynamics of the psyche. The unconscious is objective in that it is extraneous to the subjective experience of the person. The second step is based on the functionalist premise that religious beliefs fulfill other (nonreligious) needs. Freud (1907/1970) and other psychoanalytic writers (e.g., Reik, 1931) have elaborated these needs in terms of the dynamics of obsessional neurosis.

The objectification of religion in psychoanalytic theory culminates in the latent content of religious beliefs. But the latent content is excluded from the domain of experience altogether, it being merely an object for psychoanalytic theory. Thus, the objectivity sought by the psychoanalytic approach to religion may be called "theoretic objectivity." A datum is objective by virtue of conforming to theory. The possibility that there might exist religious feelings that are not motivated by nonreligious needs was incongruous with theory. Freud's defense for the exclusion of such feelings from the domain of objectivity (i.e., legitimate objects of investigation) was simply that the very idea

> sounds so strange and fits in so badly with the fabric of our psychology that one is justified in attempting to discover a psychoanalytic—that is, a genetic—explanation of such a feeling. (1930/1962, p. 12)

The "theoretic objectivity" of psychoanalysis has been an anathema to the empirically minded social psychologists. From the empiricist standpoint, the psychoanalytic formulations concerning religion are locked in a self-validating theoretic framework. It is not that testable hypotheses could not be derived from the theory, but that the tests and measuring instruments developed within the framework tend to be biased. Research on the psychoanalytic notion that religious experiences are associated with regressed ego states illustrates the point. In a review of this research, Hood (1973) demonstrated that an ego-strength scale derived from psychoanalytic concepts (Barron, 1953) correlated negatively with a measure of religiosity. However, when Barron's scale was replaced by a nonpsychodynamically oriented measured of psychological strength (Stark, 1971), positive correlation was obtained between psychological strength and religiosity. Examples of this sort abound in the literature and they have undermined the empirically oriented researchers' faith in measuring instruments derived from theory. Instead of looking to theoretic formulations, social psychologists have looked to their religious subjects for a definition of religion.

The suspicion with which social psychologists view theoretic formulations is deeply rooted in the empiricist tradition to which they claim allegiance. To the empiricist, theories are ideas in the minds of the theorists and thus fall within the category of the subjective along with religious beliefs. If the investigation is concerned with the latter, then why not deal with them directly? This is indeed what social psychologists have attempted to do. But the question as to how to render subjective beliefs accessible to research has been the enduring problem for social psychologists. Thus the task of objectification of religion has consisted of rendering public what is assumed to be essentially private—individuals' beliefs, attitudes, and experiences concerning their religion. The underlying conception of objectivity could be characterized as "methodological objectivity," in contrast with the "theoretic objectivity" of psychoanalytic theory.

Relying mainly on self-report questionnaires, social psychologists have sought to identify the various dimensions of religion. First, Glock (1962) proposed the following five dimensions: experiential, ritualistic, ideological, intellectual, and consequential. Other dimensions were soon added to the list (King & Hunt, 1972; Stark & Glock, 1968). It became evident that the list is potentially limitless and further accumulation of the dimensions would be of little help toward understanding what they were dimensions of. The essence of religiosity thus seemed to elude the multidimensional approach. A search for a unitary construct was launched by DeJong, Faulkner, and Warland (1976). The parallel between this endeavor and Spearman's search for the *g*-factor of intelligence has been pointed out by Kavanagh (1978). Indeed, a common factor, called "generic religiosity" by the authors, was found to underlie three of the six dimensions studied, namely, belief, practice, and experience. But what was "generic religiosity"? It was certainly "objective" in that it was a construct generated by publicly observable methods and not by private thoughts and experiences. But just for that reason it was as conceptually vacuous as it was devoid of experiential content. Thus social psychological research, intent upon rescuing religion from the mind of the theorist, ended up with an abstraction far removed from the mind and heart of the religious person.

INTRINSIC AND EXTRINSIC ORIENTATIONS IN RELIGION AND METHODOLOGY

The objectification of religion has left social psychological research almost empty-handed, whereas psychoanalysis has at least a theory of religion. In the true empiricist tradition, social psychologists left it up to the religious person to determine what religion is. Implicit in such a liberal gesture is the conviction that religion is *intrinsic* to the religious person in that he, rather than a preconceived theory, determines what religion means to him. The job of the empiricist is merely to observe, record, and understand—in that order—what religion means to people. Yet the methodological presuppositions of research, which are derived from Cartesian dualism, dictate that religion be investigated as *extrinsic* to the religious person, lest it remain subjective and hence unknowable. "Extrinsic" in the present context simply means "derived from a source external to the religious person." Correspondingly, "intrinsic" means "produced by the person himself." In the majority of cases, the religion that has been the object of social psychological research is not the individual's religiosity in its various experiential and behavioral manifestations, but rather the social psychologist's conceptions about the individual's religiosity. The role of the religious person in the research has been that of a collaborator who helps clarify and, most of all, reinforce, the social psychological view about religion. To put the matter differently, religion that is intrinsic to the religious person is religion as lived and experienced by him. Insofar as endorsing items in a self-report questionnaire is not the same thing as living and experiencing one's religion, it is fair to say that the religion investigated by social psychologists is *extrinsic* to the religious person.

The intrinsic-extrinsic distinction as drawn above refers to a methodological commitment on the part of the investigator. The methodological commitment to the extrinsic view is nothing but the quest for objectification as discussed earlier. Religious phenomena are made objective by turning them into objects (in the form of questionnaire items or formulations about unconscious dynamics) for the investigator as well as the religious person.

A better-known meaning of the intrinsic-extrinsic distinction refers to two types of religious orientation. In this meaning, "intrinsic" and "extrinsic" refer to the religious person's own beliefs, attitudes, and experiences rather than the methodological orientation of the investigator. As characterizations of religious orientation, these concepts were first introduced by Allport (1950) and they figure prominently in his subsequent studies on prejudice. "Extrinsic orientation" is defined by Allport and Ross (1967) as follows:

> Extrinsic values are always instrumental and utilitarian. Persons with this orientation may find religion useful in a variety of ways—to provide security, and solace, sociability and distraction, status and self-justification. The embraced creed is lightly held or else selectively shaped to fit more primary needs. In theological terms, the extrinsic type turns to God, but without turning away from self.

"Intrinsic orientation" is defined by the authors thus:

> Persons with this orientation find their master motive in religion. Other needs, strong as they may be, are regarded as of less ultimate significance,

> and they are, so far as possible, brought into harmony with the religious beliefs and prescriptions. Having embraced a creed the individual endeavors to internalize it and follow it fully. It is in this sense that he *lives* his religion.

The parallels between the two meanings of the intrinsic-extrinsic concept, namely, orientation of the religious person and methodological commitment of the investigator of religion, are obvious. Thus the extrinsic orientation in the methodologies of social psychology as well as psychoanalytic theory can be described as instrumental and utilitarian. Theories and hypothetical constructs are the creeds that the researchers hold lightly, always secondary to the concern with validity or overall theoretic coherence and fit. Research data in turn are instrumental to confirming (or disconfirming) theory or validating research tools.

The instrumental value of observation and data in research finds its most direct counterpart in the instrumental value that religious belief and practice have for the extrinsically oriented person. Could the affinity—or incompatibility—between the orientations of the religious person and the investigator of religion influence the outcome of research? I shall argue that it not only could, but necessarily does. A case in point is the fate of research into Allport's intrinsic religious orientation. Efforts to operationalize this concept have met with persistent failure (Allport & Ross, 1967; Feagin, 1964). This instructive example merits a closer look. An examination of the meaning of "intrinsic" as used by Allport will indeed show that such an orientation does not lend itself to investigation from the extrinsic perspective. In other words, the extrinsic methodological orientation introduces a bias in favor of the extrinsic religious orientation.

According to Allport, intrinsically oriented persons *live* their religion rather than use it for other ends. Religion for such people has relevance to all of life; it floods their entire life with meaning and motivation (Allport, 1959, 1966). Put differently, "intrinsic" means that the person is "in" or "part of" his religion, just as one is "in" or "part of" life itself. The point is that religion in the intrinsic orientation is not an object of contemplation, much less an object to be used for some practical gain. Religion in the intrinsic orientation is not an object at all but is the matrix within which practical concerns acquire significance and the meaning of life is contemplated (by those who are so inclined). In Allport's definition of intrinsic orientation, religion provides the ultimate criterion for significance. On the other hand, the question, What is the significance of religion itself? assumes that there is some other source of significance that is extrinsic to religion. The investigator's desire to know what religion means to people may thus introduce the extrinsic bias into his study. Consider, for example, the following item from Feagin's (1964) Intrinsic-Extrinsic Scale: "Religion is especially important to me because it answers many questions about the meaning of life." Endorsement of this item was construed as an indicator of intrinsic religiosity. The affinity between this item and Freud's definition of religion quoted earlier is ironic, in light of Freud's avowed conviction that all religion is extrinsically motivated. Small wonder that many subjects who endorsed the extrinsic items on Feagin's scale also endorsed the intrinsic ones. The subjects may not have been "indiscriminately proreligious" as Allport and Ross (1967) assume but, more likely, the scale was extrinsically biased and hence did not allow for discrimination between the two orientations.

In concluding their review of the research on the intrinsic-extrinsic concept, Hunt and King (1980) raise the following issues:

> The root meanings of [intrinsic] have been "inner," and "real," and "ultimate." All require inferences made from observed and reported behaviors that are believed to be symbols, indicators of something beyond themselves. Deciding what is "real" is a metaphysical, not empirical, operation. Deciding what is "ultimate" is even harder. Who has the perspective from which to decide what is in fact ultimate, even in the psychic economy of one person? (p. 67)

In response to these queries, we note, first, that the decisions as to what is "inner," "real," or "ultimate" are always made from some perspective or other. Second, it is not the job of the investigator of religion to engage in "metaphysical operations," but it is his responsibility to know from whose perspective decisions as to the meaning of "inner" and so on are made—his own or the religious subject's.

RADICAL EMPIRICISM: AN ALTERNATIVE APPROACH

We have seen that the bias for extrinsic religiosity is deeply entrenched in the methodology and rationale of psychological and, in general, scientific inquiry. Elimination of this bias from the investigation of religion would require nothing less than the suspension of the utilitarian, pragmatic attitude that guides and provides motivation for research. Pragmatism in science is usually understood to mean that theories, hypotheses, and methods are *tools* that are useful only to the extent that they help make sense of data. In the reality of scientific practice, as Kuhn (1971) has pointed out, this relationship tends to be reversed: Data are gathered for the purpose of confirming theory, and whether a particular data-gathering project is pursued or abandoned depends on how well it serves this purpose. Few have grasped the ethos of pragmatism in science the way William James has. And for this reason, few have surpassed him in the art of suspending the pragmatic attitude when suspension is called for. He was specifically concerned with the type of religiosity Allport has characterized as intrinsic, and he not only provided excellent phenomenological descriptions but also valuable methodological ideas for the study of this type of religiosity.

James has come down in history as the great American pragmatist. But the methodological concept of James's that is of particular concern here is radical empiricism (James, 1912/1940). What radical empiricism is and how it differs from pragmatism may be illustrated with the help of an analogy. The pragmatist is like a mapmaker who constructs a map for the purpose of guidance through an unknown territory. The radical empiricist, on the other hand, ventures into the uncharted territory armed only with curiosity and keen observation. When James claimed that the rationale for his investigation of religion is that "irrepressible curiosity imperiously leads one on" (1902/1979, p. 35), he was calling attention to the fundamental feature that distinguishes his approach from that of his pragmatically minded colleagues, namely, that his is an *intrinsically motivated* inquiry. It is not motivated by pragmatic gains such as confirmation of a hypothesis about religion or further application of a measuring instrument. Of course, James wholeheartedly welcomed benefits that might—and indeed he hoped would—accrue from his investigations. But his point is that the anticipation of pragmatic gains should be built neither into the methods of investigation nor into the definition of religion.

James proposed a working definition of religion as

> the feelings, acts, and experiences of individual men in their solitude, so far as they apprehend themselves to stand in relation to whatever they may consider the divine. (1902/1979, p. 42)

This definition is remarkable in how little it gives to the investigator and how much it leaves up to the religious individual. As such, it captures the essence of intrinsic religiosity: The individual stands "in relation to" whatever he considers the divine, rather than "outside" as a commentator on his concept of the divine or the significance of his religion. But where does it leave the investigator of religion? Insofar as religion is acknowledged to be wholly an insider's affair, the investigator remains wholly outside of it. This situation, however, is not really different from any other in which phenomena of human consciousness are being investigated. Everyone is an insider to just his own conscious experiences and an outsider to everyone else's. Radical empiricism refrains from judgments based on Cartesian-positivistic assumptions about the unknowability or unreality of the kinds of phenomena that are labeled "subjective." Rather, the method of observation may need to be adjusted so as to yield maximum information about such phenomena. In such a thoroughly empiricist conception, observation is truly an art—in which James, of course, was an unsurpassed master. The art of observation in radical empiricism consists of effortlessly shifting back and forth between the outsider's behavioral standpoint and the insider's phenomenological standpoint without, so to speak, getting caught in the middle. In the middle position, the objects of observation are neither behaviors nor directly experienced subjective phenomena but definitions and statements about these. We have seen that both psychoanalytic theory and social psychological research have dealt with such objects. In one case, they were derived from theory, in the other from empirical scales and other measuring instruments.

Much of the material presented in *Varieties* is derived from literary sources, and personal anecdotes and vignettes about well-known religious figures. James treated this material as religious behavior, that is, as expressions *of* religiosity rather than as statements *about* religion. But now the difficult question arises: If James had neither a theory nor a self-report questionnaire to guide his search for religious expression, how did he know which among the variety of phenomena to count as "religious"? It is here that James shifted to the phenomenological standpoint. He focused upon "feeling" as the chief modality of religious consciousness, not because he believed religion to be a matter of sentimental excesses but because "feeling" best captures the moment of consciousness which, in its phenomenological entirety, is far more than a mere object of cognition. He talked about the solemnity, tenderness, and rapturousness of these feelings with words that could almost evoke them in the reader. The thread that runs through and holds together the subtle nuances of feeling and the range of visions from the bizarre to the sublime is not a theory of religion nor even a definition of it but what could only be described as a sense of intimacy. Again, the question may be raised, How is it possible for an investigator who acknowledges his outsider's position with respect to the phenomena he observes to observe them with the closeness and intimacy of a participant? The astuteness of James's observations is often attributed to his considerable literary gifts. These may indeed explain how he managed to convey what he saw to his readers, but not how he saw what he did. James had an insider's access to his own consciousness, and he fully exploited this source of information. He was cognizant of the fact that, in the last analysis, subjective consciousness is the only arena in which direct observation of phenomena of consciousness can be carried out. Hence his personal explorations in this arena are consistent with radical empiricism.

Radical empiricism offers a model for a science that lacks a paradigm. Its dictates are simple: Stay close to the phenomena that are being investigated and approach them on their own terms regardless of whether they are "subjective" or "objective." James's more important and enduring message, however, is that the lack of paradigm in the investigation of religion may in fact be an advantage rather than a deficiency. For it keeps the investigator at the edge of the uncharted territory, which is what human consciousness is as much today as it was in James's time.

Recent literature shows signs of moving away from elaborate manipulations of data to a search for the experiential qualities of religious phenomena. Two types of approaches manifest this new trend. They correspond to the "insider's" and "outsider's" standpoints as discussed above. The first is a phenomenological investigation of the modalities and ramifications of some central feature of religiosity. Pruyser's (1974) work on religious "belief" and "unbelief" falls into this category. The second type of approach attempts to capture the qualities of religious experience by means of open-ended questionnaires or interviewing techniques. An excellent illustration of this approach is Hay's (1979) study. Significantly enough, Hay's solution to the dual problem of how to get at the personal experiences of the individuals who served as subjects in the study and, at the same time, how to preserve the intactness of these experiences was to use a modified Rogerian technique of interviewing. According to Hay, this allowed rapport to develop so that subjects tended to become more open. Thus, when the subject matter to be investigated lies within the innermost personal experience, there is, inevitably, a convergence of methods as well as aims of research and clinical practice.

RELIGIOUS CONSCIOUSNESS AND PSYCHOTHERAPY

Given such a convergence of methods and aims, what can psychotherapy learn from the investigation of religion and, conversely, what can the investigation of religion learn from psychotherapeutic practice? With these questions we move to a metalevel of inquiry, into the nature and general features of the two endeavors themselves. Such a metalevel inquiry comes naturally to psychoanalysts and psychotherapists who, as part of their ongoing work, reflect upon the therapeutic process and the therapist–patient relationship. But social scientists—with the exception of sociologists of knowledge—seldom reflect upon the process of their investigation, much less upon the relationship between themselves and the object of investigation. Indeed, the objectification of religion, which has been a basic methodological prescription in both psychoanalytic theory and social psychological research, can be seen as an attempt to deny the significance and even the very existence of the relationship, first between the religious individual and his religion, but ultimately between the investigator and the object of investigation.

In order, therefore, to render salient the parallels between the two endeavors, psychoanalytic psychotherapy and psychological study of religion, one needs a framework for a metalevel inquiry of the latter. A suitable framework is provided by ethnomethodology, which is an outgrowth of sociology of knowledge.

The point of departure for ethnomethodology is the observation made by contemporary sociologists of knowledge that social scientific theory construction as well as validation reflect the social conditions (i.e., commonsense notions, values, and mores) in which they arise (Berger & Luckmann, 1966; Friedrichs, 1970). Ethnomethodology extends this observation to all human thought, feeling, and action. That is, in all spheres of life, thought, feeling, and action are essentially *reflexive* (Garfinkel, 1967; Mehan & Wood, 1975). They, so to speak, mirror the situations in which they arise. In the social scientific sphere, the reflexivity or "mirroring" of thought and action is best illustrated by self-fulfilling and selfnegating prophecies. In principle, however, all research illustrates this phenomenon insofar as the purpose of research is to confirm theory. When the object of investigation is itself human thought and action that is capable of mirroring the world, as religion is, the reciprocal reflections can lead to an endless progression of ideas, like the images one sees when one looks into a mirror within a mirror. But unlike the mechanical reproductions in a mirror, the reflections of human thought and action (perhaps by virtue of their very imperfection) constitute an ever-changing, creative dialectic or "hermeneutical spiral" (Mehan & Wood, 1975, p. 199). A specific illustration of the reciprocal mirroring in the investigation of religion was provided by the "intrinsic" and "extrinsic" orientations that both the investigator and the religious person have in their respective domains of thought and action. Our earlier discussion of this case also illustrated how the dialectic of the investigation can grind to a halt because of unawareness by the investigator of his own orientation.

There has been a growing recognition among psychoanalytically oriented psychotherapists that the therapeutic situation involves a two-way interactive process, much like the creative dialectic described by ethnomethodologists. The evolution of the psychoanalytic concepts of transference and countertransference finds a striking parallel in the evolution of the social scientific understanding of the relationship between two systems of thought when one is the object of investigation for the other. This parallel has culminated in the use of the same metaphor, "mirror," to describe, in one case, the relationship between world views and, in the other, between selves. Ethnomethodologists maintain that the mirroring relationship is constitutive of the entities in the relationship. Similarly, object relations theorists (Winnicott, 1980) and self-psychologists (Goldberg, 1978; Kohut, 1971, 1977) talk about the mirroring that occurs between mother and child as part of the developmental process in which the child's self is constituted.

The psychological investigation of religion can learn from psychoanalytic psychotherapy about the pathology of the mirroring relationship and how to avoid it. When the child is made in the image of the mother, or, in Kohut's terms, when the self is usurped by the "self-object," the growth of the child is stunted. We have seen that the psychological investigation of religion has been largely an effort—ever as vigorous as Freud's—to forge religion in the image of psychological theory and methodology.

In a more positive vein, psychotherapeutic approaches have been developed for dealing with the various ways in which mirroring occurs between patient and therapist—Kohut's "narcissistic transference (1971)" or "self–object transference" (1977). These approaches can offer valuable models, and in some cases actual techniques, for the investigation of religion. I have already cited an instance (Hay, 1979) of such a promising alliance between psychotherapeutic technique and the investigation of religion. The compatibility of Rogerian client-centered therapy with Kohut's approach to self–object transference has been pointed out by Stolorow (1976). Thus Hay's use of the Rogerian technique in his study is consistent with the notion that the encounter between psychology and religion in the persons of the interviewer and the informant is essentially reflexive. The authenticity of the information gathered by the interviewer depends on his ability to empathically reflect the informant's responses without interference from his own feelings and preconceptions about the informant or his views.

But what, if anything, can psychotherapy learn from the investigation of religion? We have seen that the investigation of religion can greatly benefit from the approaches and techniques of psychoanalytic therapy. Unfortunately, the psychoanalytic theory of religion has effectively precluded the application of these approaches and techniques to religious phenomena. What psychotherapy can learn from the investigation of religion is, then, to apply its own methods to understanding religion and therewith aspects of the psyche that find expression through religion.

One cannot but be struck by the parallels between James's radical empiricism and the broad principles of psychoanalytic psychotherapy.

Thus the intrapsychic approach to therapy finds its counterpart in James's methodological commitment to the position that religion is *intrinsic* to the religious person. The same respect that James showed to his religious subject is evident in the therapist's attitude toward his patient. In both cases this respect is based on the acknowledgment that the religious person (patient) is the sole author of his productions and that the role of the investigator (therapist) is to observe and not interfere. In both cases, the art of observation (or listening) depends on the observer's ability to stand at the edge of the unknown, suspend theory, and merely observe. Both the radical empiricist and the psychoanalytically oriented therapist move back and forth across the boundary of subjectivity, relying upon introspection as much as upon observing the other. The parallels could be expanded further. Suffice it for our purposes to merely point out that psychoanalytic psychotherapy has the tools for the investigation of human consciousness. The question is, Why are these tools not applied to religion? The answer, which I have tried to substantiate in the course of this chapter, is that psychoanalytic theory, like social psychological research, has not treated religion as a phenomenon of consciousness. This is not to say that these approaches have not provided valuable insights into the unconscious or environmental determinants of some religious phenomena. But they have done little to chart the territory of religious phenomena itself.

James's enduring message is that this territory is not a faraway land of misty metaphysics, obscure theology, or plain superstition—that even when it is all these things, to a psychologist it is still of the stuff of human consciousness. To James, this territory was magnificent and potentially limitless. Neither the subterranean influences nor the neurotic constitution that he admitted may well underlie the most exalted of religious experiences (1902/1979, p. 38) could detract from its grandeur and mystery.

The phenomenological investigation of religion enhances one's aesthetic appreciation of the dimensions of human consciousness. This, I believe, is the single most important contribution of psychology of religion to psychoanalytic psychotherapy. The word "aesthetic" is used here advisedly, to mean the kind of appreciation that dwells fully in the presence of the phenomenon. Such an appreciation is the hallmark of the art of listening, which is tuning into the present with one's mind as well as all one's senses, including the eyes (Khan, 1974). Of course, such an appreciation also promotes the clarity of the psychotherapeutic mirror and thus the accuracy of the images reflected therein.

In conclusion, it is an undeniable fact that, far from declining and vanishing away, religion today is of greater concern to many than ever before—as witness the revival of indigenous religions and the spread of religions from foreign lands. Religious commitment and quest are clearly on the increase. As such, it behooves the psychoanalytic therapist to recognize and understand religious consciousness in its own right, instead of beclouding it in a welter of preconceived theories and hypotheses.

REFERENCES

Allport, G. W. *The individual and his religion.* New York: Macmillan, 1950.

Allport, G. W. Religion and prejudice. *Crane Review,* 1959, *2,* 1–10.

Allport, G. W. The religious context of prejudice. *Journal for the Scientific Study of Religion,* 1966, *5,* 447–457.

Allport, G. W., & Ross, J. M. Personal religious orientation and prejudice. *Journal of Personality and Social Psychology,* 1967, *5,* 432–443.

Barron, F. An ego-strength scale which predicts response to psychotherapy. *Journal of Consulting Psychology,* 1953, *17,* 327–333.

Becker, R. J. Religion and psychological health. In M. B. Merton (Ed.), *Research on religious development.* New York: Hawthorne, 1971.

Berger, P., & Luckmann, T. *The social construction of reality.* New York: Doubleday, 1966.

DeJong, G. F., Faulker, J. E., & Warland, R. H. Dimensions of religiosity reconsidered: Evidence from a cross-cultural study. *Social Forces,* 1976, *54,* 866–889.

Feagin, J. R. Prejudice and religious types: A focused study of Southern fundamentalists. *Journal for the Scientific Study of Religion,* 1964, *4,* 3–13.

Freud, S. *Civilization and its discontents.* New York: W. W. Norton, 1962. (Originally published, 1930)

Freud, S., *The future of an illusion.* New York: Anchor Books, 1964. (Originally published, 1927)

Freud, S. Obsessive actions and religious practices. In W. Sadler (Ed.), *Personality and religion.* New York: Harper Forum, 1970. (Originally published, 1907)

Friedrichs, R. W. *A sociology of sociology.* New York: Collier, 1970.

Friedrichs, R. W. The functionalist paradigm dominating the social-scientific study of religion and a structural alternative. *Council on the Study of Religion Bulletin,* 1982, *13*(1), 1–5.

Garfinkel, H. *Studies in ethnomethodology.* New York: Prentice-Hall, 1967.

Glock, C. On the study of religious commitment. *Research Supplement of Religious Education,* 1962, July–August, 98–110.

Goldberg, A. (Ed.). *The psychology of the self: A casebook.* New York: International Universities Press, 1978.

Hay, D. Religious experience amongst a group of post-graduate students: A qualitative study. *Journal for the Scientific Study of Religion,* 1979, *18,* 164–182.

Hood, R. W., Jr. Forms of religious commitment and intense religious experience. *Review of Religious Research,* 1973, *15,* 29–35.

Hunt, R. A., & King, M. B. The intrinsic-extrinsic concept: A review and evaluation. In J. R. Tisdale (Ed.), *Growing edges in the psychology of religion.* Chicago: Nelson-Hall, 1980.

James, W. *Essays in radical empiricism.* New York: Longmans, Green, 1940. (Originally published, 1912)

James, W. *The varieties of religious experience*. New York: Collier, 1979. (Originally published, 1902)

Kavanagh, G. *The measurement of religiosity: Problems and progress*. Unpublished manuscript: Menninger Foundation, 1978.

Khan, M. *The privacy of the self*. New York: International Universities Press, 1974.

King, M. B., & Hunt, R. A. *Measuring religious dimensions: Studies of congregational involvement* (SMU Studies in Social Science, Vol. 1). Dallas: Southern Methodist University Press, 1972.

Kohut, H., *The analysis of the self*. New York: International Universities Press, 1971.

Kohut, H. *The restoration of the self*. New York: International Universities Press, 1977.

Kuhn, T. *The structure of scientific revolutions*. Chicago: University of Chicago Press, 1971.

Mehan, H., & Wood, H., *The reality of ethnomethodology*. New York: Wiley, 1975.

Pruyser, P. *Between belief and unbelief*. New York: Harper & Row, 1974.

Reik, T. *Ritual: Psycho-analytic studies* (D. Bryan, trans). New York: W. W. Norton, 1931.

Stark, R. Psychopathology and religious commitment. *Review of Religious Research*, 1971, *12*, 165–176.

Stark, R., & Glock, C. *American piety: The nature of religious commitment*. Berkeley: University of California Press, 1968.

Stolorow, R. D. Psychoanalytic reflections on client-centered therapy in the light of modern conceptions of narcissism. *Psychotherapy: Theory, Research, and Practice*, 1976, *13*, 26–29.

Tisdale, J. R. Introduction: Definition and measurement. In J. R. Tisdale (Ed.), *Growing edges in the psychology of religion*. Chicago: Nelson-Hall, 1980.

Winnicott, D. W. *Playing and reality*. Harmondsworth, U.K.: Penguin Books, 1980.

Wuthnow, R. Two traditions in the study of religion. *Journal for the Scientific Study of Religion*, 1981, *20*, 16–32.

20

Self-Fulfilling Prophecies in Psychodynamic Practice

ROBERT H. KEISNER

That people engage in self-fulfilling prophecies is not new. What is not generally known, however, is how pervasive this phenomenon is. This chapter will focus on the identification and exploration of this process in experimental, natural, and clinical settings. A synergistic relationship between research and practice, with fantasy-created realities as a conceptual bridge, will be demonstrated. It will become clear that people in all walks of life, normal and pathological, use their own fantasies and hypotheses to create social realities.

Two models, one descriptive and one explanatory, will be presented to enhance our understanding of this phenomenon in clinical settings.

OVERVIEW

At practically identical times in history George Kelley (1955), Robert Merton (1948; 1957), and Melanie Klein (1946; 1955) made the same observation: that social relations are constructed by the beliefs of the participants. This observation was explored by Kelley (1955) within the context of his theory of personal constructs. He believed that we develop

ROBERT H. KEISNER • C. W. Post Center, Long Island University, Greenvale, New York 11548.

and test hypotheses about human behavior in order to understand, predict, and control the world around us. Sometimes our actions are designed to make our experiences confirm our "constructs." For example, one individual "constructs" another person to be hostile and then provokes that person into acting hostile. The first person then believes that he has validated his "construction" of the other person as hostile. Kelley's example of this sequence was about a man who throws rocks at his neighbor's dog, causing his neighbor to respond with angry rebukes. As a result he concludes that he has confirmed his "construct" of his neighbor as a hostile person.

At about the same time, Merton presented a sociological analysis of how stereotypes and other social perceptions are determined by self-fulfilling prophecies. According to Merton, "the self-fulfilling prophecy is a false definition of the situation evoking a new behavior which makes the originally false conception come true" (1957, p. 423).

Klein, approaching the understanding of human behavior from a psychoanalytic instinct perspective, described this process as mediated by projective identification. This is a complex process of "identification by projection that involves a combination of splitting off parts of the self and projecting them into another person" (Klein, 1955, pp. 311–312). She speculated that these processes have a fundamental effect on interpersonal relationships. That is, they allow individuals to control parts of themselves by controlling the targets of their projections. This conceptualization of how people unconsciously control their environment, by "empathically" controlling others to act, think, and feel like parts of themselves, has contributed enormously to our understanding of psychopathology and psychotherapy interactions.

The first section of this chapter will be devoted to the controlled empirical research on the social psychology of behavioral confirmation processes whereby people actually induce others to behave as expected so as to have their hypotheses about them come true. The subsequent section will summarize some of the major sociological writings stemming from Merton's concept of self-fulfilling prophecies. This literature consistently maintains that deviancy and stigmatizations are most appropriately understood as both born and nurtured by the "normal" social practices of labeling. The succeeding section will review some of the rich and complex psychoanalytic literature on the role of unconscious communication patterns in (1) pathological mother–child dyads where parental wishes are translated into childhood behaviors, and (2) patient–therapist interactions, where projective identification processes influence treatment experiences. Finally, some theoretical and clinical applications of these diverse but overlapping observations will be discussed and then illustrated by a clinical vignette.

BEHAVIORAL CONFIRMATION: A SOCIAL PSYCHOLOGICAL PERSPECTIVE

One of the most widely publicized research areas to develop out of the "skeptical" 1960s has been called the "Social Psychology of the psychology experiment" (Orne, 1969; Rosenthal, 1966; Rosenthal & Jacobson, 1968). Rosenthal's work on experimenter and teacher expectancy effects gained the greatest social attention primarily because of its implications for the education of the less socially and economically privileged. He showed that experimenter expectations of a subject's performance (based on induced experimental hypotheses) and teacher expectations regarding pupils' performance are significant determinants of respective behaviors and achievements. He demonstrated that teachers and experimenters actually elicit from their students and subjects what they want and expect to observe. Furthermore, in a comprehensive review of experimenter effects in experimental psychology, Rosenthal (1966) documented this phenomenon in the investigation of Rorschach validity, successful ESP performance, psychophysical and conditioning research, and the defensive mechanism of projection. His own research, however, is somewhat limited in scope because the persons with expectations were systematically and consciously presented with these expectations, and because their low-status positions made them especially prone to comply with the implicit demands placed on them.

The self-fulfilling nature of expectations has also been demonstrated in the study of interpersonal attraction (Jones & Panitch, 1971), interracial interactions (Ward, Zanna, & Cooper, 1974), and sex-role stereotyping (Zanna & Pack, 1975). Using a Prisoner's Dilemma game, Jones and Panitch (1971) induced beliefs in their subjects (receivers) that their partners (senders) either liked or disliked them. The receivers then chose between cooperative and competitive moves, whereupon the moves of the senders were observed as well as the attraction scores of both players. It was found that the receivers had more positive feelings for the senders in the "like" condition and that (male) receivers were relatively more cooperative in the "like" condition. Most significant, however, was the observation that male senders developed more positive feelings for their partners in a "like" condition than in the "dislike" condition. The authors hypothesize that a chain of events occurred in which the receivers, believing that the senders liked or disliked them, responded with rewards or punishments to benefit or harm the senders and then the senders actually disliked or liked the receivers because of the way they were treated. These findings represent controlled experimental support for the self-fulfilling prophecy and suggest that reciprocity is a key ingredient in this process.

Zanna and Pack (1975) examined the hypothesis that females will conform to sex-role stereotypes that desirable males have regarding so-called feminine behavior. Female subjects were given a chance to present themselves attitudinally and behaviorally to a highly desirable or not especially desirable male whose view of the ideal woman was known to closely resemble either a traditional or nontraditional stereotype. Not only did females portray themselves according to the views of others, they also adjusted their actual behaviors in such a way that those who were expected to interact with a desirable "nonsexist" male outperformed those expected to interact with a desirable "sexist" male. These findings suggest that stereotyped behaviors regarding sex differences (including "deficient" performance) may be partly a consequence of stereotyped attitudes.

A series of experiments by Ward *et al.* (1974) explored the nonverbal mediators of self-fulfilling prophecies in interracial situations. Employing naïve white job interviewers and trained white and black "applicants," it was found that black applicants received less immediacy (a measure that consisted of physical distance, forward lean, eye contact, and shoulder orientation), more speech errors, and shorter amounts of interview time. In a follow-up study designed to ascertain what effect this treatment would have on the interviewee's performance, immediacy was treated as an independent variable with the applicant's performance as a dependent measure. Applicants treated similarly to the way blacks were treated in the first study performed less well, reciprocated less immediacy, and found the interviewers to be less adequate. Apparently the self-fulfilling prophecy operates in interracial situations and is mediated by discriminatory nonverbal treatment by whites toward blacks.

Using the same Prisoner's Dilemma procedure that Jones and Panitch (1971) employed, H. H. Kelley and Stakelski (1970a, 1970b) found that some people respond to events created by their own actions and personalities. Furthermore, they observed that this contributes to a distorted perception of these individuals' worlds because they typically fail to note the causal impact of their own actions. Kelley and Stakelski found that cooperative persons believe and act as if others are heterogeneously cooperative or competitive, whereas competitive persons assume that others are homogeneously competitive. Consequently the competitive players in the Prisoners Dilemma game elicit competitive responses from others regardless of whether their partners are cooperatively or competitively oriented. Then the competitors receive feedback from the interaction that all others are competitive without recognizing that their behavior produced their partners' competitive behavior. Extending their situational findings to personality styles, the authors also note that high F-scale respondents who express negative, pessimistic, and cynical views assume that all others will do the same. As a result, without awareness, these respondents "produce" the negative reactions they expect. Miller and Holmes (1975) have argued that the implications of these observations are limited by the situational characteristics of the Prisoners Dilemma game used by Kelley and Stakelski (1970a, 1970b). For the same reason, similar findings by Jones and Panitch (1971) and Kulman and Wimberly (1976) are of limited relevance to the understanding of self-fulfilling prophecies in complex social interactions.

According to Snyder and his colleagues, self-fulfilling prophecies represent a vital link in the chain of cognitive behavioral events that result in social stereotyping (Snyder & Swann, 1978a, 1978b; Snyder, Tanke, Berschied, 1977; Swann & Snyder, 1980). Drawing on sound research evidence on the stereotyping process, they reason that the perceiver selectively notices, recalls, "sees," and reinterprets information in attempts to bolster and strengthen stereotypes. These beliefs then become "predictions" about the behavior of others, guiding and influencing the perceiver's interactions with these others. This process may then produce behaviors by stereotyped others that validate the prediction of the perceivers. The end result, they believe, is the other person's "behavioral confirmation" of the perceiver's expectations. Using stereotypes of physically attractive people, they investigated the hypothesis that differential interactions with attractive and unattractive others elicit and nurture behaviors from target persons that are consistent with the stereotypes. Therefore,

> physically attractive people actually come to behave in a friendly, likeable, sociable manner. Not because they necessarily possess these dispositions, but because the behavior of others elicits and maintains behaviors taken to be manifestations of such traits. (Snyder *et al.*, 1977, p. 659)

They had male perceivers interact with female targets whom the males believed to be physically attractive or physically unattractive. Tape recordings of each participant's conversational behavior were analyzed. These analyses revealed that target persons who were (unbeknownst to them) labeled physically attractive came to behave in a friendly, likable, and sociable manner in comparison with target persons whose perceivers regarded them as unattractive. In their search for mediators of behavioral confirmation, the authors observed that male perceivers who interacted with women whom they believed to be physically attractive were judged to be more sociable, sexually warm, humorous, interesting, outgoing, sexually permissive, bold, independent, confident, animated, and more attractive themselves than their counterparts in the unattractive target condition. The women in the attractive condition may have reciprocated these sociable overtures, whereas the women in the unattractive condition may have defensively adopted cool and aloof reactions to cope with the less than favorable treatment they received.

Having demonstrated that information about others received in social interactions is a product of the perceiver's actions toward others, Snyder and Swann (1978a) turned toward further understanding of the development and mediators of behavioral confirmation processes within the context of contemporary attribution theories. They propose that when we internalize attributions about ourselves as a result of the behavioral confirmation process

> what began in the mind of the perceiver will have become reality not only in the behavior of the target but also in the mind of the target. The target may then provide others with evidence consistent with the original perceiver's expectations and these new perceivers will treat the target accordingly. (1978a, p. 151)

To test the hypothesis that behavioral confirmation effects will be sustained across situations, the experimenters provided information to "labeling perceivers" that a target person was either a hostile or nonhostile individual. This first interaction was also structured so as to induce target persons to believe that their actions represented either personal (dispositional) responses or a response to the other person's actions (situational). With a noise weapon as a measure of hostility in the initial phase, 61% of the labeling perceivers (who expected hostile partners) used high intensities, whereas less than 28% of the labeling perceivers (who anticipated nonhostile partners) used the upper-intensity noises. The second phase occurred when targets whose perceivers expected them to be hostile delivered higher noise levels to the perceivers than did targets expected to be nonhostile persons. The third phase of this process occurred when the perceivers in the hostile label condition attributed more "aggressive natures" to the targets than did perceivers in the nonhostile label condition. Finally, preservation of behavioral confirmation over situations was confirmed under the dispositional attribution condition only. Subjects in this condition, but not the situational attribution condition, who had once been labeled as hostile continued to behave in a more hostile fashion than targets originally labeled as nonhostile. Snyder and Swann point out that self-fulfilling prophecies may not always end after the initial interaction occurs but at times may actually resemble personality traits. They argue for viewing people as constructing their own realities while unknowingly failing to take into account how hypothesis-generating and -testing procedures actually create social reality.

These hypothesis-testing procedures were examined more extensively by Snyder and Swann (1978b) in a series of four experiments designed to determine whether confirming, disconfirming, or mixed strategies were more commonly used. They asked female subjects to assess the extent to which a target person's behavior and experiences matched either a prototype extrovert or introvert, by selecting among "introverted," 'extroverted," or neutral questions. The subjects with the extrovert hypothesis chose to ask questions typically asked of people known to be extroverts, whereas those subjects armed with the notion that their partners were introverts selected questions judged as those characteristically asked of people already known to be introverts. In the second study, after subjects again selected hypothesis-confirming strategies, the targets acted accordingly. Thus the target persons in the extrovert hypothesis condition were regarded by listener-judges as more extroverted, confident, and energetic than target persons in the introvert condition. Investigations 3 and 4 were designed to explore some limits to this hypothesis-confirming behavior. Snyder and Swann were unable to deter people from using confirmatory strategies in spite of a reduced likelihood that their hypothesis would prove accurate and the offering of substantial incentives for accurate hypothesis testing.

Swann and Snyder (1980) then investigated the first link in the chain of these processes, namely, the link between the individual's beliefs about another and choice of an interaction strategy. They not only demonstrated that their subjects relied on an experimentally induced theory about the nature of abilities and translated that belief into appropriate teaching strategies, they also discovered that whether pupils actually confirmed or disconfirmed their "theories" was relatively unimportant. Subjects left the simulated training sessions believing that their initial beliefs had been confirmed—whether they actually had been or not. These results represent an important contribution to behavioral confirmation processes in that they suggest that in natural settings beliefs about the attributes of others are not only causes of how others actually behave but that other people do not even have to fulfill our prophecies for us to believe that they do. Our beliefs about the world are apparently subjectively and irrationally validated in our everyday social interactions. We select interaction strategies and process interpersonal consequences in the service of stabilizing and fantasizing about social reality.

LABELING THEORY: A SOCIOLOGICAL PERSPECTIVE

There are a number of sociological analyses in the tradition of Merton's work on self-fulfilling prophecies that focus on an understanding of deviancy, stigmatization, and mental illnesses from a labeling theory perspective (Becker, 1963; Goffman, 1963; Scheff, 1966; Schur, 1971; Scott, 1969). Applying the labeling theory of Becker to mental illness, Scheff has explored the impact of stereotypes and labeling on the launching of careers of mental illness. According to Scheff, since we are unable to attribute schizophrenia to supernatural sources such as demons and witches, our modern culture through its psychiatric representatives has developed medical and psychological models to legitimize the identification of nonconformity as mental illness. Leaving aside the political arguments, the significant principles are that stereotypes of insanity are inadvertently confirmed through social interaction, labeled deviants are rewarded or punished contingent on their playing deviant or conventional roles, and deviants are highly suggestible when they are publicly labeled during a crisis. In relation to this last point, Scheff notes that

> when societal agents and persons around the deviant react in terms of the traditional stereotypes of insanity, unstructured rulebreaking tends to crystallize in conformity to those expectations, thus becoming similar to the behavior of other deviants classified as mentally ill and a part of the deviant's orientation for guiding behavior. (1966, p. 82)

Still another example of stereotype-based labeling was offered by Scott's (1969) description of the frequently observed effects of attributions about blind people on their styles of behavior. One form of adaptation by blind people to stereotypes by the sighted is represented by a group that he calls "true believers." These people have self-concepts that embody the stereotyped attributes of helplessness, dependency, and docility that represent the core expectations by the sighted for the personalities of the blind. Scott assumes that this conformity to the expectations of the sighted is based in part on the fact that sighted people treat blind people in such a way as to preclude their acting independently and actively, therefore actualizing the stereotype beliefs. As Schur (1971) points out, definitions of deviancy, shaped primarily by stereotyped beliefs, can have such a dramatic and devastating impact that the target of these beliefs sees little alternative but to comply with the expectations. This is analogous to Milgram's (1974) explanation of why so many of his subjects exhibited obedient behavior. He proposes that they had overlearned obedient responses in their lives to such a point that when an authority figure told them to harm another person, they simply did not perceive "disobedience" as a viable alternative.

Schur (1971) describes a process of "retrospective interpretations" and "bargaining" by which observers come to see deviators or suspected deviators in a totally new light. He concludes that the perception of individuals as deviators usually comes first and then the observers recognize the apparent indicators. Garfinkel (1956) concluded that in public trials the person on trial becomes in the eyes of the condemners literally a different and new person. The individual is reconstituted, the former identity is seen as accidental, and the new identity becomes reality. Accordingly, observers now believe that what the individual is now is what, after all, he or she was all along. In addition, Goffman (1961) has observed that psychiatric case histories are developed and used by mental health professionals (or "reconstruction specialists") to substantiate the treatment and commitment of psychiatric patients.

Schur (1971) describes how bargaining and negotiations are also vital aspects of the labeling process. Psychiatrists, as doctors, can without much difficulty convert patients into accepting their diagnostic impressions and induce them to accept the illness and concurrent symptoms and traits associated with these prescribed impressions. Assuming that the therapist and patient have unequal power in the relationship (at least at the outset), the patient is likely to be extremely susceptible to the therapist's conscious and unconscious suggestions. Scheff (1966), in reanalyzing a psychiatric case history by a prominent therapist, demonstrated that through a process similar to verbal conditioning the therapist was able to shift a woman patient's complaints from centering around her husband to more self-referent complaints. Early in the interview the therapist responded to the patient's statements with flat affect, pressure to change the topic, and subtle reminders that the patient had come to see a psychiatrist (which Scheff concludes had the effect of pressuring the patient to discuss *her* problems). Eventually the patient began describing feelings of guilt, whereupon the therapist's tone and manner shifted from being distant and rejecting to warm and accepting. Through negotiation the patient and therapist reached agreement that the patient, not her husband, was responsible for their difficulties. The point is that therapists can influence, through an unwitting bargaining process, the patient's productions. They can be as responsible for their patients' interactions with them as they may assume their patients contribute to the behaviors of relatives, friends, and acquaintances.

The role of self-fulfilling prophecies in a larger institutional or organizational context is implicit in Kovel's (1970) Freudian analysis of racism. He offers a most forceful statement regarding the manner in which members of the powerful group (whites) must make members of the powerless group (blacks) "suitable to represent what has to be projected upon them" (p. 195). His model of white racism is that it consists of culturally shared, unresolved infantile conflicts regarding power and sexuality and/or dirt and property that result in some combination of two "ideal" racist types—dominative or aversive. According to Kovel aversive racists turn away from black people as if they are filth and "make" the black into the affirmation of this idea. To sustain the stereotype the black person must become dirty, sloppy, feckless, lazy, improvident, and irrational. These traits, associated with blackness and inferiority, "make" their bearer worthy of aversion. Furthermore, whites have created institutions that force blacks to live in a certain way, invariably fostering just that constellation of unworthy traits they are assumed to possess in the first place.

THE TRANSMISSION OF UNCONSCIOUS WISHES

The communication of unconscious wishes about to enter into consciousness was first described by Freud in a paper called "Dreams and Occultism" (Freud, 1933). Although Freud was discussing telepathy and thought transference, he was also planting the seed (as was often the case) for a more complete understanding of how unconscious and preconscious events are transmitted from one person to another, without either person "knowing" how the other's behavior is influenced.

Probably the most extensive clinical analysis of this type of communication influence was undertaken by Sperling (1949, 1974). She demonstrated that some mothers transmit unconscious wishes to their children through concrete signs and these are received through ordinary sensory perceptions.

> Children in this type of relationship with their mothers often make observations like paranoids who notice everything and who interpret everything in reference to themselves . . . one can, in a study which includes both the child and the mother, detect that the child is not merely reacting to a fantastic image but to a real persecutor, namely its mother. This paranoid quality is especially marked in cases where the mother rejects the child. . . . For the child in such a situation it is vitally important to observe its mother closely. . . .

> Although these children rebel in the form of behavior deviation neurosis, psychosis or psychosomatic illness—at the same time they carry out the unconscious wishes of their mother as if given a command. (1974,p. 41)

Sperling's treatment of children focuses on understanding and resolving this "sensitive synchronization." This is done by simultaneous analytic work with mother and child so that the forces to which the child is reacting and the mother's infantile strivings are identified and understood in connection with the child's illness. Among the numerous cases discussed is one where a 5-year-old girl acted provocatively and naughtily because she was used by a very inhibited "model child" mother (who unconsciously identified herself with her daughter) for vicarious satisfaction of her own sadistic impulses. "The wishes of such mothers have to be carried out because the child has not been allowed to sever its dependent ties in normal growth" (Sperling, 1974, p. 52).

PROJECTIVE IDENTIFICATION: A PSYCHOANALYTIC PERSPECTIVE

One of the most important developments in recent psychoanalytic theorizing has been the "discovery" of the concept of projective identification by Melanie Klein. In "Notes on Some Schizoid Mechanisms" Klein (1946) introduced projective identification as a

> particular form of identification which establishes the prototype of an aggressive object relation by forcefully entering the object, which gives rise to anxieties relating to the dangers threatening the subject from within the object. (pp. 8–13)

This particular understanding of projective identification is most closely represented in the current literature in Kernberg's (1975) analysis of the borderline personality organization, which involves the use of a number of defensive operations including splitting, primitive idealization, denial, omnipotence and devaluation, and projective identification. Kernberg, like Klein, understands the mechanism's purpose to be the externalization of bad, aggressive self and object images, in order to preserve the "good" self. Because of accompanying ego weakness and boundary confusion, the fear of retaliation by the object is increased. People who use this defense feel they must control parts of themselves (which have been projected excessively into another person) by controlling the other person, or, as Kernberg puts it, to "control the object in order to prevent it from attacking them under the influence of the (projected) aggressive impulses" (1975, p. 31).

Recently, projective identification has been described as a concept that can expand our understanding of countertransference processes (Boyer, 1979; Issacharoff, 1979). Boyer (1979) sees patients as using therapists as repositories or containers for object representations and attitudes that are distressful to them, believing that they have located them within the therapist. This often produces difficulty when the therapist unwittingly accepts the projections. This is accompanied by the patients' continuously scanning the analyst's behavior. If the patients observe that these projections (which they assume to be inhabiting the therapist) do not destroy or harm the therapist, they can then safely reintroject them into their personality. Boyer also considers additional possible functions of projective identifications, namely (1) to make the analyst feel and understand the patient, (2) to "blackmail" the analyst into keeping the patient out of the "mess," and (3) to reconstruct the childhood traumata so that gratification, not persecution or frustration, occurs. Bollas (1983) has warned against exaggerating the patients' wish to inform the analyst about themselves. Although Bollas concurs that "many patients convey their internal world through the establishment of an environment . . . and they necessarily manipulate the analyst through object usage into assuming different functions and roles" (pp. 129–130), it is not always clear that the effect induced is the effect intended. This is a point made in Salzman's (1980) interpretation of Sullivan's views on the obsessional personality. According to Salzman, obsessionals, because of extreme sensitivity to rejection and disapproval, actually cause targets of obsessional maneuvers to attribute hostile intentions to their obsessional operations because the targets feel hurt and anguished. Consequences do not always account for intent.

Another approach to projective identification was developed by Issacharoff (1979), who explored the role of projective and introjective identifications in the analytic relationship. A striking result of one person creating a self minus one attribute is that the other person often acts as if he or she actually possessed the projected part. In those cases either the analyst-recipient had the attribute before the interaction, or because of the patient's transferential needs the analyst was induced to have those parts. Case material related to this point will be presented later.

Issacharoff's point is that not only do patients project parental characteristics into analysts, but even more problematic is the attribution of the patients' child parts into the analyst so as to reenact, in reverse, parent–child experiences. As a result, the analyst may develop powerful childlike feelings that are highly incompatible with the analytic self. The analyst's task is made even more difficult by the need to identify and react maturely to his or her own feelings.

A more complete explanation of the stages and function of projective identification is presented by Ogden (1979). According to him, projective identification includes (1) the ridding of the self of an unwanted part that seemingly takes over the other from within, (2) the pressure exerted via the interpersonal interaction such that the "recipient" of the projection experiences pressure to think, feel, and behave in a manner congruent with the projection, and (3) recovering or reinternalizing the projected feelings after they are processed and hence modified by the recipient. These three interdependent and overlapping stages are parts of a multifunctional operation. Projective identification is identified as a defense, a mode of communication, a type of object relatedness, and a potential pathway for psychological change (if the recipient copes with the engendered feelings in a more mature fashion than the projector). Among the examples of what Ogden calls the "induction phase" is a 12-year-old patient who literally bumped into people, including her therapist, thereby putting pressure on the therapist to feel intruded upon. Another instance involved an obsessive patient who, by tormenting a therapist and causing feelings of helpless rage and retaliatory anger in him, experienced a sense of relief from knowing that his own intolerable and noxious rage and terror were alive and preserved in another. In addition, Ogden notes examples of suicidal persons' inducing others to be lacking in concern for them, family members coercing one another to comply with projected fantasies, and, finally, pressures sometimes put on infants to behave congruently with mothers' pathologies.

The psychoanalytic literature on projective identification, in particular the Kleinian and interpersonal schools, has spelled out the importance of understanding and utilizing this concept—clinically and theoretically—for the effective treatment of patients. In fact, Bion (1959) has maintained that projective identifications are the most important interaction patterns occurring between patients and therapists.

GENERAL IMPLICATIONS

The overwhelming evidence and convergent theorizing regarding the process of fantasy-constructed reality suggest unequivocally the widespread use of self-fulfilling prophecies in social relations. In freely operating clinical and natural settings as well as carefully controlled and quantifiable laboratory settings, individuals use their expectations of others to create behaviors that confirm these prior expectations. It is also clear that some, if not most, psychotherapy patients (and probably their therapists) as well as "normal" research subjects are actively involved in the inadvertent implementation and maintenance of stereotyped beliefs.

Why this process is so omnipresent is open to wide speculation. The most plausible explanation seems to be that there is some security in being able to exert control over our environment, particularly when we are unaware of the role we play in creating the environmental outcomes. This has often been understood psychodynamically as a projection, or distortion, ever since Freud first discovered the concept of transference. Only recently have we understood that this can be fact as well as fiction. Control and predictability are sought after in varying amounts by everyone. Obsessional individuals clearly need to feel in control of themselves and others, but it would be a mistake to assume that establishing and maintaining control is merely the hobgoblin of obsessional minds and behaviors.

What meaning can we derive from the overwhelming fact of belief-created situations? Indirectly, an argument can be made in defense of Sullivan's one genus postulate that "we are much more simply human than otherwise, be we happy and detached, miserable and mentally disordered or whatever" (1953, p. 16). The rationale for this lies in the consistent observation that behavioral confirmation, self-fulfilling prophecies, and projective identification are all highly similar phenomena that have been demonstrated as characterizing the interaction patterns of both disturbed and nondisturbed populations. Whether we are talking about borderline patients, individuals engaged in the everyday process of stereotyping and labeling, or subjects in psychology experiments, there is no question that this key aspect of social interaction exists in all these groups of individuals. Apparently we all have this potential to construct realities from privately held or publicly shared beliefs.

Having demonstrated the breadth of this phenomenon, it seems appropriate to conclude that clinicians, like patients, are capable of this type of behavior sequence. In fact, Kleinian analysts have been exploring the complexities of this process as a central element in countertransference reactions. Racker (1968), for example, distinguishes between "concordant countertransference" as empathic bonds based on identifications that are acceptable to analysts, and "complimentary countertransferences" that involve the analyst identifying himself with the patient's internal objects. The latter represents a neurotic response that is reactivated by the patient's conflicts. Grinberg (1979) has described the process of "projective counteridentification" where the analyst's reaction stems from the particularly intense use of projective identification by the patient. After extensively reviewing the countertransference literature, Epstein and Feiner (1979) have concluded that "we seemed to have become more receptive to the idea that in relation to our patients we are more similar than different in kind" (p. 18). This appears consistent with Racker's (1968) recommendation that psychoanalysts accept the fact that they are still children and neurotics even though they are adults and analysts. The relevance of this "similarity" assumption has been shown by Havens (1982) in an article comparing existential and interpersonal methods of psychotherapy. Arguing that clinicians cannot quantify the data of therapy and that they actually transform the data by closely observing them, Havens concludes that we must reorient ourselves toward particular critical assumptions. We will then be in a better position to make less dogmatic and more appropriately empathic and complex interventions. For example, avoiding the assumption of "psychoanalytic asymmetry," where transference is given a more prominent place than countertransference, allows clinicians to deflect facts and fantasies from the clinical encounter onto an external screen for patient and therapist to view. In this way the therapist's mistakes and misperceptions are made as obvious as the patient's, thereby inviting the patient and the therapist to make corrections. Thus, "countertransference becomes as open as transference when it is cut free of the protection of its resistances so amply provided by professional inequality" (Havens, 1982, p. 28).

Placing countertransference reactions on an equal plane with transference reactions is consistent with Sullivan's definition of mental health as the awareness of our interpersonal relationships. It is also another way of saying that therapists and patients are much more like one another than they are different. However, since this question is truly unanswerable in any given transaction, it seems most sensible to conclude that clinicians need at least to remain open to the real probability that they are more like their patients—in terms of such phenomena as defensive operations, modes of communication, developmental levels, life stressors, and so on—than they would prefer to believe. Cooper and Witenberg (1983) propose that analysts' "reactions may be similar to the patient's—or they may differ—but either way the awareness of these differences and/or similarities will allow the analyst to understand the patient" (p. 250).

CONCLUSIONS

Self-fulfilling prophecies in clinical encounters can be organized into two paradigms—one descriptive, the other explanatory. The first model focuses on the stages that characterize the hypothesis-confirming sequence. The second represents an attempt to locate the motivation for behavioral confirmation in the individual's effort to transform anxiety into fear. These two levels of understanding will be explored using a clinical example.

During the course of almost 2 years of therapy, a 35-year-old male patient would threaten to stop treatment in response to failures in empathy on my part. He would accuse me of being selfish, treating him without compassion, and being "just like everyone else." I considered this to be a "blackmailing" effort on his part. I saw him as attempting to coerce me into telling him how much I did care, how much I needed him, and how special he was to me. Forcing these reactions out of me was an unsuccessful attempt to restore his damaged self-esteem. Although this pattern became less frequent, I became aware of disliking him more and more whenever he became abusive. In fact, I began to entertain wishes that he would follow through on his threats to end our work together. In other words he was successfully inducing me to feel toward him the way he expected me to feel. In fact, after one of these sessions he confided in me his belief that in spite of all the positive experiences he had during the course of therapy, he always expected that someday I would hate and reject him. The fascinating part of this patient's therapy was that his hypothesis had actually come true, largely because of his own actions.

This process involves several overlapping stages. First, a hypothesis is established. The universality of this approach to events was described by G. A. Kelley (1955) in his personality theory of personal constructs. He said that we use our own "theories" and hypotheses about human behavior to make judgments and interpretations of our experiences. Undoubtedly these hypotheses stem from prior experiences. When these hypotheses are actualized in the treatment exchange they are often understood as transference reactions. The second stage coincides with the commencement of the relationship between the patient and therapist. Both act like "scientists," looking for data to confirm their hypotheses. Selective attention, perception, and evaluation prevail. As Swann and Snyder (1980) observed, research subjects believed their beliefs were confirmed whether they actually had been or not. In the third stage of this process the patient and therapist manipulate one another into behaving the way they are "supposed" to. This stage in the hypothesis-confirming sequence was observed by Snyder *et al.* (1977), who found that "attractive" others became more likable because they were treated more positively, and by Snyder and Swann (1978a), who observed that "hostile" others became aggressive after being treated more aggressively.

Why does this process occur? What motivates people, particularly psychotherapy patients, to fulfill their prophecies? Classical psychoanalytic theory has offered the repetition compulsion as an explanation. Social psychology focuses on such issues as control and predictability as explanations. The interpersonal psychoanalytic theory of Sullivan, brought up-to-date by Chrzanowski (1977), offers another framework within which to explain this phenomenon. Patients with hypotheses that involve expectations of victimization are likely to experience excessive anxiety states. They are made anxious by their own hypotheses since they "know" they are going to experience reduced feelings of self-worth. Anxiety is obscure and intangible since it excludes the factor that creates the discomfort. By "forcing" therapists to dislike them, these patients are transforming anxiety into fear. If the therapist actually dislikes and victimizes them, a real external threat exists. Now an internal, undefined, diffuse danger becomes an external, concretely identified danger that the patient can cope with in one form or another (Chrzanowski, 1977, p. 62). Unless a therapeutic response is forthcoming, however, the patient is likely to win the battle but continue to lose the war.

Psychotherapists need to be aware of how often and how creatively patients will attempt to fulfill their interpersonal predictions. Being "used" in this manner is likely to be a disconcerting experience. Bollas's (1983) recent article on handling one's countertransference reactions offers therapists an opportunity to turn an obstacle into an advantage. He believes that therapists must allow patients to affect them and to "disturb" them. In other words, therapists must be prepared on occasion to become "situationally ill" so that their patients' unconscious emotional states can become conscious through the therapist's own feeling state. Nevertheless, simply knowing that patients engage in self-fulfilling prophecies, as well as understanding why they feel compelled to do so, can be anxiety reducing. If the therapist is less anxious, skillful interventions are more likely to follow.

ACKNOWLEDGMENTS

I would like to thank Bonnie Keisner and Jerome Tognoli for their helpful comments on an earlier version of this chapter.

REFERENCES

Becker, H. *The outsiders.* New York: Free Press, 1963.

Bion, W. *Experiences in groups.* New York: Basic Books, 1959.

Bollas, C. Expressive uses of the countertransference. *Contemporary Psychoanalysis,* 1983, *19,* 1–34.

Boyer, L. B. Countertransference with severely repressed patients. In L. Epstein & A. Feiner (Eds.), *Countertransference*. New York: Jason Aronson, 1979.
Chrzanowski, G. *Interpersonal approach to psychoanalysis*. New York: Gardner Press, 1977.
Cooper, A., & Witenberg, E. The stimulation of curiosity in the supervisory process. *Contemporary Psychoanalysis*, 1983, *19*, 248–264.
Epstein, L., & Feiner, A. (Eds.). *Countertransference*. New York: Jason Aronson, 1979.
Freud, S. *New introductory lectures on psychoanalysis*. New York: W. W. Norton, 1933.
Garfinkel, H. Conditions of successful degradation ceremonies. *American Journal of Sociology*, 1956, *61*, 420–424.
Goffman, E. *Asylums*. Garden City, N.Y.: Doubleday/Anchor, 1961.
Goffman, E. *Stigma*. Englewood Cliffs, N.J.: Prentice-Hall, 1963.
Grinberg, L. Projective counteridentification and countertransference. In E. Epstein & A. Feiner (Eds.), *Countertransference*. New York: Jason Aronson, 1979.
Havens, L. The choice of clinical methods. *Contemporary Psychoanalysis*, 1982, *18*, 16–42.
Issacharoff, A. Barriers to knowing. In L. Epstein & A. Feiner (Eds.), *Countertransference*. New York: Jason Aronson, 1979.
Jones, S., & Panitch, D. The self-fulfilling prophecy and interpersonal attraction. Journal of Experimental Social Psychology, 1971, *7*, 356–366.
Kelley, G. A. *The psychology of personal constructs*. New York: W. W. Norton, 1955.
Kelley, H. H., & Stakelski, A. J. Errors in perception of intentions in a mixed-motive game. *Journal of Experimental Psychology*, 1970, *6*, 401–419. (a)
Kelley, H. H., & Stakelski, A. J. The social interaction basis of cooperator's and competitor's beliefs about others. *Journal of Personality and Social Psychology*, 1970, *16*, 66–91. (b)
Kernberg, O. *Borderline conditions and pathological narcissism*. New York: Jason Aronson, 1975.
Klein, M. Notes on some schizoid mechanisms. *International Journal of Psycho-analysis*, 1946, *27*, 99–110.
Klein, M. On identification. In M. Klein, P. Heinmann, & R. Money-Kyrle (Eds.), *New directions in psychoanalysis*. New York: Basic Books, 1955.
Kovel, J. *White racism: A psychohistory*. New York: Pantheon, 1970.
Kulman, D. M., & Wimberley, D. H. Expectations of choice behavior held by cooperators, competitors, and individualists across four classes of experimental game. *Journal of Personality and Social Psychology*, 1976, *34*, 69–81.
Merton, R. K. The self-fulfilling prophecy. *Antioch Review*, 1948, *8*, 193–210.
Merton, R. K. *Social theory and social structure*. New York: Free Press, 1957.
Milgram, S. *Obedience to authority: An experimental view*. New York: Harper & Row, 1974.
Miller, J., & Holmes, J. The role of situational restrictiveness on self-fulfilling prophecies: A theoretical and empirical extension of Kelley and Stakelski's Triangle hypothesis. *Journal of Personality and Social Psychology*, 1975, *31*, 661–673.
Ogden, T. On projective identification. *International Journal of Psychoanalysis*, 1979, *60*, 357–373.
Orne, M. J. Demand characteristics and the concept of quasi-controls. In R. Rosenthal & R. L. Rosnow (Eds.), *Artifact in behavioral research*. New York: Academic Press, 1969.
Racker, H. *Transference and countertransference*. New York: International Universities Press, 1968.
Rosenthal, R. *Experimenter effects in behavioral research*. New York: Appleton-Century-Crofts, 1966.
Rosenthal, R., & Jacobson, L. *Pygmalion in the classroom: Teachers' expectations and pupils' intellectual development*. New York: Holt, Rinehart and Winston, 1968.

Salzman, L. Sullivan's views on the obsessional states. *Contemporary Psychoanalysis*, 1980, *16*, 271–286.
Scheff, J. *Being mentally ill*. Chicago: Aldine, 1966.
Schur, E. *Labeling deviant behavior*. New York: Harper & Row, 1971.
Scott, R. *The making of blind men*. New York: Russell Sage, 1969.
Snyder, M., & Swann, W. Behavioral confirmation in social interaction: From social perception to social reality. *Journal of Experimental Social Psychology*, 1978, *14*, 148–162. (a)
Snyder, M., & Swann, W. Hypothesis-testing processes in social interaction. *Journal of Personality and Social Psychology*, 1978, *36*, 1202–1212. (b)
Snyder, M., Tanke, E. D., & Berschied, E. Social perception and interpersonal behavior: On the self-fulfilling nature of social stereotypes. *Journal of Personality and Social Psychology*, 1977, *35*, 656–666.
Sperling, M. The role of mother in psychosomatic disorders in children. *Psychosomatic Medicine*, 1949, *11*, 377–385.
Sperling, M. *The major neuroses and behavior disorders in children*. New York: Jason Aronson, 1974.
Sullivan, H. S. *Conceptions of modern psychiatry*. New York: W. W. Norton, 1953.
Swann, W., & Snyder, M. On translating beliefs into action: Theories of ability and their application in an instructional setting. *Journal of Personality and Social Psychology*, 1980, *38*, 879–888.
Ward, C., Zanna, M., & Cooper, J. The nonverbal mediation of self-fulfilling prophecies in interractial interaction. *Journal of Experimental Social Psychology*, 1974, *10*, 109–120.
Zanna, M., & Pack, S. J. On the self-fulfilling nature of apparent sex differences in behavior. *Journal of Experimental Social Psychology*, 1975, *11*, 583–591.

Index

Abreaction, 124
Academic achievement, 276
Adolescence, 313–325
  biology and, 313–314
  loneliness and, 195
  milieu of, 321–324
  normality and, 316–321
  problems in reporting on, 313–316
  psychopathology and, 323–324
  See also Children and childhood
Adult functioning, 265
Aestheticism, 424
Affect
  charisma and, 82–83, 85
  cognition and, 280
  fantasy and, 390
  fathers and, 275
  infants and, 251–252, 256
  memory and, 17–18
  metacognition and, 272–273
  peer relations and, 276
  psycholinguistics and, 403
  recall distortion and, 29–30
  religion and, 413–414, 420
  social exchange and, 79
Aggression, 113–135
  control and, 120–121
  excitation-transfer theory, 121–122
  frustration–aggression theory, 115–119
  instinct theory, 113–114
  psychoanalysis and, 123–124
  psychotherapy and, 122–123
  research/practice, 114–115
Aggression (*Cont.*)
  social learning theory and, 119–120
  social psychology of, 115–122
  therapeutic relationship and, 118–119
Aging, 328
  See also Midlife change
American Psychiatric Association, 301
Anorexia nervosa, 256
Anthropology, 90–91
Anxiety
  memory and, 17–18
  negative cuing effect, 26
  nonverbal communication, 93
  sense distortion recall, 30
Associative organizers concept, 355–357
  See also Shielding
Attachment behavior, 250
Attachment figures, 273–279
Attention deployment, 16
Attitude change theory, 139–153
  clinical practice and, 150–152
  cognitive dissonance theory, 144–145
  cognitive response theory, 146–147
  interaction theory, 147–148
  learning theory, 143–148
  psychotherapy and, 139–142, 148–150
  reasoned action theory, 146
  self-perception theory, 145
  social psychology and, 142
Attitudes
  interpersonal attraction and, 184
  sex role stereotyping and, 165
  toward research, 7–8, 9–13
Attraction. *See* Interpersonal attraction

Attribution theory
loneliness, 198
relationships and, 190–191
Authoritarian parenting, 269
Autism, 359, 408

Battered families. *See* Domestic violence
Behavior
attribution theory, 190–191
fantasy and, 390–391
gender role disturbances, 296–298
infants, 249
leader–follower relationship, 76
leader situation and, 75
leader style, 71
mental health policy and, 52–54
psychotherapy and, 15
reasoned action theory and, 146
self-fulfilling prophecy and, 428
sex role and, 161, 162
situation and, 73
social exchange and, 79
stereotyping and, 159
Behaviorism
attitude change and, 141
gender disturbances and, 298
leadership studies and, 75–76
power relationships in, 80
research/practice, 16
Behavior modification
research and, 4
research/practice, 12–13
Behavior therapists and therapy
leader style, 72
psychoanalysis and, 130
research/practice, 12–13
See also Cognitive behavior therapy
Beliefs, 8–9
Biofeedback, 387
Biology
adolescence and, 313–314
fantasy and, 384–385
gender role development and, 288–290, 293
midlife change and, 337, 339–340
Bisexuality, 289
Blindness, 434–435
Body
ego and, 258
shielding and, 358–359, 360
Body communication. *See* Nonverbal communication
Brain, 289
Breakdown, 407

Cardiovascular disease, 53
Cartesian dualism, 413, 416, 419–420
Castration anxiety, 290, 291
Catharsis
aggression and, 116, 121, 125–126
psychoanalysis and, 124
social learning theory and, 120
Channel discrepancy, 93
Character. *See* Personality
Charisma
leadership and, 82–84
therapeutic relationship and, 84–85
Childhood development (early), 265–281
childrearing styles and, 266–270
metacognition, 270–273
multiple attachment figures, 273–279
psychotherapy and, 265–266
research implications of, 279–281
Childrearing
adolescence and, 215–216
gender role development and, 292–293
gender role disturbances and, 297–298
peer groups and, 277
research on, 266–270
Children and childhood
adulthood and, 327–328
aggression and, 119
context and, 21
environment and, 53
injustice and, 228–229
just world needs and, 221
loneliness and, 195
nonverbal memory codes, 21
psycholinguistics and, 404–405, 406–407
shielding and, 364, 365
unconscious wish transmission, 436–437
victimization and, 231
See also Adolescence
Clinical practice
memory and, 15–42
See also Practice; Research/practice

Clinical psychology
origins of, xi
psychotherapy and, xii
recent developments in, xxi–xiii
Cognition
affect and, 280
aggression and, 121–122
charisma and, 82
childhood development, 270–273
environment and, 53
fantasy and, 377
gender role differences and, 293
gender role disturbances and, 302
infants, 251, 253, 254
injustice and, 228
metacognition, 270–273
religion and, 413
self-fulfilling prophecy and, 431
stereotyping and, 158, 159
Cognitive behavior therapy, 12, 16
attitude change and, 140, 141–142
frequency information distortion, 40–41
memory and, 18
See also Behavior therapists and therapy
Cognitive developmental theory, 294–296
Cognitive dissonance theory
attitude change and, 144–145
effort justification hypothesis, 148
Cognitive function, 16–17
Cognitive response theory, 146–147
Communication
associative organizers, 356
cognitive response theory, 146–147
fantasy and, 392
infants and, 249–250, 255
relationships, 186
unconscious wishes, 436–437
Companionate love, 200–201
Compliance, 85
Conflict, 190–191
Context
memory and, 20–21
nonverbal communication and, 93
Context-dependent learning, 24–25
Contingency models, 74
Control
aggression and, 120–121, 129, 130
injustice and, 222–224
Control (*Cont.*)
self-blame and, 226
Controllable-world needs, 228–229
Coping strategies, 16
Cost–benefit analysis
environment and, 53–54
mental health policy, 50–52
Counseling psychology
nonverbal communication, 93–94
sex role stereotyping and, 165–166
Countertransference
gender role disturbances, 307–308
injustice and, 233–234, 235
metacognition and, 273
midlife change and, 347
projective identification and, 438, 440–441
social exchange and, 79
See also Therapeutic relationship; Transference
Crime
adolescence and, 314
aggression and, 129
controllable-world needs and, 223
Cross-cultural studies, 90–91
Cross-gender identification, 296–297, 298
Cues, 34–37
Culture, 303

Daydreams, 275, 276
See also Fantasy
Deep structure, 73
Deindividualization, 126–127
Deinstitutionalization, 46, 54–55
Demography, 324
Dependency
fear of breakdown, 407
psycholinguistics, 408
violence, 129
Depression
aggression and, 120
helplessness and, 40–41
midlife change, 345
nonverbal communication, 93
Developmental psychology, 243–244
adolescence, 313–325
childhood development (early), 265–281
gender role development, 287–309
harmdoers, 231
infant development research, 245–260

Developmental psychology (*Cont.*)
  injustice and, 229
  midlife change, 327–348
  movement studies, 96
  nonverbal communication, 91
  shielding and, 363–364
  victims and, 232
Diagnosis
  differential diagnosis, 9–10
  gender role development, 302–304
  movement studies and, 98–99
  nonverbal communication and, 103–105
Diagnostic and Statistical Manual of Mental Disorders, 246
Differential diagnosis, 9–10
Discipline, 268–269
Disincentives, 54–55
Dismantling, 408
Displacement, 116–117
Dissonance theory. *See* Cognitive dissonance theory
Distortion
  childhood development and, 279–280
  harmdoers, 220
  infant research, 256
  injustice and, 228, 233
  See also Recall distortion
Divorce, 192
Domestic violence
  aggression and, 128, 129
  injustice and, 235–237
  self-blame and, 225
  See also Violence
Drive states. *See* Instinct theory
Duration (of therapy), 75

Education, 7–8
Effort justification hypothesis, 148
Ego
  associative organizers and, 356
  infant research and, 258
  metacognition and, 272
  shielding and, 366
  therapeutic mode and, 279
  therapeutic relationship and, 268
Ego theory, 254
Elaboration. *See* Inferential elaboration
Elderly
  loneliness, 195
  See also Aging; Midlife change
Emotion. *See* Affect
Empiricism, 414–415, 418–421, 423–424
Environment
  aggression and, 121
  injustice and, 228
  mental health policy and, 52–54
  shielding and, 370
Epidemiology, 47
Equity, 76–77
Erikson, E., 330
Ethnomethodology, 422–423
Everyday exchanges, 189–190
Excitation-transfer theory, 121–122
Eye movement, 385–386

Facial expression
  infant research, 254
  See also Nonverbal communication
Family
  adolescence and, 322–324
  psycholinguistics and, 404–405, 406–407
  self-fulfilling prophecy, 428
Fantasy, 375–396
  biology and, 382–383, 384–385
  capacity for, 387
  clinical practice and, 389–392
  communication and, 392
  definition of, 375–377
  eye movement and, 385–386
  historical perspective on, 377–379
  individual styles of, 384
  information load and, 385
  normative aspect of, 380–381
  operant thought and, 386–387
  posture and, 383–384
  psycholinguistics and, 404
  schizophrenia and, 381–382
  self-awareness and, 388
  self-fulfilling prophecy and, 427
  sexual fantasy, 388–389
  therapeutic models utilizing, 393–396
Fathers
  childrearing and, 273–275
  sex role development and, 267, 275, 292, 293
Fear of breakdown, 407
Feeling of knowing phenomenon, 26–27
Feminism, 303
Field dependency, 362–363
Frequency information distortion, 37–41

Freud, Anna, 317
Freud, Sigmund, xi, 5, 91, 95, 113, 114, 245, 258, 274, 288–289, 290, 291, 327–328, 353–355, 371, 378, 402, 413, 414, 418, 436
Friendship, 275–276
Frustration-aggression theory, 115–119, 120

Gender. *See entries under* Sex
Goals, 7–8
Grant Study of Adult Development, 333–334
Guided memory, 21–22
Guilt, 220

Harmdoers, 220, 230–231, 232–237
Health maintenance organizations, 55–56
Hearing, 248–249
Helplessness, 40–41
Homosexuality, 39–40
  fantasy and, 388–389
  psychopathology contrasted, 301
Hormones, 289
Housing, 50–51

Identification
  change and, 85
  projective identification, 437–439
  self-fulfilling prophecy and, 428
  therapeutic relationship and, 131
Illusory correlation phenomenon, 38–39, 40
Imagery, 375
  See also Fantasy
Imagination, 376, 378
  See also Fantasy
Implosion therapy, 148–149
Individuality, 224
Infancy
  gender role development, 293
  parent–child relationship, 267
  psychoanalysis and, 245–246
  shielding and, 363
Infant development research, 245–260
  basic competence, 247–252
  infant intervention, 259–260
  practice implications of, 257–260
  theory update, 252–256
Inferential elaboration, 30–32

Injustice, 217–238
  controllable-world needs, 222–224, 228–229
  definitions, 218–220
  harmdoers, 220
  individual differences and, 224
  just-world needs, 221–222
  observers in, 221
  perceptions of, 217–218
  psychoanalysis and, 228, 229–232
  response to, 227–238
  therapeutic relationship and, 232–237
  treatment and, 237–238
  victims of, 224–227
Inpatient care, 54–55
Instinct theory
  aggression, 113–114, 115
  infants, 251
  injustice and, 229
Institutionalization, 54–55
Interaction theory, 147–148
Interdependence, 189
Internalization, 85
Interpersonal attraction, 179–208
  conflict and, 190–191
  control and, 207–208
  ending relationships, 191–194
  everyday exchanges, 189–190
  initial attraction, 180–185
  initiating relationships, 185–186
  intimacy and, 204–207
  loneliness and, 194–198
  love concerns, 199–201
  physical attractiveness, 182–183
  proximity, 180–182
  relationships growth, 186–187
  self-disclosure in, 187–189
  self-fulfilling prophecy and, 429–430, 431–432
  sex concerns, 201–204
  sex differences in, 198–208
  similarity in, 183–185
  See also Relationships
Intimacy
  conflict and, 190
  fantasy and, 396
  interpersonal relations and, 204–207
  self-disclosure and, 187
  sexual activity and, 203–204

Jung, C., 329

Just-world hypothesis, 221–222

Labeling, 428, 432–433
Labeling theory
  self-fulfilling prophecy and, 434–436
  See also Stereotyping
Language
  associative organizers and, 356
  shielding and, 361
speech and, 402
See also Psycholinguistics; Shielding
Leader–follower relationship
  situation and, 74
  social exchange perspective on, 76
  therapeutic relationship contrasted, 68–69
Leadership
  charisma and, 82–84
  leader attributes analysis, 69–70
  leader style analysis, 71–73
  power and, 79–82
  relationships and, 67
  situation and, 73–75
  social exchange approaches to, 75–77
Learned helplessness, 120
Learning
  context-dependent, 24–25
  environment and, 53
  memory and, 19
  state-dependent, 24–25
Learning theory. *See* Social learning theory
Linguistics. *See* Psycholinguistics
Loneliness, 194–198
  coping with, 198
  defined, 194–195
  demography of, 195
  experience of, 195–196
  precipitating factors of, 196–198
Long-term memory, 19
Love
  self-disclosure and, 188
  sex differences and, 199–201

Magical beliefs, xi
Marriage, 184–185
Media
  aggression and, 120
  nonverbal communication, 91
Medicare/Medicaid, 54
Memory
  affect and, 17–18
  clinical practice and, 15–42
  cognition contrasted, 271
  cognitive behavior therapy and, 18
  fantasy and, 383
  guided memory, 21–22
  psychodynamic therapy, 17, 18
  recall distortions, 27–41
  recall failure, 19–27
  repression and, 23
  theories of, 18–19
  See also Recall
Menopause, 339–340
Mental blocks, 391
Mental health policy
  cost-benefit analysis, 50–52
  defined, 45–46
  deinstitutionalization, 46
  environmental/behavioral factors in, 52–54
  marketplace incentives, 54–57
  meta-analysis in, 49–50
  modes of practice, 57
  needs analysis in, 47
  patient education and, 57–58
  physical health and, 58–59
  practitioners in, 46–47
  research and, 45–59
  service maximization, 47–48
  social networks/volunteer groups and, 54
Meta-analysis
  injustice, 223
  mental health policy, 49–50
Metacognition, 280
Midlife change, 327–348
  childhood and, 327–328
  clinical implications of, 343–347
  cross-sectional research in, 334–337
  definition and, 331
  individual research, 337–343
  interest in, 328–329
  longitudinal research in, 331–334
  research suggestions in, 347–348
  sociological perspective on, 327
  theories of adult development, 329–331
Midlife crisis, 330–331
Mirroring, 423
  See also Posture mirroring

Misattribution theory, 127
Misleading cues biasing, 34–37
Modeling
  fantasy and, 391–392
  gender role behavior and, 301–302
  psychoanalysis and, 130–131
Money exchange, 74–75
Mortality, 342
Movement, 95–99

Narcissistic disorders, 268
  midlife change and, 345–346
  religion and, 423
National Institute of Mental Health, 46
Negative cuing effect, 25–26
Nonverbal behavior, 89–108
Nonverbal codes, 21–22
Nonverbal communication
  historical perspective on, 89–95
  infants, 249–250
  objectivity/subjectivity, 94–95
  personality and, 95–99
  posture mirroring, 94–95
  practice and, 106–108
  psychology education and, 106–108
  relationship messages, 99–106
  specificity in, 92
  therapeutic relationship and, 91–92, 93–95, 97–99, 100–106
  verbalization and, 97–98
Norms, 316–321

Oakland Growth Study, 332
Objectivity
  assumptions about, 10–11
  nonverbal communication, 94–95
  religion and, 412, 414, 415
Object relations
  childhood development and, 273–279
  gender role development and, 290–291
  gender role disturbances and, 300
  infant research, 253
  psycholinguistics, 405, 408
  research and, 280–281
  therapeutic relationship and, 268
Observers, 221, 222, 229–230
Obsessive-compulsive disorders
  metacognition and, 273
  psycholinguistics and, 403
Oedipus complex, 290, 291
Operant thought, 386–387
Outpatient care, 46, 54–55

Paradoxical directive technique, 147
Paranoia, 273
Parenting, 277
Patient–therapist relationship. *See* Therapeutic relationship
Peer groups
  adolescence and, 315, 321, 322
  child development and, 275–276
  gender role disturbances, 297, 298, 301
  parents and, 277
Penis envy, 290
Permissive parenting, 269
Personality
  aggression and, 116
  childrearing and, 268–269
  nonverbal communication and, 95–99
  relationships and, 187
  stereotyping and, 158
Phenomenology, 424
Physical attractiveness, 182–183
Physical health, 58–59
Play behavior, 276
  fathers and, 293
  peer groups and, 277
Policy. *See* Mental health policy
Pornography, 201–202
Posture, 383–384
Posture mirroring, 94–95
Power, 79–82
Practice
  abnormal populations and, 11–12
  idiographic assumption, 12
  research and, 3–13
  subjectivity and, 10
  See also Research/practice
Pragmatism, 418–419
Prejudice, 416
Pre-oedipal phase
  gender role disturbances, 300
  gender role identity, 291–292
  infant research, 252–255
President's Commission on Mental Health, 47, 54
Professionalization, 9
Professional service review organizations, 46

Projection, 428
Projective identification, 437–439
Proximity, 180–182, 184
Psychiatrists, 46
Psychoanalysis
  adolescence and, 317, 318
  aggression and, 123–124
  attitude change and, 140
  behavior therapy and, 130
  cognitive development theory and, 295
  fathers and, 274
  gender role development and, 290–292
  gender role disturbance and, 298, 299–300
  infancy and, 245–246, 254–255, 256
  injustice and, 228, 229–232
  midlife change and, 328, 329, 330
  organ affect, 252
  projective identification concept, 437–443
  psycholinguistics and, 402
  psychopathology and, 279
  religious experience and, 411–425
  self-fulfilling prophecy and, 428
  sex of therapist, 170–173
  stereotyping and, 157
  victimized/harmdoing patients, 232–237
Psycholinguistics, 401–409
Psychologists, 46
Psychopathology
  adolescence and, 319–321, 323–324
  childhood development and, 278–279
  fantasy and, 378, 381–382
  gender role disturbance and, 302–303
  homosexuality contrasted, 301
  labeling theory and, 434
  psycholinguistics and, 401–409
  research/practice and, 11–12
  sexual fantasy and, 389
  shielding and, 357–358, 359–362, 366, 369
Psychosis, 404
  See also Psychopathology
Psychotherapy
  aggression and, 114, 122–123
  attitude change and, 139–142, 148–150
  behavior and, 15
Psychotherapy (*Cont.*)
  childhood development and, 265–266, 280
  cognitive function and, 16–17
  definitions of, 140
  fantasy and, 389–392
  frequency information distortion, 40–41
  gender role disturbances, 299–300, 304–308
  infant research and, 257–259
  inferential elaboration and, 32
  interaction theory and, 147
  loneliness and, 198
  meta-analysis and, 49, 50
  metacognition and, 270–273
  midlife change and, 338, 342, 343–347
  misleading cues biasing, 36–37
  outcome analysis, 56–57
  power relationships in, 80
  recall distortion and, 28
  religion and, 421–425
  sense distortion recall, 29–30
  sex bias in, 303
  sex role stereotyping and, 163–170
  shielding and, 367–371
  social exchange approach to, 77–79
  social psychology and, 63
  spread of, xii
  therapeutic relationship in, 67–68
  verbal label biasing, 34
  See also Psychoanalysis; Therapeutic relationship
Public policy. *See* Mental health policy

Quality of life, 53–53

Race differences
  nonverbal communication, 105
  self-fulfilling prophecy, 429, 430, 436
Racial stereotyping, 158
Radical behaviorism, 16
Radical empiricism. *See* Empiricism
Rational-emotive therapy, 140
Reality, 228
Reasoned action theory, 146
Recall
  failure of, 19–27
  process of, 19
  See also Memory

Recall distortion, 27–41
  defined, 27
  frequency of information, 37–41
  inferential elaboration, 30–32
  misleading cues biasing, 34–37
  sense distortion, 28–30
  verbal label biasing, 32–34
Recall failure
  context-dependent learning, 24–25
  decision criterion, 25
  encoding inadequacy, 20–21
  "feeling of knowing" phenomenon, 26–27
  negative cuing effect, 25–26
  verbal code types, 22–23
  verbal versus nonverbal codes, 21–22
Reciprocity, 430
Referrals, 299–302
Reinforcement
  aggression and, 119, 120
  gender disturbances and, 298
  psychoanalysis and, 130–131, 132–133
Reinforcement schedules, 16
Relationships
  leadership and, 67
  nonverbal communication and, 100–106
  See also Interpersonal attraction; Therapeutic relationship
Religion, 411–425
  defined, 411, 413, 418
  intrinsic/extrinsic orientations in, 415–418
  objectification of, 412–415
  psychotherapy and, 421–425
  radical empiricism and, 418–421
Repression, 20, 21, 23, 25
Research
  attitudes toward, 7–8, 9–13
  beliefs and, 9
  definitions of, 5, 6
  goals of, 7–8
  memory and, 15–42
  mental health policy and, 45–59
  methodological variety in, 5
  nomothetic assumption, 12
  normal populations and, 11–12
  objectivity and, 10
  practice and, 3–13 (*See also* Research/practice)
  self-fulfilling prophecy, 429
Research (*Cont.*)
  utilization of, 6
Research/practice
  aggression and, 114–115
  attitude changes and, 9–13, 150–152
  behaviorists and, 12–13, 16
  beliefs and, 8–9
  definitions of, 5–6
  dichotomy between, 3–4
  Freud and, 114
  goals and, 7–8
  infant research, 255, 256
  nonverbal communication, 106–108
  religion and, 412
  self-fulfilling prophecy and, 427
Resistance, 403
Responsibility, 220
Revivification procedures (memory), 22, 25
Rogers, Carl, xii, 72, 73, 267–268
Romantic love. *See* Interpersonal attraction; Love

Schizophrenia
  family relations and, 41
  fantasy and, 381–382
  psycholinguistics and, 401–409
  shielding and, 360, 361–362, 363
Science
  error types in, 7
  magical beliefs and, xi
  methodological variety in, 5
  religion and, 411, 418
Self-blame, 225, 226
Self-concept
  fathers and, 275
  midlife change and, 335–336
  sex role and, 160
  victims, 225–226, 231, 232
Self-disclosure, 187–189
Self-esteem
  breakups and, 192
  loneliness and, 198
Self-fulfilling prophecy, 427–443
  implications of, 439–441
  labeling theory and, 434–436
  overview of, 427–429
  projective identification, 437–439
  social psychology and, 429–434
  unconscious wishes and, 436–437
Self-perception theory, 144–145

Semantic attributes, 22–23
Sensorimotor matrix, 355–356
Sex differences
 adolescent research and, 314–315
 control and, 207–208
 fantasy and, 384–385
 gender role development, 293
 injustice and, 226–227
 interpersonal relations and, 192–193, 198–208
 intimacy and, 205–207
 leadership attributes and, 70
 love concerns and, 199–201
 midlife change and, 335, 341
 self-concept and, 160
 self-disclosure and, 188
 sexual concerns and, 201–204
 sexual experience and, 203
Sexism, 303
 psychotherapy and, 163–164
 self-fulfilling prophecy and, 430
Sex role development, 287–309
 biological factors in, 288–290
 children and, 289–290
 cognitive developmental theory, 294–296
 countertransferences, 307–308
 diagnostic issues in, 302–304
 disturbances in, 296–298
 fathers and, 275
 life span perspective on, 295, 301
 parent–child relationship and, 267
 psychoanalysis and, 290–292
 referrals process and, 299–302
 social learning theory and, 292–294
 transference and, 304–307
 treatment goals and, 304
Sex-role stereotyping
 discussion of, 170–175
 psychotherapy and, 163–170
 self-fulfilling prophecy and, 429, 430
Sexual fantasy, 388–389, 393
Sexuality, 201–204
Shielding, 353–371
 associative organizers concept and, 355–357
 clinical practice and, 368–371
 concept of, 353–355
 development and, 363–364
 field dependency and, 362–363
 naive observation phase, 357–359
Shielding (*Cont.*)
 psycholinguistics and, 404
 psychopathology and, 359–362, 366
 regulation and compensation, 364–366
 treatment process and, 367–368
Short-term memory, 18–19
 See also Memory; Recall
Simmel, Georg, 67–68
Situations
 aggression theory, 115
 leadership and, 73–75
Social exchange, 77–79
Socialization, 293, 295, 307–308
Social learning theory
 aggression and, 119–120
 attitude change and, 143–144
 cognitive development theory and, 295
 gender role development and, 292–294
 gender role disturbance and, 301–302
 psychoanalysis and, 130–131
Social networks, 54
Social psychology
 aggression and, 113–135
 attitude change theory, 139–153
 injustice and, 217–238
 interpersonal attraction, 179–208
 leadership studies, 67–86
 nonverbal behavior and, 89–108
 psychotherapy and, 63
 religion and, 412–415
 self-fulfilling prophecy and, 428, 429–434
 social stereotypes, 157–177
Socioeconomic class, 315
Specialization, 9–10
Speech, 402
Splitting, 272–273
Stanford Cardiovascular Project, 53
State-dependent learning, 24–25
Stereotyping, 9, 157–177
 controllable world needs and, 223
 discussion of, 170–175
 minimization of, 157
 overview of, 158–159
 physical attractiveness, 183
 psychotherapy and, 163–170
 self-fulfilling prophecy and, 428, 430–431

Stereotyping (*Cont.*)
  sex differences, 208
  sex roles, 159–163, 303
Stream of consciousness, 379, 380–381
Stress, 54
Structure
  leader social exchange, 76
  leader style, 72–73
  social exchange and, 78
Students, 7–8
Student–teacher relationship, 68
Subjectivism and subjectivity
  assumptions about, 10–11
  nonverbal communication, 270–271
  religion and, 414–415
Suicide, 270–271

Task performance, 78
Teachers
  gender role development, 293–294
  student–teacher relationship, 68
Terman Gifted Group Study, 332–333
Therapeutic relationship
  age of therapist and, 346–347
  aggression and, 118–119
  charisma and, 84–85
  childrearing compared, 266, 269
  gender role disturbances, 304–308
  injustice and, 228, 232–237
  interaction theory and, 147
  labeling process in, 435–436
  leader attributes and, 69–70
  leadership relationship contrasted, 68–69
  leader situation and, 74–75
  leader social exchange, 76–77
  leader style and, 72–73
  learning theory and, 130–132
  movement studies and, 97
  nonverbal communication, 91–92, 93–95, 97–98, 100–106
  power and, 80–82
  psychotherapy and, 67–68
  self-disclosure and, 188
  self-fulfilling prophecy and, 428
  sex of therapist, 170–173, 298, 305
  social exchange and, 77–79
  stereotyping and, 157
  warmth in, 268
  See also Psychoanalysis; Psychotherapy
Tip-of-the-tongue phenomenon, 26–27
Transference
  adolescence and, 321
  aggression and, 127
  charisma and, 85
  childrearing and, 266
  gender role development and, 304–307
  injustice and, 234, 235
  leader style and, 73
  metacognition and, 272
  modeling and, 131
  movement studies and, 97
  power relationship and, 81–82
  religion and, 423
  shielding and, 367
  See also Countertransference
Transsexuals, 297–298
Trauma, 246, 369

Ucs, 420, 403
Unconscious
  psycholinguistics and, 402–403
  self-fulfilling prophecy and, 428, 436–437
Utilization, 5, 6

Verbal codes, 21–23
Verbalization, 97–98
Verbal label biasing, 32–34
Victims, 224–227
  blaming of, 227–228, 230
  harmdoers and, 220
  just-world hypothesis and, 222
  observers and, 221
  psychoanalysis and, 231–232
  therapeutic relationship and, 232–237
Violence
  aggression and, 124, 128
  fantasy and, 381–382
  See also Domestic violence
Visual experience, 247, 248
Visual forms, 32–33
Volunteer groups, 54

Writer's block, 391

Yale Communication and Attitude Change Program, 143–444
Youth. *See* Children

GPSR Compliance

The European Union's (EU) General Product Safety Regulation (GPSR) is a set of rules that requires consumer products to be safe and our obligations to ensure this.

If you have any concerns about our products, you can contact us on ProductSafety@springernature.com

In case Publisher is established outside the EU, the EU authorized representative is:

Springer Nature Customer Service Center GmbH
Europaplatz 3
69115 Heidelberg, Germany

Batch number: 10844655

Printed by Printforce, the Netherlands